Protein–Carbohydrate Interactions in Infectious Diseases

RSC Biomolecular Sciences

Editorial Board

This series is devoted to coverage of the interface between the chemical and biological sciences, especially structural biology, chemical biology, bio- and chemo-informatics, drug discovery and development, chemical enzymology and biophysical chemistry.

Ideal as reference and state-of-the-art guides at the graduate and post-graduate level.

Titles in the series:

Visit our website on www.rsc.org/biomolecularsciences

For further information please contact:
Sales and Customer Care, Royal Society of Chemistry, Thomas Graham House, Science Park, Milton Road, Cambridge, CB4 0WF, UK
Telephone: +44 (0)1223 432360, Fax: +44 (0)1223 426017, Email: sales@rsc.org

Protein–Carbohydrate Interactions in Infectious Diseases

Edited by

Carole A. Bewley
National Institutes of Health, Bethesda, Maryland, USA

RSCPublishing

ISBN-10: 0-85404-802-2
ISBN-13: 978-0-85404-802-1

A catalogue record for this book is available from the British Library

Published by The Royal Society of Chemistry,
Thomas Graham House, Science Park, Milton Road,
Cambridge CB4 0WF, UK

Registered Charity Number 207890

For further information see our web site at www.rsc.org

Typeset by Macmillan India Ltd, Bangalore, India
Printed by Henry Ling Ltd, Dorchester, Dorset, UK

Preface

Protein–carbohydrate interactions are ubiquitous in nature. Though sometimes underappreciated, they play pivotal roles in countless and diverse biological processes ranging from mediating leukocyte rolling during the course of inflammation, gamete–gamete interactions that initiate fertilization, and recognition by and attachment to host cells by pathogenic organisms. During the past two decades, great strides have been made toward better understanding the chemical and molecular basis for protein–carbohydrate interactions and how they manifest in biology. In concert with many elegant biological studies, much of this progress can be attributed to advances in synthetic, organic, or carbohydrate chemistry; in the development of powerful tools that allow for robust, rapid, and diverse screening of protein–carbohydrate interactions such as carbohydrate and protein microarrays; in instrumentation used to make biophysical measurements such as isothermal titration calorimetry and surface plasmon resonance; in the development of carbohydrate-based conjugate vaccines; and in high-resolution structural analyses by X-ray crystallography and NMR. Perhaps the well-studied aspect of protein–carbohydrate interactions in biology pertains to their involvement in infectious diseases. Indeed, the outer surfaces of all cells of most pathogens are decorated with a distinct set of complex carbohydrate structures, usually in the context of glycolipids or glycoproteins; distinct carbohydrate-binding proteins (typically lectins or toxins) that recognize with specificity particular carbohydrate structures; or both. Consequently, the initial contact between two cells or a cell and pathogen almost certainly boils down to protein–carbohydrate interactions. Depending on the biological system, this initial event is one and the same with cellular recognition, attachment, adhesion, or a combination of the three. Given the interdisciplinary nature of the fields of carbohydrate chemistry and glycobiology, the study of protein–carbohydrate interactions in infectious diseases presents a field that is wide open for exploration.

A number of truly outstanding text and reference books giving comprehensive coverage to individual areas such as organic or enzymatic synthesis of carbohydrates, saccharide biosynthesis, and glycobiology, to name a few, are available. It is my hope that collectively, the set of chapters presented here, will educate and inform graduate students, postdoctoral fellows, and non-specialist readers alike, on the central role that protein–carbohydrate interactions play in mediating and initiating most infectious diseases; and the indispensable roles that chemistry and structural and cell biology play in studying these interactions. An introductory chapter outlining the atomic features behind carbohydrate recognition is presented by Sharon. Blaauw and Appelmelk describe carbohydrate structures displayed from glycolipids that are unique to mycobacterial species and account for virulence, while Imberty *et al.* and Sharon and Ofek describe co-crystal structures and biological studies that reveal the basis for carbohydrate recognition by surface lectins of *Pseudomonas aeruginosa* and diverse enterobacterial species such as *Escherichia coli*, *Klebsiella pneumoniae* and *Salmonellae* species, and introduce the notion of anti-adhesive therapies. Using high-resolution NMR techniques, Bernardi, Podlipnik, and Jiménez-Barbero describe the structural basis for bacterial enterotoxins binding to gangliosides as well as synthetic multivalent GM-1 mimetics. Chapters by Lehrer and van Kooyk and colleagues provide an opposing view of carbohydrate recognition, that is from the host cell's side, and review the defensive and immunologic roles of primate and human lectins known as theta defensins and C-type lectins, respectively. In a chapter covering sialic acid metabolism, Vimr and Steenbergen review the prevalence of sialic acid in microbial cell walls and touch on a topic on everyone's mind, namely that of targeting carbohydrate structures and metabolism, or alternatively carbohydrate receptors for chemotherapy. Chapters by Hölemann and Seeberger, and Kováč on carbohydrate-based conjugate vaccines for malaria and cholera, respectively, fill two essential aspects of the book's subject matter, namely elegant synthetic routes to assemble complex and homogeneous carbohydrate structures and protein conjugates on a fairly large scale, and current progress on the efficacy of carbohydrate-based vaccines. In the spirit of the 'omics' revolution, the final chapter by Shin describes the construction and use of carbohydrate or glycan microarrays that can provide initial information on carbohydrate specificity of lectins, toxins, enzymes, or whole cells. For those seeking more information, the book contains nearly 1000 references to the chemistry and glycobiology literature, and will hopefully give the reader an overall view of how carbohydrates and their protein partners mediate infection.

Contents

CHAPTER 1

Atomic Basis of Protein–Carbohydrate Interactions: An Overview

NATHAN SHARON

Department of Biological Chemistry, The Weizmann Institute of Science, Rehovot 76100, Israel

1 Introduction

Protein–carbohydrate interactions are the basis of numerous biological processes, both normal and pathological ones. They include the enzymatic synthesis and degradation of oligo- and polysaccharides, intracellular sorting of glycoconjugates, transport of carbohydrates into living cells and of their derivatives into subcellular organelles, the immunological response to carbohydrate antigens, and migration of leukocytes to sites of inflammation. These interactions also play a key role in a variety of cell adhesion phenomena, among them the attachment of parasites, fungi, bacteria, and viruses to host cells, the first step in the initiation of infection.[1,2] The high selectivity required for this attachment, as well as the binding of a variety of microbial toxins to cells, is provided by a specific stereochemical fit between complementary molecules, one a carrier of biological information (such as complex carbohydrates) and the other capable of decoding such information (carbohydrate-binding proteins, belonging to the class of lectins).[3] This concept has its origins in the lock-and-key hypothesis, introduced by Emil Fisher at the end of the 19th century to explain the specificity of interactions between enzymes and their substrates, *i.e.*, between molecules in solution; it was subsequently extended to describe the interactions of cells with soluble molecules and with other cells.

Complex carbohydrates are commonly found at the cell surface, where they are positioned to interact with suitable proteinaceous receptors, primarily lectins, in solution or on the surfaces of other cells. These proteins, originally identified as sugar-specific hemagglutinins, are currently known as cell-recognition molecules.[4] They are geared to distinguish between different oligosaccharides, whether as such or as part of glycoconjugates (primarily glycoproteins and glycolipids). Like other carbohydrate-binding proteins, whether enzymes, anticarbohydrate antibodies, or sugar transporters, lectins are structurally diverse, differing markedly in size, tertiary and quaternary structure, as well as the structure of their combining sites.[5]

2   Combining Sites

A major source of information about the combining sites of lectins is the X-ray crystallography study of their complexes with ligands. By now, the structures of close to 200 lectins, and over 300 of their complexes with carbohydrates, have been solved largely by this method (www.cermav.cnrs.fr/lectines/); most of those from bacterial or viral sources are listed in Table 1. Other inputs include binding experiments with sugars and their derivatives, site-directed mutagenesis, and, to a limited extent, also NMR experiments and molecular modeling. Such studies have shown that like the lectins themselves, the sites are diverse, even when their specificity is the same, although within a given lectin family the sites may be similar.[5,6] The sites appear to be preformed, since few conformational changes occur upon ligand binding. They are mostly in the form of shallow depressions on the surface of the protein, where typically only one or two edges or faces of the ligand are bound, and are thus similar

Table 1 *Complexes of bacterial and viral lectins of known three-dimensional structure*[*]

Source and name of lectin	Ligand	Resolution (Å)	PDB code
Clostridium botulinum			
Botulinum neurotoxin B	NeuAcα2,3Galβ4Glc	2.60	1F31
Clostridium tetani			
Tetanus toxin	Galβ4Glc	1.80	1DLL
	NeuAc	2.60	1DFQ
Escherichia coli			
Heat-labile enterotoxin	*m*NPαGal[†]	1.60	1EFI
FimH	Man	2.79	1KLF
F17-G	GlcNAc	1.65	1O9W
PapG	GalNAcβ3Galα4-Galβ4-GlcβCer	1.80	1J8R
Verotoxin-1 (shiga-like)	Galα4Galβ4Glc	2.80	1BOS
Influenza virus			
Hemagglutinin	NeuAcα2,3βGalβ3-GlcNAcβ3Galβ4Glc	2.50	IRV0
Pseudomonas aeruginosa			
PA-IL	Gal	1.60	1OKO
PA-IIL	L-Fuc	1.30	1GZT
Ralstonia solanacearum			
RS-IIL	Man	1.70	1UQX
Staphylococcus aureus			
Enterotoxin B	Galβ4Glc	1.90	1SE4
Vibrio cholerae			
Cholera toxin	NPαGal[†]	2.00	1EEI

[*]For references, see www.cermav.cnrs.fr/lectines.
[†]*m*NP, *m*-nitrophenyl.

to those of anticarbohydrate antibodies or glycosidases.[7–9] In a few lectins, the combining sites are in the form of deep clefts.

The participation of a particular amino acid of a lectin and of a specific group of the carbohydrate ligand in the interaction between the two can be assessed by different methods. In the case of the amino acids, it is done mainly by site-directed mutagenesis, as mentioned earlier. This technique, combined with ligand-binding experiments, also provides information on the relative contribution of individual residues to the protein–carbohydrate interaction. As to the carbohydrate ligand, its hydrogen bonding pattern in the complex can be mapped by studies with deoxy-, methoxy-, and deoxyfluoro sugar analogs.[10,11] It is only X-ray crystallography, however, that provides detailed information at the atomic level on the interplay between the two molecules. Still, it should be kept in mind that a crystal represents a frozen constellation often obtained from a concentrated solution in a nonphysiological milieu, without any indication on the status of the involved molecules prior to complex formation.[12] The use of nonphysiological crystallization conditions may induce conformational changes in the protein and alter the mode of binding. Therefore, although the value of X-ray crystallography is not disputable, it is essential to complement the data obtained by this method with information from other sources (*e.g.*, NMR) on the solution structure of the lectin–ligand complex.

3 Lectin–Carbohydrate Bonds

The bonds involved in the formation of lectin–carbohydrate complexes are in principle not different from those involved in the formation of the corresponding complexes of other carbohydrate-binding proteins. Lectins combine with their ligands primarily by a network of hydrogen bonds and hydrophobic interactions; in rare cases, electrostatic interactions (or ion pairing) and coordination with metal ions also play a role.[13–15] Bonding is sometimes mediated by one or more water molecules (explained later). Although in a single lectin a limited set of residues contribute to the interactions with the ligand, on the whole almost all kinds of amino acids participate in ligand binding.[15]

3.1 Hydrogen Bonds

Hydrogen bonds that are directional are heavily involved in conferring specificity to protein–carbohydrate interactions, as well as contributing to their affinity.[8,9] They depend largely on interactions between the hydroxyls of the carbohydrate and the amino acid side chains of the protein, most frequently of aspartic acid, aparagine, glutamic acid, glutamine, arginine, and serine residues. Main chain NH and CO groups also contribute to hydrogen bonding, but to a lesser extent.

A sugar hydroxyl has the capacity to interact with a protein both as a hydrogen bond donor and as an acceptor, by way of the lone electron pairs. Moreover, as donor, the hydroxyl possesses the added advantage of having rotational freedom about the C–OH torsional angle, thus enabling it to attain the best possible linear bond with an acceptor group, which is important in imparting specificity. When each of two adjacent hydroxyls of a monosaccharide interacts with different atoms of the same amino acid (*e.g.* the two oxygens of the carboxylate of glutamic or aspartic acid), they form bidentate hydrogen bonds.[8,9] Another kind of hydrogen bond characteristic of protein–carbohydrate complexes is the cooperative bond, in which the hydroxyl group acts simultaneously as donor and acceptor.

3.2   Hydrophobic Interactions

Even though carbohydrates are highly polar molecules, the steric disposition of the hydroxyl groups creates hydrophobic patches on their surfaces, which can form contacts with hydrophobic side chains of the protein.[16,17] One widely occurring interaction of this kind is the stacking of a monosaccharide on a side chain of an aromatic amino acid. It is the result of the presence of partial positive charges on the aliphatic protons on one face of a hexopyranose ring and a partial negative charge from the π-electrons of the aromatic system.[18] Such stacking is found in the sugar complexes of almost all legume lectins, of the galectins and some C-type lectins, as well as of bacterial toxins (*e.g. E. coli* lytic toxin and tetanus toxin). In addition, the methyl moiety of the acetamide of acetamido sugars often interacts with aromatic residues of the lectins (*e.g.* WGA and influenza virus hemagglutinin that are specific for such sugars), as does the methyl group of fucose. Hydrophobic contacts also occur with side chains of aliphatic amino acids, such as valine or leucine.

3.3   Other Interactions

Most saccharides are uncharged, and therefore ionic, *i.e.*, charge–charge, interactions do not commonly participate in the formation of protein–carbohydrate complexes; such interactions, however, occur in complexes with sialic acids, *e.g.*, influenza virus hemagglutinin,[19] and are common in complexes with glycoaminoglycans. An unusual type of bond, found only in the C-type lectins, is between the protein-bound Ca^{2+} and certain hydroxyls of the ligand.[20]

4   Role of Water

As mentioned, contacts between the protein and its ligands are sometimes mediated by water bridges.[21–23] Water acts as a molecular mortar, its small size and ability to serve as both a hydrogen bond donor and acceptor endowing it with ideal properties for this purpose. Such bridges, which consist of a single water molecule or chains of several water molecules, may be important for ligand recognition.

  Both the protein and the ligand in aqueous solutions are normally bonded to water molecules by hydrogen bonding. Therefore, the overall process of binding between the two involves the meeting of a solvated polyhydroxylated glycan and a solvated protein combining site. In the process of complexation, the protein–water and ligand–water hydrogen bonds are replaced by protein–ligand bonds and the released water returns to the bulk solvent.[11] When the complex finally forms, it presents a new surface to the surrounding medium, which is also hydrated. Solvation–desolvation energies are very large, due to entropy and cannot be reliably calculated for hydrophilic compounds such as sugars. Thus, even though the energetic contributions of van der Waals and hydrogen bonding interactions in the combining site can be estimated, errors in the estimation of the attendant solvation energy changes can be much larger, making the overall calculations of binding energy difficult.[24] Comparison of the structures of unligated lectins from the same family with those of their complexes with the same or different sugars has shown that certain ordered water molecules are conserved in all such structures.[21]

References

1. K.A. Karlsson, *Biochem. Soc. Trans.*, 1999, **27**, 471.
2. I. Ofek, D.L. Hasty and R.J. Doyle, *Bacterial Adhesion to Animal Cells and Tissues*, ASM Press, Washington, DC, 2003, 416.
3. N. Sharon and H. Lis, *Science*, 1989, **216**, 227.
4. N. Sharon and H. Lis, *Glycobiology*, 2004. **14**, 53R.
5. N. Sharon, *Trends Biochem. Sci.*, 1993, **18**, 221.
6. N. Sharon and H. Lis, *Lectins*, 2nd edn, Chapters 5 and 6, Kluwer Academic Publishers/Springer, Amsterdam, 2003.
7. E.A. Kabat, *J. Supramol. Struct.*, 1978, **8**, 79.
8. F.A. Quiocho, *Biochem. Soc. Trans.*, 1993, **21**, 445.
9. D. Bundle, in *Carbohydrates*, S. Hecht (ed), Oxford University Press, Oxford, 1998, 370.
10. G. Magnusson, S.J. Hultgren and J. Kihlberg, *Methods Enzymol.*, 1995, **253**, 105.
11. R.U. Lemieux, *Acc. Chem. Res.*, 1996, **29**, 373.
12. H. Rüdiger, *Acta Anat.*, 1998, **161**, 130.
13. C. Cambillau, in *Glycoproteins*, J. Montreuil, H. Schachter and J.F.G. Vliegenthart (eds), Elsevier, Amsterdam, 1995, 29.
14. W.I. Weis and K. Drickamer, *Ann. Rev. Biochem.*, 1996, **65**, 441.
15. H. Lis and N. Sharon, *Chem. Rev.*, 1998, **98**, 637.
16. Y. Bourne, P. Rouge and C. Cambillau, *J. Biol. Chem.*, 1992, **267**, 97.
17. J. Bouckaert, J. Berglund, M. Schembri, E. De Genst, L. Cools, M. Wuhrer, C.S. Hung, J. Pinkner, R. Slattegard, A. Zavialov, D. Choudhury, S. Langermann, S.J. Hultgren, L. Wyns, P. Klemm, S. Oscarson, S.D. Knight, H. De Greve, *Mol. Microbiol.*, 2005, **55**, 441.
18. M. Levitt and M.F. Perutz, *J. Mol. Biol.*, 1988, **201**, 751.
19. S. Kelm, J.C. Paulson, U. Rose, R. Brossmer, W. Schmid, B.P. Bandgar, E. Schreiner, M. Hartmann, E. Zbiral, *Eur. J. Biochem.*, 1992, **205**, 147.
20. W.I. Weis, K. Drickamer and W.A. Hendrickson, *Nature*, 1992, **360**, 122.
21. R. Loris, P.P.G. Stas and L. Wyns, *J. Biol. Chem.*, 1994, **269**, 26722.
22. E.J. Toone, *Curr. Opin. Struct. Biol.*, 1994, **4**, 719.
23. P. Adhikari, K. Bacchawat-Sidker, C.J. Thomas, R. Ravishankar, A.A. Jeyaprakash, V. Sharma, M. Vijayan, A. Surolia, *J. Biol. Chem.*, 2001, **276**, 40734.
24. J. Janin, *Structure*, 1997, **3**, 473.

CHAPTER 2

Mycobacterial Glycolipids and the Host: Role of Phenolic Glycolipid and Lipoarabinomannan

G.J. BLAAUW AND B.J. APPELMELK

Department of Medical Microbiology, VUmc Vrije Universiteit Medical Center, Van der Boechorststraat 7, 1081 BT Amsterdam, The Netherlands

1 Introduction

Among mycobacterial infectious diseases, tuberculosis and leprosy are the best known. Their causative agents are *Mycobacterium tuberculosis* and *Mycobacterium leprae*, respectively. They have in common the staining characteristic referred to as acid fastness, which is due to the presence of a highly characteristic cell wall that includes the presence of very large covalently attached lipids known as mycolic acids. Leprosy, also referred to as Hansen's disease,[1] is a chronic infection of the skin and peripheral nerves. It is not very contagious and is transmitted via droplets from the nose and mouth during close and frequent contact. It is a mutilating disease because of the progressive and permanent damage to the skin, nerves, limbs and eyes. It mainly occurs in third world countries such as India and Brazil. It is estimated that 2 million people worldwide are disabled because of leprosy[2] and between 500,000 and 1,000,000 cases are detected each year. Leprosy in fact presents itself by a spectrum of manifestations: on one end the tuberculoid type, which is a non-progressive disease, mainly affecting the peripheral nerves, resulting in loss of sensation and muscle atrophy, whereas on the other end the lepromatous type, which is the progressive form of leprosy, mainly affects the skin, mutilating the face, fingers, and toes. Progression in the latter type is due to the inability of the host immune response to inhibit bacillary multiplication. Leprosy can be treated and should be detected as early as possible, because of the irreversible damage it causes.

Tuberculosis, also referred to as Koch's disease,[3] is mainly a pulmonary disease, but other sites of infection are also possible, owing to early bacteremia after initial infection. It is caused by *M. tuberculosis* or, less frequently, by other mycobacterial species within the *M. tuberculosis* complex, such as *M. bovis, M. africanum*, and *M. microti*. Tuberculosis is currently one of the leading causes of death worldwide, responsible for 50 million new infections and claiming 2–3 million lives per year.[4] Infection is acquired through inhalation of aerosols that contain infectious bacilli. However, only 10% of those who are infected actually get the disease. The other 90% is able to contain the infection with the help of an adequate immune response, resulting in the formation of granulomas in which (infected) macrophages and lymphocytes are predominant. The tubercle bacilli enter into a state of dormancy in which they do not replicate or disseminate. Dormancy reflects a state of equilibrium, which is beneficial to both the host and the infectious agent. The former is able to contain the infection, while the latter is able to survive. The hypoxic milieu within granulomas seems to be a major factor in induction of this state of dormancy. It is estimated that nearly one-third of the world population is in this way latently infected.[5] The moment this equilibrium is disturbed, the bacilli can be reactivated to cause disease. Therefore host factors and especially the state of the immune system, such as the number of CD4 T-cells, are crucial determinants for disease progress after initial infection or, alternatively, for containment or even elimination. This chapter will focus on two glycolipid components of these pathogenic mycobacteria, lipoarabinomannan (LAM) and phenolic glycolipids (PGLs). These glycolipids play, through interactions with host receptors such as lectins, a crucial role in the pathogenesis of tuberculosis and leprosy. A brief history of these compounds, their structure, biosynthesis and role in pathogenicity and immune modulation will be discussed in this chapter.

2 Phenolic Glycolipids

2.1 Phenolic Glycolipids History

Characterization by infrared (IR) spectroscopy of antigen fractions from mycobacteria signified the beginning of phenolic glycolipids (PGLs) history.[6] These fractions contained lipids, characteristic for a given mycobacterial species. In 1960 these lipids were named mycosides and were defined as "glycolipids or glycolipid peptides limited in distribution to a single species of mycobacteria." Not long afterwards, the presence of mycosides A, B, and G in lipid extracts of *M. kansasii, M. bovis*[7] and *M. marinum*[8] was also demonstrated. They were classified as members of the PGL family, because IR spectra indicated the presence of an aromatic group, differentiating them from other mycobacterial cell wall glycolipids.[6] Analytical tools, such as gas–liquid chromatography, NMR- and mass spectrometry, made it possible to elucidate the chemical structure of these different PGLs. The structures of PGLs derived from different mycobacterial species are principally the same in their lipid core and are attached to mycocerosic acids (Figure 1), but they vary in the composition of oligosaccharide side chain (Figure 2) attached to the phenol moiety.

A: phenolic glycolipid (PGL)

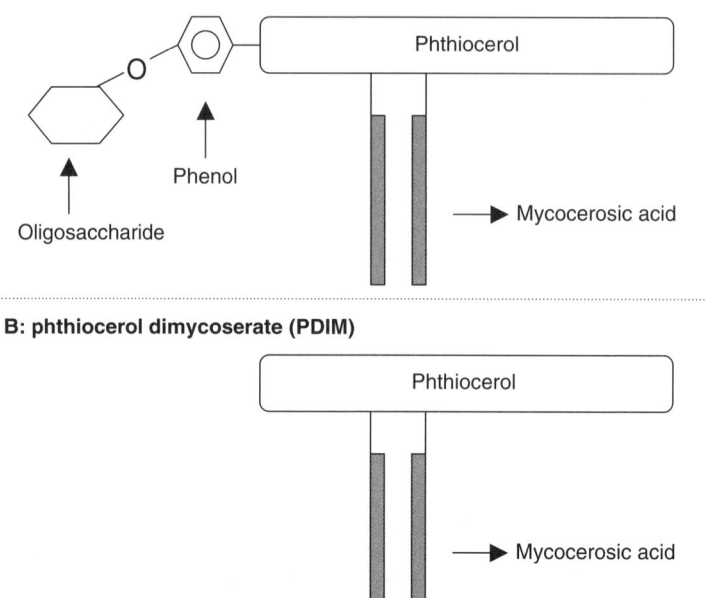

B: phthiocerol dimycoserate (PDIM)

Figure 1 *Schematic structures of PGL (A) and phthiocerol dimycocerosate (B). They have in common a lipid core with two mycocerosic tails. PGL has an additional glycosylated phenol group attached to the lipid core. This oligosaccharide can be a mono-, di-, tri- or tetrasaccharide, depending on the mycobacterial species from which it is isolated*

After a period of negligence, interest in PGLs was revived when a *M. leprae*-specific antigen (PGL-I) (Figure 3) was isolated from *M. leprae*-infected liver tissue of armadillos. This antigen was also classified as a member of the PGL family.[9,10] Owing to its high antigenicity, PGL-I could be used in serodiagnosis of leprosy. With respect to the diagnosis of leprosy, the synthesis of neoglycoproteins of PGL-I, the major PGL of *M. leprae*, has been a major leap forward.[11,12] Antibodies against PGL-I can now be detected in the sera of leprosy patients with high sensitivity and specificity.

In addition to its high antigenicity and the use for serodiagnosis, the role of PGL-I in pathogenicity of *M. leprae* and immune modulation of the host was investigated.[13,14,15] In analogy with *M. leprae*, a PGL (PGL-Tb 1) (Figure 3B) was extracted from *M. tuberculosis* (strain Canetti).[16] In contrast to leprosy, PGL of *M. tuberculosis* cannot be used for serodiagnosis of tuberculosis: a too large variation is observed in the anti-PGL antibody responses in the sera of tuberculosis patients.[17,18,19] Nevertheless, the production of PGLs by specific strains of *M. tuberculosis* has recently been recognized as a virulence factor of major importance and we foresee major future interest in studies on the interaction of PGL with host receptors such as immune cell-surface lectins and signaling pathways.

A
CH3O
O
O
R′
CH3
OH
O
O
OCH3
CH3
O
O
OCH3
CH3
CH3O
CH3O
O
O
OCH3
H3C
OH
B
OH
O
O
R′
CH3
OH
OCH3
OH
O
O
R′
C
CH3
OCH3
OH

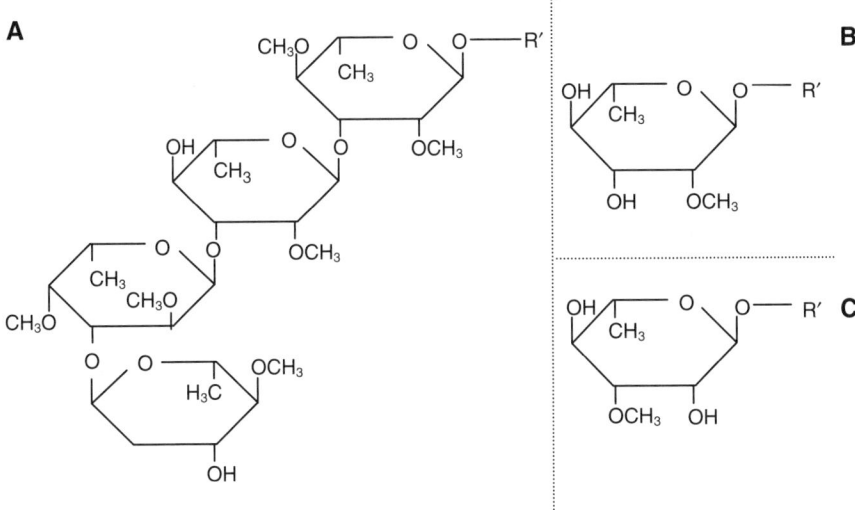

Figure 2 *Structures of the oligosaccharide moieties of, respectively, the major PGL of M. kansasii, also known as mycoside A (A); mycoside B, the major PGL of M. bovis (B); and mycoside G, the major PGL of M. marinum (C) (R′: phenolphthiocerol)*

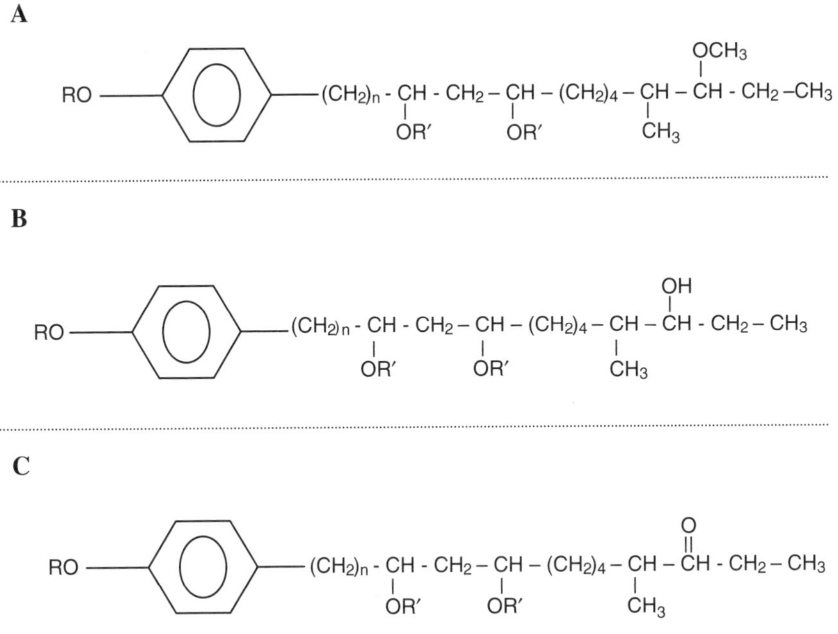

A
OCH3
RO
(CH2)n - CH - CH2 – CH – (CH2)4 – CH – CH - CH2 –CH3
OR′
OR′
CH3
B
OH
RO
(CH2)n - CH - CH2 – CH – (CH2)4 – CH – CH - CH2 – CH3
OR′
OR′
CH3
C
O
RO
(CH2)n - CH - CH2 – CH – (CH2)4 – CH – CH - CH2 – CH3
OR′
OR′
CH3

Figure 3 *Variation in lipid cores within a single M. marinum strain (MNC 842): phenolphthiocerol (A), phenolphthiotriol (B), and phenolphthiodiolone (C) (R: oligosaccharide moiety, R′: mycocerosic acid)*

2.2   Structure and Biosynthesis of Phenolic Glycolipids

PGL is a collective term for phenol phthiocerol and related compounds, which contain the following two distinctive chemical elements (Figure 1): a long-chain β-diol, naturally occurring as diesters of polymethyl-branched fatty acids and the mycocerosic acids. The lipid core is terminated by an aromatic nucleus, usually glycosylated by a mono- or oligo- (up to tetra-) saccharide that is linked through the OH group of the phenol moiety. The presence of this glycosylated phenol moiety differentiates it from PDIM (Figure 1B). Both PDIM and PGL are unique to pathogenic mycobacteria.[20,21] In contrast to PDIM, PGL is not present in all strains within the *M. tuberculosis* complex[22] and in fact many *M. tuberculosis*, strains do not produce PGL. Variability in number and kind of saccharide units attached to the aromatic nucleus, is mainly species specific, but different PGLs can be found within a single strain. So far the structures of seven different PGLs from *M. kansasii* have been elucidated[23–26] (Table 1). The major PGL Phe Gl K-I is also known as mycoside A. Some of the PGLs in *M. kansasii* are also found in *M. gastri*.[27] In *M. bovis*, four different PGLs (Phe Gl B [mycoside B], B-1, B-2, B-3) have been identified.

They differ not only in sugar moiety but in lipid core as well. Phe Gl B and B-3 contain a phenolphthiocerol lipid core, while Phe Gl B-1 and B-2 have a phenolphthiodiolone lipid core.[28] The latter has a keto instead of a methoxy group. Phenolphthiocerol, phenolphthiodiolone and phenolphthiotriol are the principal lipid cores of the PGLs in *M. marinum* (Figure 4). Two strains of this species have been characterized. A variable length of the lipid core is observed: C_{26} dimethyl and C_{27} and C_{29} trimethyl in MNC 170, whereas in strain MNC 842 the C_{27} trimethyl acid is the major component. Besides, strain MNC 170 only produces phenolphthiocerols, whereas strain MNC 842 also had phenolphthiodiolones and phenolphthiotriols. Phe Gl M (mycoside G) in *M. marinum* is almost identical to Phe Gl B (mycoside B), the major PGL variant found in *M. bovis* (Figure 2). They differ only in the location of the OCH_3 group in the monosaccharide residue.[29]

Of the two most pathogenic mycobacterial species, all strains of *M. leprae* tested produce PGL. So far, three different PGLs have been isolated.[10,30] In contrast, the presence of PGL in *M. tuberculosis* is strain specific. As in *M. leprae*, different PGLs can be found within the same strain of *M. tuberculosis*. For instance, *M. tuberculosis* strain Canetti harbors a major triglycosyl variant (PGL-Tb-1), a minor triglycosyl variant and two minor monoglycosyl variants[16] (Table 1).

Other strains contain, apart from the major triglycosyl variant, a phenolphthiotriol, in which the methoxy group of the lipid core is replaced by a hydroxyl group. However, strains expressing this mixture of lipid cores can also vary in the triglycosyl moiety and in some PGLs a methyl group is missing in the fucopyranosyl part of the sugar.[31] In a single PGL variant, even the number of carbon atoms of the mycocerosic acid side chains can vary. Variation also exists in the branching pattern of the mycocerosic acids. Apart from the aromatic nucleus with attached oligosaccharide moiety, PDIMs and PGLs are identical. Therefore, they also share a common biosynthetic pathway with two different precursors, either C22–C24 fatty acids or *p*-hydroxyphenylalkanoic acid. These two precursors are elongated to a common lipid core by three malonyl CoA and two methylmalonyl CoA units.[32] The genetic organization of

Table 1 *Oligosaccharide moieties of PGL in different mycobacterial species*

PGL	Oligosaccharide moiety
M. tuberculosis	
Tb 1/Tb-O	2,3,4-Tri-*O*-methyl-fucopyranosyl-(1→3)-rhamnopyranosyl-(1→3)-2-*O*-methyl-rhamnopyranosyl-(1→)[16,31]
*	2,3,4-Tri-*O*-methyl-fucopyranosyl-(1→3)-rhamnopyranosyl-(1→3)-rhamnopyranosyl-(1→)[16]
*	2-*O*-Methyl-rhamnopyranosyl-(1→)[16]
*	Rhamnopyranosyl-(1→)[29]
Tb-K	2,4-Di-*O*-methyl-fucopyranosyl-(1→3)-rhamnopyranosyl-(1→3)-2-*O*-methyl-rhamnopyranosyl-(1→)[31]
M. leprae	
PGL-I	3,6-Di-*O*-methyl-glycopyranosyl-(1→4)-2,3-di-*O*-methyl-rhamnosyl-(1→2)-3-*O*-methyl-rhamnopyranosyl-(1→)[29]
PGL-II	3,6-Di-*O*-methyl-glycopyranosyl-(1→4)-3-*O*-methyl-rhamnosyl-(1→2)-3-*O*-methyl-rhamnopyranosyl-(1→)[29]
PGL-III	6-*O*-Methyl-glycopyranosyl-(1→4)-2,3-di-*O*-methyl-rhamnosyl-(1→2)-3-*O*-methyl-rhamnopyranosyl-(1→)[29]
M. kansasii	
K-I	2,6-Dideoxy-4-*O*-methyl-arabinohexopyranosyl(1→3)-4-*O*-acetyl-2-*O*-methyl-fucopyranosyl(1→3)-2-*O*-methyl-rhamnopyranosyl(1→3)-2,4-di-*O*-methyl-rhamnopyranosyl(1→)[26]
K-II	2,4-Di-*O*-methyl-mannopyranosyl(1→3)-4-*O*-acetyl-2-*O*-methyl-fucopyranosyl(1→3)-2-*O*-methyl-rhamnopyranosyl(1→3)-2,4-di-*O*-methyl-rhamnopyranosyl(1→)[23]
K-IV	2-*O*-Methyl-mannopyranosyl(1→3)-4-*O*-acetyl-2-*O*-methyl-fucopyranosyl(1→3)-2-*O*-methyl-rhamnopyranosyl(1→3)-2,4-di-*O*-methyl-rhamnopyranosyl(1→)[23]
K-5	2,6-Dideoxy-4-*O*-methyl-arabinohexapyranosyl(1→3)-4-*O*-propionyl -2-*O*-methyl-fucopyranosyl(1→3)-2-*O*-methyl-rhamnopyranosyl(1→3)-2,4-di-*O*-methyl-rhamnopyranosyl(1→)[23]
K-6	2,6-Dideoxy-4-*O*-methyl-arabinohexapyranosyl(1→3)-2-*O*-methyl-fucopyranosyl(1→3)-2-*O*-methyl-rhamnopyranosyl(1→3)-2,4-di-*O*-methyl-rhamnopyranosyl(1→)[23]
K-7	2-*O*-Methyl-fucopyranosyl(1→3)-2-*O*-methyl-rhamnopyranosyl(1→3)-2,4-di-*O*-methyl-rhamnopyranosyl(1→)[23]
K-8	2-*O*-Methyl-fucopyranosyl(1→3)-2-*O*-methyl-rhamnopyranosyl(1→3)-2-*O*-methyl-rhamnopyranosyl(1→)[23]
M. bovis	
B	2-*O*-Methyl-rhamnopyranosyl (lipidcore: *phenolphthiocerol*)[28]
B-1	2-*O*-Methyl-rhamnopyranosyl (lipidcore: *phenolphtiodiolone*)[28]
B-2	Rhamnopyranosyl (lipidcore: *phenolphtiodiolone*)[28]
B-3	Rhamnopyranosyl-2-*O*-methyl-rhamnopyranosyl (lipidcore: *phenolphthiocerol*)[28]

the genes involved in this pathway comprises a transcriptionally coupled cluster of *pps* genes (*pps A-E*) that encodes a set of enzymes that catalyze the different biosynthesis steps.[32,33] The biosynthetic pathway of both mycocerosic side chains and their subsequent *trans*-esterification onto the lipid core of (phenol)phthiocerol has also been

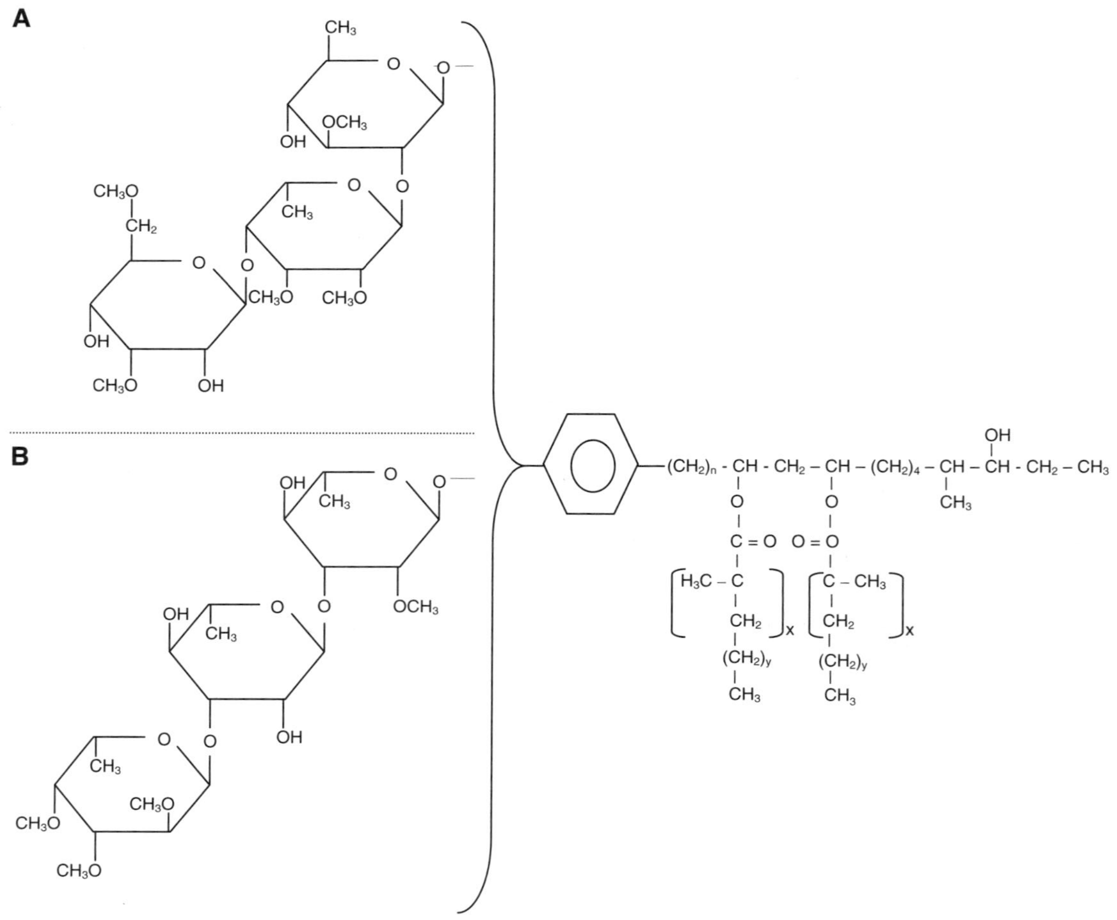

Figure 4 *Structures of PGL-I: M. leprae (A) and PGL-Tb1: M. tuberculosis (B) (Rʼ: phenolphthiocerol)*

elucidated.[34,35]The biosynthesis of *p*-hydroxyphenylalkanoic acid involves the elongation of *p*-hydroxybenzoic acid I and II (*p*-HBAD) with eight to nine series of malonyl CoA units (Figure 5). *p*-HBAD I is the precursor molecule of mycoside B (monosaccharide) and *p*-HBAD II is the precursor molecule of PGL-Tb1 (trisaccharide).

On the genetic level this process is encoded by the *pks15/1* gene. *M. tuberculosis* strains that do not produce PGL are usually affected by a seven base pair deletion in this gene. The deletion causes a frameshift in the *pks15/1* gene and therefore results in a disrupted open reading frame (ORF) (Figure 6). These strains secrete glycosylated derivates of *p*-HBAD into the culture media, indicative of the fact that elongation to *p*-hydroxyphenylalkanoic acid has not taken place.[22] In *M. bovis* BCG, a deletion of six base pairs is observed, which does not result in a frameshift and therefore leaves the ORF intact[22] (Figure 6).

The formation of the carbohydrate domain of PGL-Tb and *p*-HBAD II requires the sequential transfer of three monosaccharides. The genes involved, *Rv2962c, Rv2958c,* and *Rv2957,* encode three putative glycosyl transferases.[36] These enzymes transfer sugar residues both onto *p*-HBAD and phenolphthiocerol. Although the role of *p*-HBAD as a precursor molecule for the synthesis of PGL has been made plausible,[22,37] the source of this molecule in mycobacteria has still not been identified and further research needs to be done. Contradictory reports have been published on the timing of glycosylation within the process of PGL biosynthesis. Some reports show that attachment of the glycosyl moiety takes place on the phenolphthiocerol molecule, indicating it to be the final step.[38] Others report the presence of glycosylated

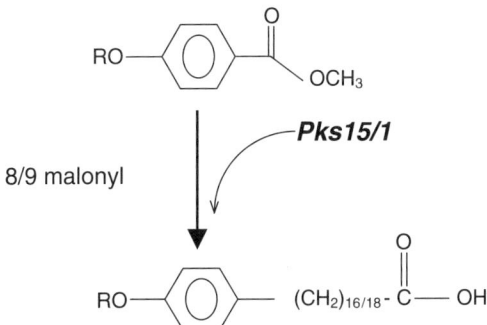

Figure 5 *Elongation of p-HBAD (I or II) to p-hydroxyphenylalkanoic acid by pks15/1 as a determining step in the biosynthesis of PGL*

```
M. tub./Beijing    GAGGCGAGCGAAAGCACCGGGGGCCGCGGGCCGCGGCCGTCGATGGTGCCG
M. tub./Canetti    GAGGCGAGCGAAAGCACCGGGGGCCGCGGGCCGCGGCCGTCGATGGTGCCG
M. bovis           GAGGCGAGCGAAAGCACCGGGGG------GCCGCGGCCGTCGATGGTGCCG
M. tub.            GAGGCGAGCGAAAGCACCGGGGG-------CCGCGGCCGTCGATGGTGCCG
M. tub.            GAGGCGAGCGAAAGCACCGGGGG-------CCGCGGCCGTCGATGGTGCCG
```

Figure 6 *Sequence alignment of a part of pks15/1 of various strains of M. tuberculosis and M. bovis*

HBAD in the media of cultured *M. tuberculosis*, which means that glycosylation takes place early in the biosynthesis of PGL.[22]

PGL secretion is blocked by the inactivation of a transmembrane protein MmpL7, which has been shown to also facilitate PDIM transport.[20,39] However, PGL synthesis was not affected in these mutant strains. This indicates that glycosylation of PGL occurs at the inside of the bacterial cell membrane, before transport across the cell membrane takes place. This chapter will mainly focus on the oligosaccharide moiety of the PGL molecule, since this seems to be essential for its immunomodulatory properties.

2.3   Role of PGL in Pathogenesis

2.3.1   *Leprosy*

M. leprae was the first pathogenic mycobacterial species in which the role of PGL was recognized.[9] *M. leprae* is unique among other mycobacterial species in its tropism for neural tissue. Schwann cells, the myelinating glial cells of the peripheral nervous system, are the principal targets of *M. leprae*. Destruction of the myelin sheath of the peripheral nerve causes progressive loss of sensitivity downstream of the affected nerve. To enter the Schwann cell *M. leprae* binds to laminin-2, a glycoprotein consisting of three polypeptide chains (α, β, and γ). This protein then forms a bridge between the bacilli and the laminin receptor on the surface of the Schwann cell.[13,14] The molecule responsible for binding to laminin-2 is PGL-I, to be more precise, the trisaccharide moiety of PGL-I. The oligosaccharide moieties of other PGLs cannot bind to laminin, which explains why only *M. leprae* enters the Schwann cell.[40] Leprosy presents itself with a broad spectrum of clinical manifestations, depending on the patient's immune reaction.[41] On one end is the lepromatous form of leprosy with a strong antibody response and a poor cellular response, resulting in abundant bacilli in macrophages and in the Schwann cells of the peripheral nerves. On the other end of the spectrum is tuberculoid leprosy, which exhibits a weak humoral antibody response and a strong cellular response, resulting in lymphocyte proliferation and granuloma formation with relatively few bacilli present. Contradicting reports have been published on whether PGL-I has an inhibiting effect on the lymphocyte proliferative response in lepromatous leprosy only or in both lepromatous and tuberculoid leprosy.[15,42] Since PGL-I is present in all strains of *M. leprae*, the paper of Mehra *et al.*[15] suggested the presence of specific T-lymphocyte suppressor cells in patients who suffer from lepromatous leprosy. Evidence was provided that these suppressor T-cells recognize the terminal trisaccharide moiety of PGL-I. In contrast, the study by Prasad *et al.*[42] denied that PGL-I has a central role in determining the spectrum of leprosy. The role of PGL-I in inhibition of lymphocyte proliferation and suppression of the IFN-γ response after infection with *M. leprae* was confirmed in another study.[43] Lymphocyte proliferation and IFN-γ production became significantly induced after PGL-I, presented on the surface of *M. leprae*-infected dendritic cells (DCs), was masked by a PGL-specific monoclonal antibody; this Mab recognizes the oligosaccharide unit of *M. leprae* PGL. These data suggest that *M. leprae* PGL suppresses T-cell function and that recognition of the trisaccharide by a not-yet-identified host lectin seems mandatory for this downregulation.

A striking characteristic of pathogenic mycobacteria is that they can survive and even multiply inside the human macrophage. Phagocytic cells contain several oxygen-dependent and independent antimicrobial systems. The former uses toxic oxygen metabolites, such as superoxide (O_2^-), hydrogen peroxide (H_2O_2), hydroxyl radical ($OH^·$) and possibly singlet oxygen ($^1O_2$) after the phagocytosis-induced respiratory burst. This respiratory burst is induced after activation of macrophages by IFN-γ. Some data suggest that PGL-I plays a role as a scavenger of the above-mentioned toxic substances, thereby protecting intracellular bacilli against their harmful effects.[44]

Cytokines in tuberculoid leprosy and lepromatous leprosy are differentially expressed.[45.] In tuberculoid leprosy, IFN-γ response is induced thereby upregulating IL-12 production and downregulating IL-10, which directs the immune response in a T_H1 manner (cellular immune response). However, in lepromatous leprosy, IFN-γ response is reduced thereby downregulating IL-12 and upregulating IL-10 production, which directs the immune response in a T_H2 manner (humoral immune response). In contrast to the cellular immune response, the humoral immune response cannot deal with intracellular pathogens, so that lepromatous leprosy manifests itself by disseminated disease. As noted earlier in this chapter, PGL-I plays a role in inhibition of lymphocyte proliferation and in suppression of IFN-γ response. In theory this could contribute to the downregulation of IL-12 and upregulation of IL-10, which results in an anti-inflammatory pathway of the immune response. The former perfectly matches the immune response in lepromatous leprosy. However, it is not very likely that PGL plays a decisive role in directing the host-immune response, because it is present in both tuberculoid and lepromatous leprosy. Other strain-related factors in *M. leprae*, yet unknown, might play a role. However, considerable evidence is present that polymorphisms in the host-immune genes direct whether lepromatous or, alternatively, tuberculoid disease develops in leprosy patients.[46,47]

2.3.2 Tuberculosis

M. tuberculosis strain Canetti was the first strain in which PGL (PGL-Tb1) was found to be present.[16] In aforementioned studies, the presence (and structure) of PGL was investigated by time-consuming chemical techniques. The classical way to determine whether isolated mycobacterial strains produce PGL or not is by lipid extraction from a fresh culture and chemical analysis of this extract by thin layer chromatography (TLC).[48] Elucidation of the structure of PGL is done by NMR- and mass spectrometry. For large-scale epidemiological studies, this method is very time consuming. Therefore, lipid extracts of *M. tuberculosis* were tested with dot-blot enzyme-linked immunosorbent assay (ELISA), in which monoclonal antibodies (MabIII604) binding to the oligosaccharide moiety of PGL of *M. tuberculosis* detected the presence of PGL. In this assay, 64% of the strains tested, scored positive. In contrast, TLC of lipid extracts yielded evidence for the presence of PGL in only 16% of these strains.[49] Since MabIII604 binds to the oligosaccharide moiety of PGL, dot-spot analysis can be false positive, owing to the presence of HBAD II, a precursor molecule of PGL present in most *M. tuberculosis* strains (Figure 5).[22]

As in the case of leprosy, the synthesis of a neoglycoprotein, analogous to PGL Tb1, offered the opportunity to examine the sera of patients for the presence of antibodies against PGL-Tb1. Unfortunately the seroreactivity of the tested sera was very low: 24 of 119 sera tested positive.[17] In comparable studies, with purified PGL as target, the presence of anti-PGL antibodies in sera of tuberculosis patients was investigated and found to vary widely: from 20% to over 95%.[17–19,50] Altogether several findings indicated that many *M. tuberculosis* strains do not express PGL, which made PGL antibody detection unsuitable as a diagnostic tool for tuberculosis. Consequently, the role of PGL in pathogenesis of tuberculosis was thought to be less significant as compared to its role in leprosy. Very recent studies have changed this view, however. It was shown that the great majority of *M. tuberculosis* strains that do produce PGL belong to the so-called Beijing genotype.[51–53] Molecular typing techniques, such as IS 6110 restriction fragment length polymorphism and spoligotyping, have made it possible to differentiate between strains of *M. tuberculosis*. Several genotypes can be recognized and their prevalence established. The Beijing genotype of *M. tuberculosis* has proved to be very successful, since it is highly prevalent in Asia[51,54] and has been responsible for major outbreaks throughout the world;[52] some of the genotypes have been associated with multiple drug resistance.[55,56] As can be anticipated, these strains have an intact ORF in the *pks15/1* gene.[39,53] Strains belonging to this family are also hypervirulent in murine disease models.[57,58] In addition, *in vitro* experiments showed that cytokine expression profiles in human monocytes, after infection with these strains or after exposure to their lipid extracts, exhibited a T_H2 immune response with inhibition of TNF-α, IL-10, and IL-12, as compared to non-Beijing strains, which elicited cytokine responses toward a T_H1 response.[59,58] Recent experiments showed that Beijing strain HN 878, disrupted in the *pks15/1* gene by insertion of a hygromycin cassette, could no longer synthesize PGL. Mice infected with this knockout strain survived longer, as compared to the wild-type strain. *In vitro* release of proinflammatory cytokines, such as TNF-α, IL-6, and IL-12, by bone marrow-derived macrophages was, as compared to macrophages infected with the mutant strain, inhibited in those infected by the wild-type strain. When this mutant was complemented with an intact *pks15/1* gene, PGL synthesis was restored and production of proinflammatory cytokines by macrophages decreased by 2–2.5 times.[39] From these experiments, it appears that PGL is an important factor in the pathogenesis of the W/Beijing family of *M. tuberculosis* and acts by modulating the innate immune response. Interestingly, strain NHN 5, a non-Beijing strain,[59] has an intact *pks15/1* gene but does not show detectable PGL synthesis, as determined by TLC analysis of crude lipid extract. This means that the particular deletion of seven base pairs is not the only mutation responsible for the lack of PGL synthesis. It would have been interesting to sequence the whole *pks15/1* gene of this particular strain and analyze it in comparison with others. It is not unlikely that a single nucleotide polymorphism is responsible for a disruption in its ORF. As indicated already by strain NHN 5[39] and strain Canetti,[22] an intact ORF in the *pks15/1* gene is not an exclusive property of the W/Beijing strain. In one of the aforementioned reports, 26 strains of *M. tuberculosis* were tested for the presence of an intact ORF in *pks15/1*.[53] It appeared that these strains all belonged to the defined genetic group 1, which is characterized by not having a mutation in *katG*[463] CTG.[60]

Among these strains were also strains that have the TbD1 region present such as strain Canetti (Figure 7). The TbD1 deletion marks the difference between ancestral and modern *M. tuberculosis* strains[53,61] (Figure 7). It was shown that considerable differences in virulence exist among strains within the same genotype.[57,58] Among these strains, the Beijing genotype was represented.

Survival rates of mice infected with the Beijing genotype strains were significantly lower, as compared to other genotypes, such as Haarlem and Somali. Macro- and microscopic inspection of lungs, liver, kidneys and spleen revealed much more pathology in samples infected with the Beijing genotype. Infection experiments in a rabbit/meningitis model showed more or less the same results.[62]

2.4 Evaluation

The role of PGL in pathogenesis of *M. leprae* is well established. In addition, for at least certain strains of *M. tuberculosis*, its role as a mediator of virulence through modulation of the immune response is established. The receptors of PGL have not been identified yet, nor is the molecular mechanism of action clarified. There is strong support for the hypothesis that the saccharide moiety is an essential part of the mole- cule. It is very likely to be the binding site recognized by the presumed receptors. In the 1970s, a study in India indicated that *M. tuberculosis* strains attenuated in guinea pigs were characterized by the presence of a phenolic PDIM without the oligosac- charide moiety, named attenuation indicator lipid.[63] As the name suggests, this lipid has been used as an indicator for reduced virulence. Since it only differs in absence of the trisaccharide part as compared to PGL, it supports a role of this oligosaccha- ride domain in PGL activity. Later, more substantial proof for the specific role of the sugar moiety was provided by showing that spleen cell proliferation in response to ConA induction was reduced in the presence of PGL, which suggests that PGL has a

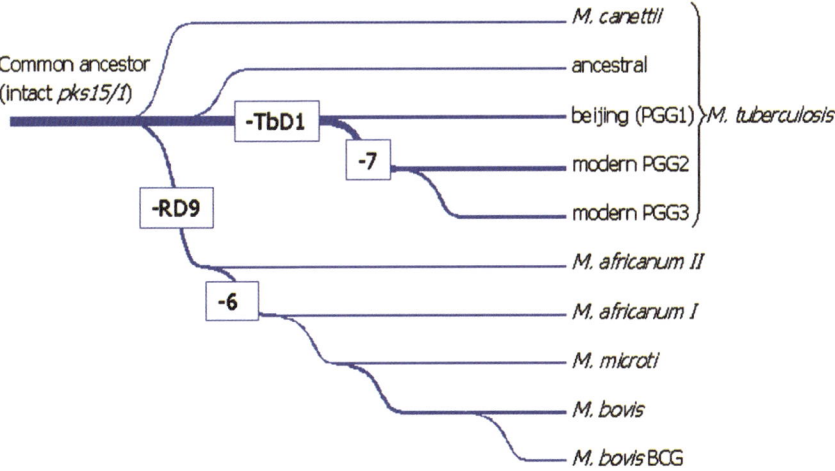

Figure 7 *Evolutionary scheme after Brosch et al.[61] and in analogy with Marmiesse et al.[53] in relation to the pks15/1 polymorphism*

direct effect on T-lymphocytes and their proliferation[15] (Figure 8B). Modifications of the sugar moiety of PGL, such as mono-deglycosylation and demethylation, reversed this effect, as was also true for blocking PGL with monoclonal antibodies directed to the oligosaccharide part. Meanwhile, removal of a mycocerosic acid side chain from the lipid core did not have an effect on the level of PGL-mediated suppression. In 2002, it was reported that monocyte-derived DCs, infected with *M. leprae*, presented PGL on their cell surface. These infected cells were able to suppress production of IFN-γ by T-lymphocytes as well as their proliferation (Figure 8C). The level of suppression of IFN-γ production and lymphocyte proliferation was dependent on the

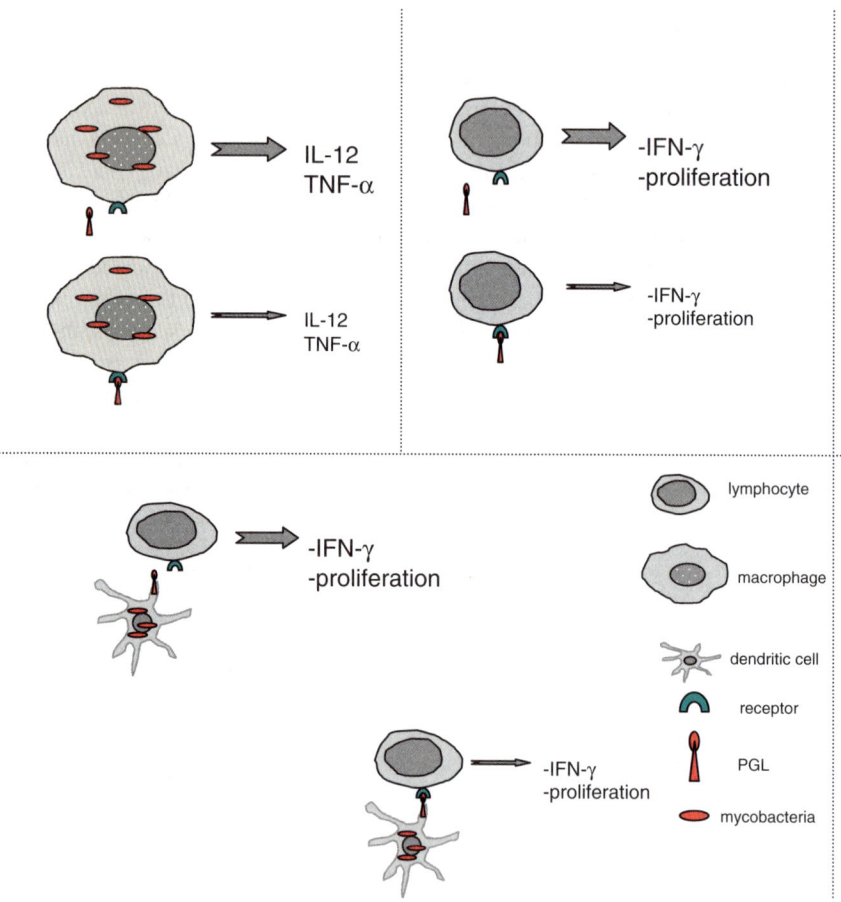

Figure 8 *Immune modulation of PGL. Three models: (A) Binding of PGL to a, as yet unidentified, (lectin) receptor on the surface of a macrophage results in the reduced production of IL-12 and TNF-α. (B) Binding of PGL to T-lymphocytes results in reduced production of IFN-γ and inhibition of lymphocyte proliferation. (C) PGL, presented on the surface of infected DCs or macrophages, binds to a receptor on the surface of T-lymphocytes. Subsequently, IFN- γ production is reduced and lymphocyte proliferation inhibited*

mycobacterial species the DCs were infected with. *M. avium* induced a strong IFN-γ response, *M. bovis* BCG partially suppressed this, and *M. leprae* completely blocked it. When PGL on the surface of DCs was blocked by Mabs directed against PGL, T-cell proliferation and IFN-γ response were restored subsequently.[43] These data would suggest a model where PGL is presented on the surface of DCs and that the oligosaccharide part is recognized by unidentified T-cell lectins. In another *in vitro* model, PGL was added to murine macrophages and was found to have a direct effect on the cytokine expression profile of these macrophages. Production of proinflammatory cytokines IL-6 and TNF-α was inhibited in a dose-dependent manner (Figure 8A). These studies suggest that macrophages express a PGL receptor, too. In addition, PGL of *M. tuberculosis* had a stronger suppressive effect than PGL of *M. bovis* BCG,[39] which confirms the fact that the trisaccharide moiety determines the inhibitory effect of PGL. Hence, one can conclude that PGL is important as a virulence factor within a number of mycobacterial species.

Within the species of *M. leprae*, the role of PGL is well established. So far, however, its role in differentiating the immune response in tuberculoid leprosy and lepramotous leprosy is not very likely, since it is present in all strains of *M. leprae*. Other strain-related factors and host factors might be of more importance in this respect. Within the *M. tuberculosis* complex, there is strong evidence that PGL accounts for a differential course in infection, both in experimental animals[39,62] and in diseased human TB patients.[64] Since PGL is also present in less virulent representatives of *M. tuberculosis*, such as strain Canetti and the Somali genotype,[58,57] one can conclude that PGL in the context of the Beijing strains is a major virulence factor. In the context of other strains, its effect might be obscured and/or attenuated by other features of that particular genotype, which results in a less virulent phenotype, despite the presence of PGL.

2.5 Future Prospects

1. So far, the effects of PGL *in vivo* and *in vitro* have been studied extensively, but its mode of action and signaling pathway is still a black box. The single most important problem to be solved is the nature of the receptors on the cell surface of host cells. Are Toll, C-type lectin, or G-protein coupled receptors involved? Do macrophages and T-cells express one and the same receptor or, alternatively, is there more than one? The type of receptor might give an indication for further research on the intracellular molecular signaling after initiation by PGL–cell surface interaction.

2. Apart from studying the effect of PGL on host cells and cytokine secretion, another line of research could be which factors influence the production of PGL by mycobacteria. Several studies indicate that the effect of PGL is dose dependent.[15,39,43] Which factors, be it in host or pathogen, determine how much PGL is actually secreted?

3. Is there some species specificity in PGL recognition? Are the PGL saccharide units recognized best by those hosts for which that particular mycobacterial species is a natural pathogen? For instance, is PGL of *M. leprae* recognized better by human cells as compared to murine cells? Does PGL of *M. bovis*, known

to affect human cells to only a limited degree, have a larger effect on immune cells of cattle?

4. Production of PGL is attributed to the presence of an intact ORF in the *pks15/1* gene. Yet, strain NHN 5 has an intact gene but does not produce PGL.[39] Sequence analysis of the *pks15/1* gene of this strain might reveal relevant single nucleotide polymorphisms, responsible for disrupting the ORF. Alternatively, expression of the *pks15/1* gene might also be influenced by other not-yet-identified gene products.

3 Lipoarabinomannan

3.1 Introduction

The mycobacterial surface glycolipid lipoarabinomannan (LAM; Figure 9) consists of a glycosyl phosphatidyl inositol (GPI) anchor, a mannan domain (an α1.6-linked main chain with α1.2- or sometimes1.3-mannose side chains) and a complex, branched α-arabinan domain ending in a single β-arabinose. This reducing β-arabinose is not substituted in *M. chelonae*; this form of LAM is called araLAM. In *M. smegmatis*, β-arabinose is nonstoichiometrically substituted with phospho-*myo*-inositol; this form is called PI-LAM. In various species (*M. tuberculosis, M. bovis, M. leprae, M. avium* subsp. *paratuberculosis, M. marinum, M. kansasii*), it is substituted (again nonstoichiometrically) with mannose residues also called caps, this form is called manLAM, and mannose cap length ranges from one to three residues. In some species, the majority of the caps consist of one or at most two mannoses (*e.g., M. marinum*), while in other species, such as *M. tuberculosis*, the cap is longer, that is, often two or three mannoses. The Man–Ara linkage is α1.5, and the Man–Man linkage in the cap is α1.2.

Several aspects of LAM have been studied extensively, such as structure analysis, genetics, enzymology, biosynthesis and organic synthesis of substructures. During

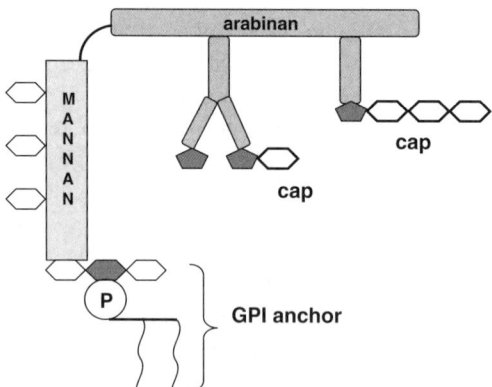

Figure 9 *Structure of LAM. LAM consists of a mannosylated (◇ mannose) GPI anchor containing inositol (◆). The arabinan domain ends in a β-arabinose ◆ . The mannose cap, composed of α1-2 interlinked mannose residues, is indicated with ◇*

the last years, in particular, the interaction between LAM and the host is being studied. The single most important conclusion drawn from LAM–host interaction studies is that manLAM contributes to persistence of mycobacteria by suppressing host innate immunity and that this effect is due to the mannose caps: manLAM but not PI-LAM was shown to downmodulate IL-12 and upmodulate IL-10 production.[65–67] In addition, two research groups have shown that manLAM but not araLAM inhibit phagolysosome fusion.[68,69] The state of the art of LAM research *anno* 2004 has been excellently reviewed by Briken *et al.*,[70] and hence the focus of this review will be on a very recent development in the field, that is, the use of synthetic LAM neoglyco-conjugates to analyze the nature of LAM–host interaction on the molecular level.

3.2 Synthetic LAM Oligosaccharides

The synthesis of the following LAM oligosaccharides has been shown in Figure 10.[71] These five compounds represent both the arabinan domain and the mannose caps of manLAM. Note also the amino-octyl spacer at the nonreducing end. This allows the

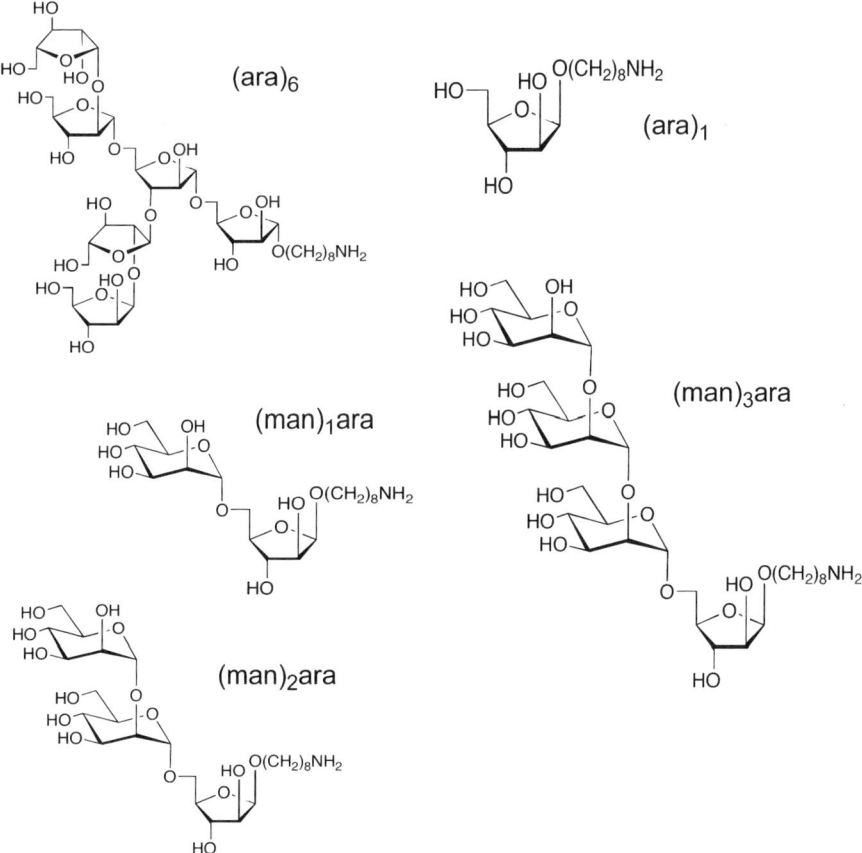

Figure 10 *Five LAM oligosaccharides that were synthesized*

incorporation of these oligosaccharides into neoglycoconjugates, for instance, coupling to bovine serum albumin or incorporation into biotin-labeled polyacrylamide (PAA) poly-mers. The synthesis route for the latter is rather simple and was pioneered by N. Bovin:[72] in a one-pot synthesis poly (*p*-nitrophenylacrylate) is mixed with the amino-octyl-oligosaccharides and with amino-alkyl-spacered biotin. This reaction drives to comple-tion, yielding polyvalent biotin-labeled neoglycopolymers with a molecular weight of 30 kDa and with 50–70 glycan chains/polymer (Figure 11). This type of conjugate can be used in various assay formats to study LAM–host interaction, such as fluorescence-acti-vated cell sorting (FACS) analysis and ELISA. Subsequently, these neoglycoconjugates were used to identify the molecular specificity of host lectins interacting with manLAM, as well as to study the specificity of anti-LAM monoclonal antibodies.

3.2.1   Synthetic LAM Oligosaccharides: ManLAM–Lectin Interactions

In our initial studies we found that manLAM- (but not PI-LAM-) modulated LPS-stimulated human DCs to increase IL-10 secretion. We wondered which part of manLAM was able to bind to the DCs. To answer this, DCs were incubated first with the biotin-labeled PAA neoglycoconjugates and then with a fluorescent streptavidin conjugate, after which FACS analysis took place.[73] The data (Figure 12) show that specifically the mannose cap is responsible for binding of manLAM to DCs. The evident question is the nature of lectins involved in the cap–DC interactions. The lectin DC-SIGN (DC-specific intercellular adhesion molecule nonintegrin, CD 209) is known to bind HIV gp120 via its high-mannose glycans chains (Figure 13). Strikingly, high-mannose chains carry a manα1.2manα1.2man trimannoside (D1 branch), which is a structure identical to the mannose cap of mycobacteria.

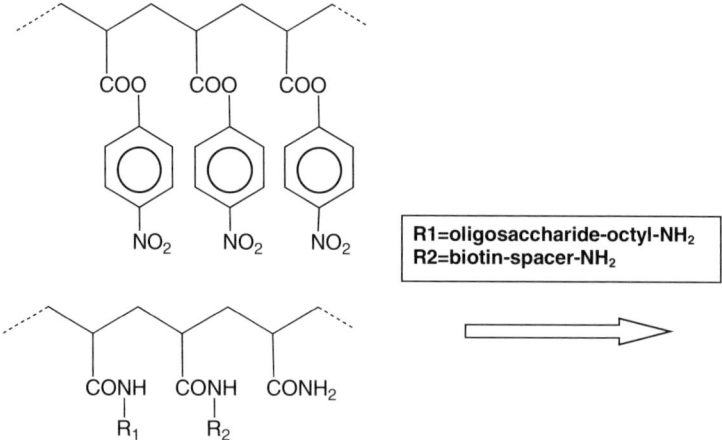

Figure 11 *Synthesis of polyacrylamide neoglycoconjugates. Poly (p-nitrophenylacrylate) (upper panel) is reacted with octylamino-spacered LAM oligosaccharide and amino-derivatized biotin. This results in a biotinylated polyacrylamide neogly-copolymer (lower panel)*

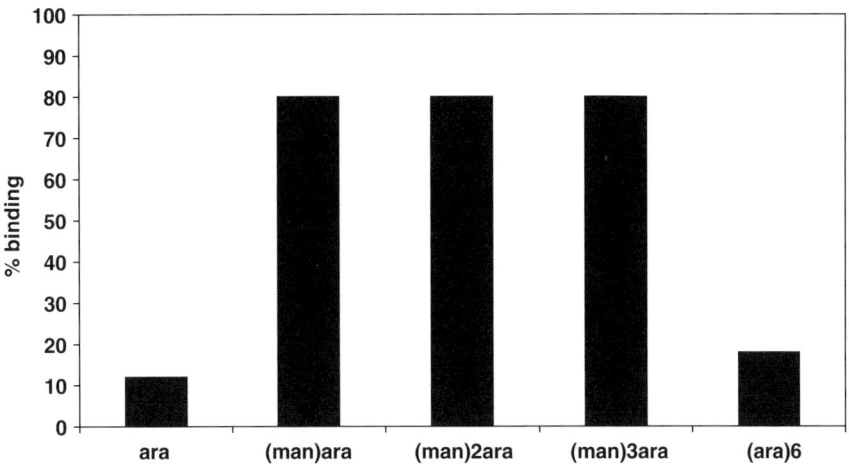

Figure 12 *Binding of biotinylated LAM PAA neoglycoconjugates to human DCs in FACS analysis. Vertical axis represents the percentage of DCs that binds to the LAM probe indicated on the horizontal axis*

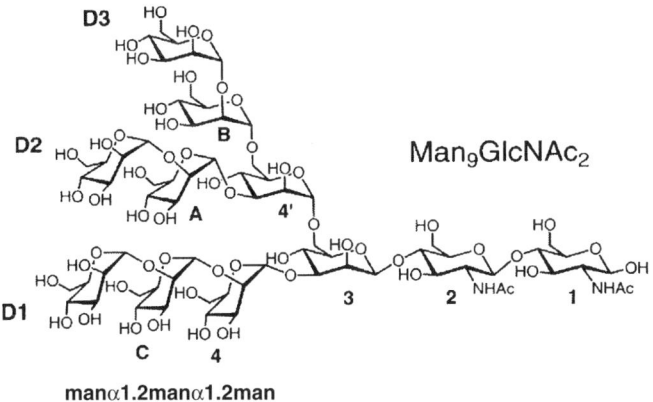

Figure 13 *Covalent structure of a high-mannose glycan*

Evidence in favor of the hypothesis that DC-SIGN is the DC lectin interacting with the cap was obtained in ELISA studies with synthetic cap PAA neoglycopolymers as coating and a DC-SIGN-Fc construct as a probe (Figure 14). Figure 14 shows that DC-SIGN binds better to a longer cap, which is in agreement with DC-SIGN binding studies to manLAM from a diversity of mycobacterial species and a diversity of various cap length.[74] However, also within a single strain, the natural LAM are all chemically diverse with regard to cap length, and these data show how synthetic LAM neoglycoconjugates are indispensable for defining LAM–lectin interactions on the molecular level.

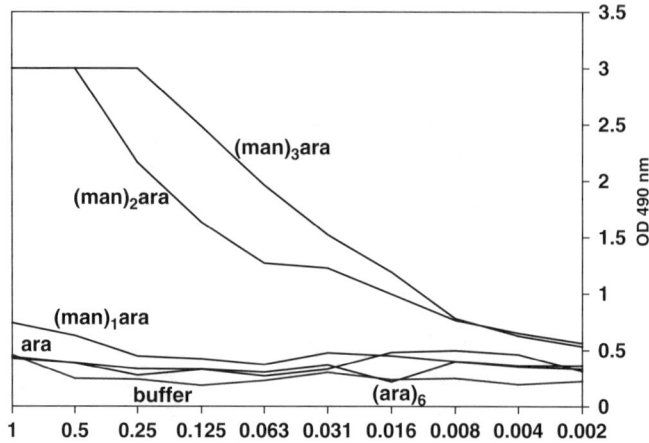

Figure 14 *Binding of DC-SIGN to synthetic LAM neoglycoconjugates in ELISA*

3.2.2   Synthetic LAM Oligosaccharides: ManLAM–Antibody Interactions

Protection against mycobacterial disease is commonly held to require an adequate T-cell response. However, a few past[75] and also more recent studies suggest that antibodies to mycobacterial surface glycans can protect, too. In 2003, Svenson and co-workers[76] demonstrated that an arabinomannan–protein conjugate was an effective anti-TB vaccine, and this vaccine together with a novel adjuvant, suitable for human use, is currently tested further and commercialized (http://www.eurocine.se). Subsequently, it was shown that passive protection with an anti-LAM monoclonal antibody yielded partial protection against experimental murine tuberculosis.[77] The molecular specificity of the protective Mab has not been described yet. As an example, we will exemplify below how we have used the synthetic LAM neoglycoconjugates to determine the epitope specificity of anti-LAM Mabs. Libraries of Mabs, obtained from mice, immunized with mycobacterial antigen were tested for reactivity in ELISA against our panel of neoglycoconjugates. The reactivity of one such Mab (CS-35) is shown in Table 2, which shows that the Mab is specific for the $(ara)_6$ motif and does not react with either monoarabinoside or the mannose cap. It binds to LAM and bacteria of both *M. tuberculosis* and *M. smegmatis*, which demonstrates that the capping is not at all complete and nonsubstituted arabinan remains available for interaction with anti-ara Mabs. Similar data for Mab cs-35 were described by Kaur *et al.*[78]

3.3   Future Prospects

Most studies on the effect of LAM on phagosome maturation or immunosuppression have compared PI-LAM (immunostimulatory, does not inhibit phagolysosome fusion) with manLAM (Immunosuppressive, inhibits phagolysosome fusion). Classically, differences between these two forms of LAM have been attributed to the

Table 2 *Reactivity of anti-LAM Mab CS-35 with natural LAMs and synthetic LAM neoglycoconjugates*

Antigen tested	Reactivity (ELISA data OD_{492})
ManLAM from *M. tuberculosis*	>2.5
PI-LAM from *M. smegmatis*	>2.5
M. tuberculosis cells	>2.5
M. smegmatis cells	>2.5
M. marinum cells	>2.5
Ara-PAA	<0.4
Ara-BSA	<0.4
$(Ara)_6$-PAA	>2.5
$(Ara)_6$-BSA	>2.5
Man-Ara-PAA	<0.4
Man-Ara-BSA	<0.4
$(Man)_2$-Ara-PAA	<0.4
$(Man)_2$-Ara-BSA	<0.4
$(Man)_3$-Ara-PAA	<0.4
$(Man)_3$-Ara-BSA	<0.4
$(Man)_2$-Ara-PAA	<0.4

mannose cap. Recent studies show that this is incorrect: inositol phosphate itself seems to be immunostimulatory and these studies should be repeated, but now with true araLAM (from *M. chelonae*).[79] However, even the use of *M. chelonae* LAM might not suffice as differences between manLAM and araLAM are also found in the mannan domain and the GPI anchor. For full evaluation of the role of the mannose cap, capless mutants should be generated.

It would be of interest to determine the epitope specificities of protective anti-LAM Mabs and potentially synthetic LAM vaccines yielding high titers of protective antibodies could be designed.

Although in human DCs, DC-SIGN is the dominant lectin that recognizes the mannose cap, for other human host cells (*e.g.*, macrophages or Langerhans cells) or equivalent host cells in other animal species of interest for tuberculosis research (mouse, Zebra fish), lectins other than DC-SIGN might be important. DC-SIGN-independent DC downregulation has also been described;[80] which receptors are involved is not known. The neoglycoconjugates described could be of value in finding novel lectins: after binding of these cap conjugates to the host cell of choice, lectin–cap complexes could be isolated with magnetic streptavidin beads and identified by mass spectrometry after sodium dodecyl sulphate-polyacrylamide gel electrophoresis (SDS-PAGE).

Acknowledgments

We thank A. Kolk Amsterdam and J. Belisle, Colorado State University, Fort Collins, CO, and the National Institutes of Health, Bethesda, MD (contract NO1 AI-75320) for providing mycobacterial antigens and Mabs. We also thank T. Lowary (Edmonton, Alberta, Canada) for oligosaccharide synthesis and Nicolai

Bovin (Moscow, Russia) and Ron Hokke (Leiden, Netherlands) for synthesis of neoglycoconjugates. We thank Wilbert Bitter for critically reading the manuscript.

References

1. G.A. Hansen, *Nor. Mag. Laegevidensk.*, 1874, **4**, 1.
2. D.N. Lockwood and S. Suneetha, *Bull. World Health Organ.*, 2005, **83**, 230.
3. R. Koch, *Berl. Klin. Wochenschr.*, 1882, **19**, 221.
4. P. Tiruviluamala and L.B. Reichman, *Annu. Rev. Publ. Health*, 2002, **23**, 403.
5. C. Dye, S. Scheele, P. Dolin, V. Pathania and M.C. Raviglione, *JAMA*, 1999, **282**, 677.
6. D.W. Smith, W.K. Harrell and H.M. Randall, *Am. Rev. Tuberc.*, 1954, **69**, 505.
7. D.W. Smith, H.M. Randall, A.P. Maclennan and E. Lederer, *Nature*, 1960, **186**, 887.
8. R.G. Navalkar, E. Wiegeshaus, E. Kondo, H.K. Kim and D.W. Smith, *J. Bacteriol.*, 1965, **90**, 262.
9. S.W. Hunter and P.J. Brennan, *J. Bacteriol.*, 1981, **147**, 728.
10. S.W. Hunter, T. Fujiwara and P.J. Brennan, *J. Biol. Chem.*, 1982, **257**, 15072.
11. X. Wu, J.R. Marino-Albernas, F.I. Auzanneau, V. Verez-Bencomo and B.M. Pinto, *Carbohydr. Res.*, 1998, **306**, 493.
12. T. Fujiwara, S.W. Hunter, S.N. Cho, G.O. Aspinall and P.J. Brennan, *Infect. Immun.*, 1984, **43**, 245.
13. A. Rambukkana, J.L. Salzer, P.D. Yurchenco and E.I. Tuomanen, *Cell*, 1997, **88**, 811.
14. A. Rambukkana, H. Yamada, G. Zanazzi, T. Mathus, J.L. Salzer, P.D. Yurchenco, K.P. Campbell and V.A. Fischetti, *Science*, 1998, **282**, 2076.
15. V. Mehra, P.J. Brennan, E. Rada, J. Convit and B.R. Bloom, *Nature*, 1984, **308**, 194.
16. M. Daffe, C. Lacave, M.A. Laneelle and G. Laneelle, *Eur. J. Biochem.*, 1987, **167**, 155.
17. M. Daffe, S.N. Cho, D. Chatterjee and P.J. Brennan, *J. Infect. Dis.*, 1991, **163**, 161.
18. J. Torgal-Garcia, H.L. David and F. Papa, *Ann. Inst. Pasteur. Microbiol.*, 1988, **139**, 289.
19. J. Torgal-Garcia, F. Papa and H.L. David, *Acta. Leprol.*, 1989, **7**(Suppl 1), 136.
20. J.S. Cox, B. Chen, M. McNeil and W.R. Jacobs Jr., *Nature*, 1999, **402**, 79.
21. M. Daffe and M.A. Laneelle, *J. Gen. Microbiol.*, 1988, **134**, 2049.
22. P. Constant, E. Perez, W. Malaga, M.A. Laneelle, O. Saurel, M. Daffe and C. Guilhot, *J. Biol. Chem.*, 2002, **277**, 38148.
23. M. Watanabe, Y. Aoyagi and A. Ohta, *Eur. J. Biochem.*, 1997, **248**, 93.
24. J.J. Fournie, M. Riviere and G. Puzo, *Eur. J. Biochem.*, 1987, **168**, 181.
25. J.J. Fournie, M. Riviere, F. Papa and G. Puzo, *J. Biol. Chem.*, 1987, **262**, 3180.
26. J.J. Fournie, M. Riviere and G. Puzo, *J. Biol. Chem.*, 1987, **262**, 3174.
27. M. Gilleron, A. Venisse, J.J. Fournie, M. Riviere, M.A. Dupont, N. Gas and G. Puzo, *Eur. J. Biochem.*, 1990, **189**, 167.
28. A. Vercellone and G. Puzo, *J. Biol. Chem.*, 1989, **264**, 7447.

29. G. Puzo, *Crit. Rev. Microbiol.*, 1990, **17**, 305.

30. S.W. Hunter and P.J. Brennan, *J. Biol. Chem.*, 1983, **258**, 7556.

31. M. Watanabe, Y. Yamada, K. Iguchi and D.E. Minnikin, *Biochim. Biphys. Acta.* 1994, **1210**, 174.

32. A.K. Azad, T.D. Sirakova, N.D. Fernandes and P.E. Kolattukudy, *J. Biol. Chem.*, 1997, **272**, 16741.

33. L.R. Camacho, P. Constant, C. Raynaud, M.A. Laneelle, J.A. Triccas, B. Gicquel, M. Daffe and C. Guilhot, *J. Biol. Chem.*, 2001, **276**, 19845.

34. O.A. Trivedi, P. Arora, A. Vats, M.Z. Ansari, R. Tickoo, V. Sridharan, D. Mohanty and R.S. Gokhale, *Mol. Cell.*, 2005, **17**, 631.

35. A.K. Azad, T.D. Sirakova, L.M. Rogers and P.E. Kolattukudy, *Proc. Natl. Acad. Sci. USA*, 1996, **93**, 4787.

36. E. Perez, P. Constant, A. Lemassu, F. Laval, M. Daffe and C. Guilhot, *J. Biol. Chem.*, 2004, **279**, 42574.

37. M. Daffe and P. Draper, *Adv. Microb. Physiol.*, 1998, **39**, 131.

38. P.F. Thurman, W. Chai, J.R. Rosankiewicz, H.J. Rogers, A.M. Lawson and P. Draper, *Eur. J. Biochem.*, 1993, **212**, 705.

39. M.B. Reed, P. Domenech, C. Manca, H. Su, A.K. Barczak, B.N. Kreiswirth, G. Kaplan and C.E. Barry, III, *Nature*, 2004, **431**, 84.

40. V. Ng, G. Zanazzi, R. Timpl, J.F. Talts, J.L. Salzer, P.J. Brennan and A. Rambukkana, *Cell*, 2000, **103**, 511.

41. D.S. Ridley and W.H. Jopling, *Int. J. Lepr. Other Mycobact. Dis.*, 1966, **34**, 255.

42. H.K. Prasad, R.S. Mishra and I. Nath, *J. Exp. Med.*, 1987, **165**, 239.

43. K. Hashimoto, Y. Maeda, H. Kimura, K. Suzuki, A. Masuda, M. Matsuoka and M. Makino, *Infect. Immun.*, 2002, **70**, 5167.

44. M.A. Neill and S.J. Klebanoff, *J. Exp. Med.*, 1988, **167**, 30.

45. M. Yamamura, K. Uyemura, R.J. Deans, K. Weinberg, T.H. Rea, B.R. Bloom and R. L. Modlin, *Science*, 1991, **254**, 277.

46. T.J. Kang, S.B. Lee and G.T. Chae, *Cytokine*, 2002, **20**, 56.

47. A.R. Santos, P.N. Suffys, P.R. Vanderborght, M.O. Moraes, L.M. Vieira, P.H. Cabello, A.M. Bakker, H.J. Matos, T.W. Huizinga, T.H. Ottenhoff, E.P. Sampaio and E.N. Sarno, *J. Infect. Dis.*, 2002, **186**, 1687.

48. E.G. Bligh and W.J. Dyer, *Can. J. Biochem. Physiol.*, 1959, **37**, 911.

49. S.N. Cho, J.S. Shin, M. Daffe, Y. Chong, S.K. Kim and J.D. Kim, *J. Clin. Microbiol.*, 1992, **30**, 3065.

50. N.M. Casabona, T.G. Fuente, L.A. Arce, J.O. Entraigas and R.V. Pla, *Acta Leprol.*, 1989, **7**, 89.

51. D. van Soolingen, L. Qian, P.E. de Haas, J.T. Douglas, H. Traore, F. Portaels, H.Z. Qing, D. Enkhsaikan, P. Nymadawa and J.D. van Embden, *J. Clin. Microbiol.*, 1995, **33**, 3234.

52. P.J. Bifani, B. Mathema, N.E. Kurepina and B.N. Kreiswirth, *Trends Microbiol.*, 2002, **10**, 45.

53. M. Marmiesse, P. Brodin, C. Buchrieser, C. Gutierrez, N. Simoes, V. Vincent, P. Glaser, S.T. Cole and R. Brosch, *Microbiology*, 2004, **150**, 483.

54. M.Y. Chan, M. Borgdorff, C.W. Yip, P.E. de Haas, W.S. Wong, K.M. Kam and D. van Soolingen, *Epidemiol. Infect.*, 2001, **127**, 169.

55. D.D. Anh, M.W. Borgdorff, L.N. Van, N.T. Lan, T. van Gorkom, K. Kremer and D. van Soolingen, *Emerg. Infect. Dis.*, 2000, **6**, 302.

56. P.J. Bifani, B.B. Plikaytis, V. Kapur, K. Stockbauer, X. Pan, M.L. Lutfey, S.L. Moghazeh, W. Eisner, T.M. Daniel, M.H. Kaplan, J.T. Crawford, J.M. Musser and B.N. Kreiswirth, *JAMA*, 1996, **275**, 452.

57. J. Dormans, M. Burger, D. Aguilar, R. Hernandez-Pando, K. Kremer, P. Roholl, S.M. Arend and D. van Soolingen, *Clin. Exp. Immunol.*, 2004, **137**, 460.

58. B. Lopez, D. Aguilar, H. Orozco, M. Burger, C. Espitia, V. Ritacco, L. Barrera, K. Kremer, R. Hernandez-Pando, K. Huygen and D. van Soolingen, *Clin. Exp. Immunol.*, 2003, **133**, 30.

59. C. Manca, L. Tsenova, A. Bergtold, S. Freeman, M. Tovey, J.M. Musser, C.E. Barry III, V.H. Freedman and G. Kaplan, *Proc. Natl. Acad. Sci. USA*, 2001, **98**, 5752.

60. S. Sreevatsan, X. Pan, K.E. Stockbauer, N.D. Connell, B.N. Kreiswirth, T.S. Whittam and J.M. Musser, *Proc. Natl. Acad. Sci. USA*, 1997, **94**, 9869.

61. R. Brosch, S.V. Gordon, M. Marmiesse, P. Brodin, C. Buchrieser, K. Eiglmeier, T. Garnier, C. Gutierrez, G. Hewinson, K. Kremer, L.M. Parsons, A.S. Pym, S. Samper, D. van Soolingen and S.T. Cole, *Proc. Natl. Acad. Sci. USA*, 2002, **99**, 3684.

62. L. Tsenova, E. Ellison, R. Harbacheuski, A.L. Moreira, N. Kurepina, M.B. Reed, B. Mathema, C.E. Barry, III and G. Kaplan, *J. Infect. Dis.*, 2005, **192**, 98.

63. M.B. Goren, O. Brokl and W.B. Schaefer, *Infect. Immun.*, 1974, **9**, 150.

64. F. Drobniewski, Y. Balabanova, V. Nikolayevsky, M. Ruddy, S. Kuznetzov, S. Zakharova, A. Melentyev and I. Fedorin, *JAMA*, 2005, **293**, 2726.

65. J. Nigou, C. Zelle-Rieser, M. Gilleron, M. Thurnher and G. Puzo, *J. Immunol.*, 2001, **166**, 7477.

66. L. Tailleux, O. Schwartz, J.L. Herrmann, E. Pivert, M. Jackson, A. Amara, L. Legres, D. Dreher, L.P. Nicod, J.C. Gluckman, P.H. Lagrange, B. Gicquel and O. Neyrolles, *J. Exp. Med.*, 2003, **197**, 121.

67. T.B. Geijtenbeek, S.J. Van Vliet, E.A. Koppel, M. Sanchez-Hernandez, C.M. Vandenbroucke-Grauls, B. Appelmelk and Y. Van Kooyk, *J. Exp. Med.*, 2003, **197**, 7.

68. I. Vergne, J. Chua, S.B. Singh and V. Deretic, *Annu. Rev. Cell. Dev. Biol.*, 2004, **20**, 367.

69. Z. Hmama, K. Sendide, A. Talal, R. Garcia, K. Dobos and N.E. Reiner, *J. Cell. Sci.*, 2004, **117**, 2131.

70. V. Briken, S.A. Porcelli, G.S. Besra and L. Kremer, *Mol. Microbiol.*, 2004, **53**, 391.

71. R.R. Gadikota, C.S. Callam, B.J. Appelmelk and T.L. Lowary, *J. Carbohydr. Chem.*, 2003, **22**, 459.

72. N.V. Bovin, E.Y. Korchagina, T.V. Zemlyanukhina, N.E. Byramova, O.E. Galanina, A.E. Zemlyakov, A.E. Ivanov, V.P. Zubov and L.V. Mochalova, *Glycoconj. J.*, 1993, **10**, 142.

73. E.A. Koppel, I.S. Ludwig, M.S. Hernandez, T.L. Lowary, R.R. Gadikota, A.B. Tuzikov, C.M. Vandenbroucke-Grauls, Y. Van Kooyk, B.J. Appelmelk and T.B. Geijtenbeek, *Immunobiology*, 2004, **209**, 117.

74. N. Maeda, J. Nigou, J.L. Herrmann, M. Jackson, A. Amara, P.H. Lagrange, G. Puzo, B. Gicquel and O. Neyrolles, *J. Biol. Chem.*, 2003, **278**, 5513.

75. R. Teitelbaum, A. Glatman-Freedman, B. Chen, J.B. Robbins, E. Unanue, A. Casadevall and B.R. Bloom, *Proc. Natl. Acad. Sci. USA*, 1998, **95**, 15688.

76. B. Hamasur, M. Haile, A. Pawlowski, U. Schroder, A. Williams, G. Hatch, G. Hall, P. Marsh, G. Kallenius and S.B. Svenson, *Vaccine*, 2003, **21**, 4081.

77. B. Hamasur, M. Haile, A. Pawlowski, U. Schroder, G. Kallenius and S.B. Svenson, *Clin. Exp. Immunol.*, 2004, **138**, 30.

78. D. Kaur, T.L. Lowary, V.D. Vissa, D.C. Crick and P.J. Brennan, *Microbiology*, 2002, **148**, 3049.

79. C. Vignal, Y. Guerardel, L. Kremer, M. Masson, D. Legrand, J. Mazurier and E. Elass, *J. Immunol.*, 2003, **171**, 2014.

80. M.C. Gagliardi, R. Teloni, F. Giannoni, M. Pardini, V. Sargentini, L. Brunori, L. Fattorini and R. Nisini, *J. Leukoc. Biol.*, 2005, **78**, 106.

CHAPTER 3

Structures and Roles of *Pseudomonas aeruginosa* Lectins

ANNE IMBERTY[1], MICHAELA WIMMEROVÁ[2], CHARLES
SABIN[1,3] AND EDWARD P. MITCHELL[3]

[1]CERMAV-CNRS (affiliated with Université Joseph Fourier), Grenoble BP53,
F-38041 Grenoble cedex 09, France
[2]National Centre for Biomolecular Research and Department of Biochemistry,
Masaryk University, Kotlarska 2, 611 37 Brno, Czech Republic
[3]E.S.R.F. Experiments Division, BP-220, F-38043 Grenoble cedex 09, France

Pseudomonas aeruginosa is a gram-negative bacterium, which is found in various environments including soil, water, and vegetation. It is also an opportunistic pathogen, responsible for numerous nosocomial infections in immunocompromised patients, for whom it may cause a wide number of diseases such as septicaemia, urinary tract infections, pancreatitis and dermatitis.[1] The bacteria colonise patients with chronic lung diseases as well as those under mechanical ventilation. These infections are fatal in cystic fibrosis patients. In addition, *P. aeruginosa* is capable of infecting many other organisms such as *Caenorhabditis elegans* and *Drosophila melanogaster*,[2] and even plants such as *Arabidopsis thaliana*.

For *P. aeruginosa* and many other pathogenic organisms, the ability to adhere to host tissues is essential to initiate an infection. Oligosaccharide-mediated recognition and adhesion are key points in the early steps of *P. aeruginosa* pathogenesis and are followed by processes such as chronic colonisation and biofilm formation together with alginate production under the control of quorum sensing.[3] These last adaptive changes of the bacteria make their eradication difficult with the bacteria becoming resistant to antibiotic drugs when in the biofilm phenotype.[4]

Carbohydrates play a central role in bacterial virulence. They are major components of lipopolysaccharide (LPS) – the element characteristic of gram-negative bacterial envelopes, which protects them against the diffusion of antibiotics. They also form the mucoid exopolysaccharide, composed of mannuronic and guluronic acid, which *P. aeruginosa* starts to produce in long-lasting infections. These carbohydrates can also be specific targets in the environment (mucins in the lumen lining mucosa) or on mucosal surfaces (receptors at the host cell surface) that the bacterium selects for binding through soluble virulence factors (soluble lectins) and adhesins present on fimbriae.

P. aeruginosa produces a wide variety of carbohydrate-binding proteins, but their role in recognition and adhesion is far from being elucidated (Figure 1). Several types of receptors have been identified: the unique flagellum of *P. aeruginosa* is involved in lung infection and it has been demonstrated that two of its proteins, flagellin, and flagellar cap protein, recognise mucin oligosaccharides.[5] Adhesins, located on pili, also play a role in carbohydrate-mediated adhesion. Pili or fimbriae are often involved in attachment to host surface.[6] Type IV pili (general secretion pathway) have been shown to recognise asialo-GM1 and -GM2 glycolipids[7] and to be involved in adhesion of *P. aeruginosa* to host.[8] Other types of pili (chaperone–usher pathway, Cup) are known to be involved in host glycan binding in a variety of pathogens through a lectin domain located at the tip of the pili.[9] Although not yet characterised in *P. aeruginosa*, sequence analyses indicate the presence of several *cup* fimbrial gene clusters in the *P. aeruginosa* genome,[10] but nothing is yet known about the carbohydrate-binding properties of the corresponding pili. The soluble lectins PA-IL (gene *lecA)* and PA-IIL (gene *lecB*), specific for galactose and fucose, respectively, have been characterised originally by Prof. Gilboa-Garber[11] and are the subject of the present review.

1 PA-IL: A Calcium-Mediated Galactose-Binding Lectin

1.1 Characteristics and Specificity

The galactophilic lectin PA-IL was the first *P. aeruginosa* lectin to be isolated by affinity chromatography using a sepharose column.[12] The *lecA* gene coding for PA-IL was cloned and the translation product was shown to consist of 121 amino acids (excluding initiator methionine) with a molecular weight of 12.75 kDa.[13] The localisation of the lectin is predominantly intracellular although the presence of the lectin on the outer membrane enabled intact cells of *P. aeruginosa* to agglutinate papain-treated human erythrocytes.[14]

This lectin has a narrow specificity spectrum for D-galactose-containing molecules and the sugar binding activity is dependent upon divalent cations.[15] The affinity for galactose is in the medium range with an association constant (K_a) of 3.4 $\times$ 10^4 M^{-1} as reported from an equilibrium dialysis study.[16] PA-IL has a preference for the α-anomer of galactose and binds to disaccharide in a decreasing order as follows: αGal1-6Gal (melibiose) > αGal1-4Gal (galabiose) > αGal1-3Gal[17] (Table 1). The lectin has been demonstrated to bind strongly to glycosphingolipids containing terminal and nonsubstituted αGal1-3Gal or αGal1-4Gal structures.[18] It also recognises the most common human antigens: I, B and especially Pk which contains the αGal1-4Gal moiety. This specificity makes it an interesting tool in the field of xenotransplantation since it can label the αGal1-3Gal epitope, *i.e.*, the xeno-antigen responsible for hyperacute rejection when grafting pig organs into human or apes.[19] The fact that PA-IL also displays binding activity with micromolar affinity towards hydrophobic ligands containing aromatic rings was demonstrated by Stoitsova *et al.*[20]

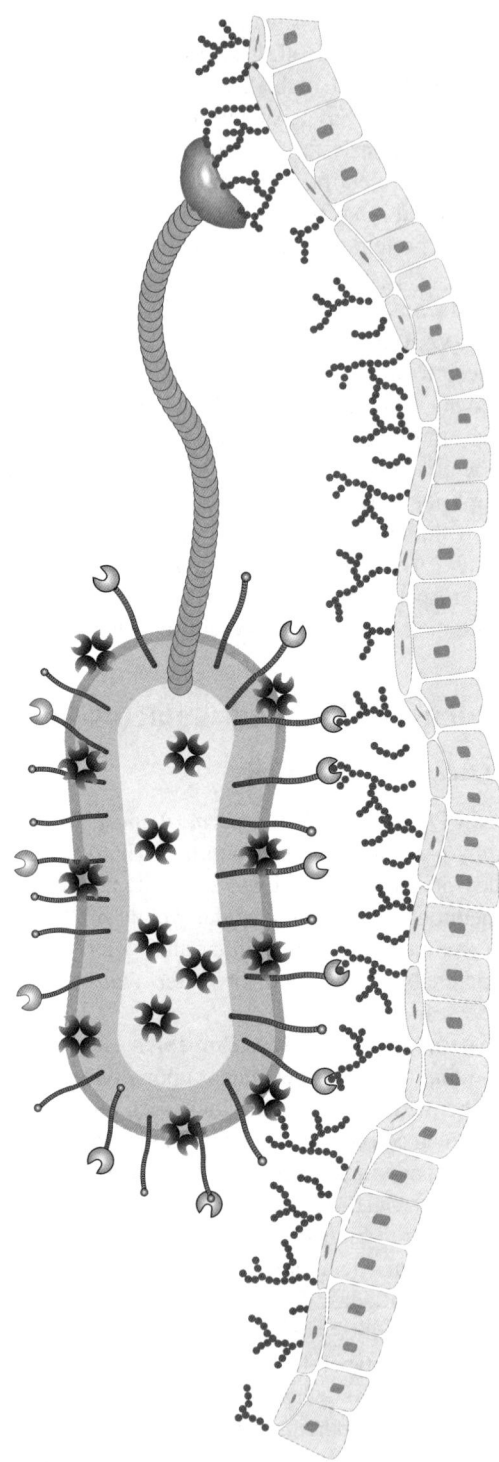

Figure 1 *Schematic representation of one P. aeruginosa bacterium with different carbohydrate-binding proteins interacting with glycoconjugates on a host cell surface*

Table 1 *Carbohydrate specificity of PA-IL.[16,17] Reference value is fixed for galactose (1.0 for a IC_{50} of 40 nM for inhibition of PA-IL binding to hydatid cyst glycoprotein)*

Inhibitor	Potency
Phenyl-β-Gal	57.1
αGal1-6Glc (melibiose)	13.3
αGal1-3αGal-*O*-Me	4.7
αGal-*O*-Me	2.7
βGal-*O*-Me	2.2
αGal1-4Gal	1.8
D-galactose	1.0
αGal1-3Gal	0.8
D-GalNAc	0.5
βGal1-4Glc	0.5
D-fucose	0.02

with adenine and naphthalene derivatives. The binding site, with a stochiometry of one molecule per PA-IL monomer, is independent from the galactose-binding site and can also accommodate acyl homoserine lactones,[21] which are the quorum-sensing signal molecules responsible for cell–cell communication in bacteria. Such a dual function for a lectin with a hydrophobic site able to bind phytohormones, independent from the carbohydrate-binding site, has been shown previously for plant lectins, *e.g.*, the *Dolichos biflorus* lectins.[22]

1.2 Structure of PA-IL

PA-IL has been crystallised in the presence of calcium,[23] and the structure, solved at 1.5-Å resolution, is deposited in the Protein Data Bank[24] with code 1L7L. The structure consists of a tetramer containing four identical monomers assembled by 222 symmetry (Figure 2). Each monomer adopts a small jellyroll-type β-sandwich fold, consisting of two curved sheets, each of four antiparallel β-strands. Tetramerisation occurs by interaction between the largest sheets for one interface and by contacts between C-terminus moieties for the other interface. Each monomer contains one calcium ion located in the region 100–108 of the amino acid sequence, which forms a loop and a short one-turn α-helix. The side chains of Asp100, Asn107, and Asn108 participate in the coordination of calcium together with the main chain carbonyl of Thr104. A fifth contact is established by the carbonyl main chain group of Tyr36, which is located on a neighbouring loop.

The structure of the complex of PA-IL with galactose has been solved at 1.6 Å resolution (PDB code 1OKO) and showed the presence of one calcium ion and one galactose ligand in the same binding site (Figure 2).[25] Two oxygen atoms of galactose, *i.e.*, O3 and O4 are involved directly in the coordination sphere of the calcium ion. From the calcium-binding loop (100–108), Asp100 establishes contact with O4 of galactose and Asn107 forms hydrogen bonds with O2 and O3, while Val101 and

A   B   C

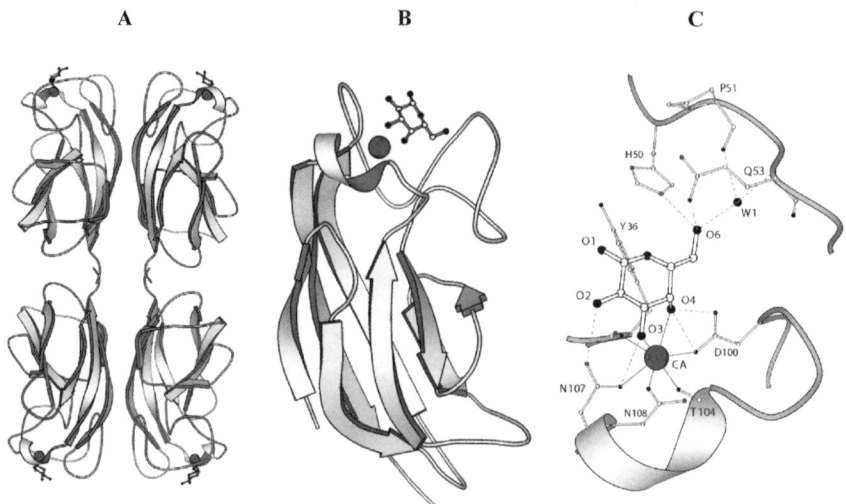

Figure 2 *(A) Ribbon representation of the tetrameric PA-IL/galactose complex with stick repre-*
sentation of sugar and cpk representation of the calcium ions (PDB code 1OKO).[25] (B)
Representation of one monomer. (C) Ligand-binding site: coordination contacts are
shown as solid lines and hydrogen bonds as dashed lines. Graphical representations
were prepared with the use of MOLSCRIPT program[74]

Thr104 are involved in hydrophobic contacts. Tyr36, also participating in coordination of the calcium ion through its carbonyl backbone oxygen, makes a
hydrophobic contact with C2 of galactose. Another part of the same long loop
(36–64) interacts specifically with the O6 hydroxymethyl group that establishes
hydrogen bonds with His50, Gln53, and one water molecule making a bridge to the
protein backbone.

1.3   Comparison with other Proteins

A search for sequences similar to PA-IL yields only one significant hit in the sequence
genome of *Pseudorhabdus luminescens*, which is an insect pathogen. The unknown
hypothetical protein Plu2096 displays 36% identity and 48% similarity with PA-IL
and all amino acids involved in calcium binding are conserved (Figure 3). At three-
dimensional level, PA-IL belongs to the "Galactose-binding domain-like" fold family as classified in the SCOP database.[26] This family contains a large number of
small nine-stranded β-sandwich proteins, most of them being involved in carbohydrate recognition. With the exception of the fucose-binding lectin from eel, all of
these carbohydrate-binding β-sandwiches are not lectins, but carbohydrate-binding
modules (CBMs) that are usually domains of larger proteins with glycosylhydrolase
activities. One of these CBMs shows interesting similarities with PA-IL since it
binds xylo-oligosaccharide through contacts with calcium.[27] The binding site is
located on the same apex of the β-sandwich as for PA-IL, but there are no similarities in the loops and sequences involved (Figure 4). Another example of carbohydrate

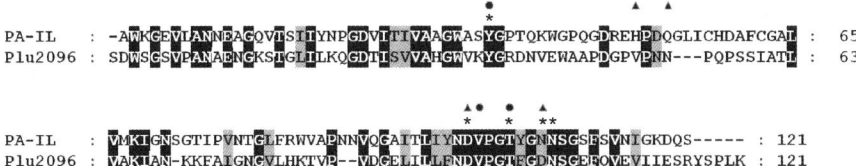

```
                                         •             ▲   ▲
                                         *
PA-IL    : -AWKGEVLANNEAGQVLSILYNPGDVITTVAAGWASYGPTQKWGPQGDREHPDQGLICHDAFCGAL :  65
Plu2096  : SDWSCSVPANAENGKSIGLILKQGDTISVVAHGWVKYGRDNVEWAAPDGPVPNN---PQPSSIATL :  63


                                       ▲•    •    ▲
                                       *     *    **
PA-IL    : VMKIGNSGTIPVNTGLFRWVAPNNVQGAITLIYNDVPGTYGNNSGSFSVNIGKDQS----- : 121
Plu2096  : VAKIAN-KKFAIGNGVLHKTVP--VDGELILLFNDVPGTFGDNSGEFQVEVIIESRYSPLK : 121
```

Figure 3 *Sequence alignment of PA-IL with a similar sequence from the genome of Photorhabus luminescens. Strictly conserved amino acids are shown with a black background and similar ones with a grey background. Amino acids involved in calcium binding are indicated by stars. Amino acids interacting with the galactose ligand by hydrogen bonds and hydrophobic contacts are indicated by triangles and circles, respectively*

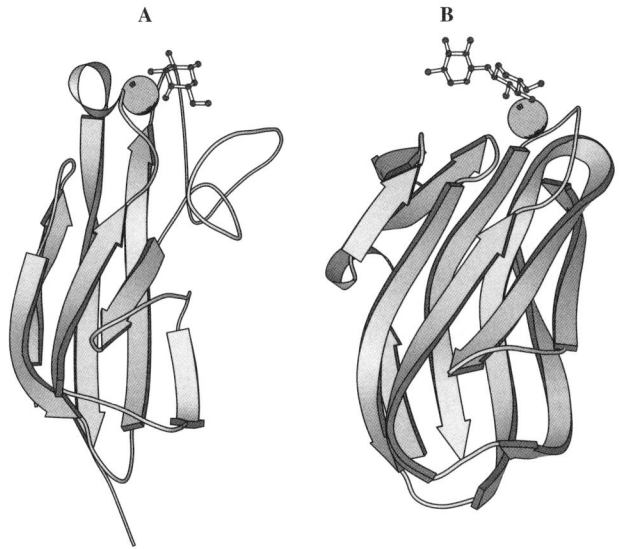

Figure 4 *Comparison of two calcium-dependant carbohydrate-binding proteins with a β-sandwich fold. (A) PA-IL complexed with calcium and galactose (PDB code 1OKO).[25] (B) CBM36 from Paenibacillus polymyxa xylanase (PDB code1UX7)[27] complexed with calcium and xylotriose*

binding mediated by one calcium ion is the C-type lectins, a family of proteins that are widely distributed in animal kingdom, but not in yeast.[28] The carbohydrate-binding domain (CRD) of C-type lectins shows specific calcium-dependent binding of various monosaccharides with dissociation constants in the millimolar range. The majority of C-type lectins can be divided into D-mannose-binding (also recognising D-glucose and L-fucose) and D-galactose-binding ones. Despite the lack of similarity at the sequence or at the fold level, the galactose-binding mode of PA-IL shows a striking resemblance to the ones observed in the crystal structure of rat mannose-binding protein A (APDWG mutant) and TC14 from the tunicate *Polyandrocarpa misakiensis*, which have both been crystallised with galactose.[29,30]

1.4   Roles of PA-IL in Bacterial Infection

The PA-IL lectin possesses specific characteristics that allow it to be described as a virulence factor. The expression of many virulence genes in *P. aeruginosa* depends on a system termed "quorum sensing," which is utilised for cell–cell communications in many bacteria.[31] Small molecules, termed autoinducers, are produced by the bacterial cell and accumulate in the environment at a high population density, until a threshold level is reached. In *P. aeruginosa*, quorum sensing has been shown to regulate the production of virulence factors, to be involved in biofilm formation and development and to be implicated in antibiotic resistance.[32] In this bacterium, two separate quorum-sensing circuits, named *las* and *rhl*, modulate gene transcription in response to two different molecules of the *N*-acylhomoserine lactone family (AHL). The production of PA-IL lectin was found to be directly dependent on the *rhl* locus and expression of the *lecA* gene coding for PA-IL can be activated by the corresponding AHL.[33] These results were confirmed with the recent analysis of the *P. aeruginosa* transcriptome. Since the sequence of the *P. aeruginosa* strain PAO1 genome was published,[34] several teams have analysed the transcriptional responses of *P. aeruginosa* genes to the presence of AHL signalling components using DNA microarray chips. Two independent studies indicated that the expression of the *lecA* gene is enhanced by 36- to 200-fold when the quorum sensing system is activated.[35,36]

The question whether PA-IL could act directly as a virulence factor was assessed on respiratory epithelial cells in primary culture.[37] The lectin inhibited the growth of respiratory cells and decreased the percentage of activated ciliated cells. It could therefore be demonstrated that PA-IL has a dose-dependent cytotoxic effect and could directly damage airway epithelia during respiratory infection. In a mouse model of gut-derived sepsis, the injection of *P. aeruginosa* strains lacking functional PA-IL were nonlethal, whereas the wild-type bacteria were lethal.[38] Using the same model, it could be demonstrated that PA-IL injected into the mouse cecum does not induce mortality, whereas the same lectin associated with other toxins such as exotoxin A and elastase result in a high yield of mortality.[39,40] This effect can be inhibited by the presence of GalNAc, a monosaccharide that binds to PA-IL almost as effectively as galactose. It has been concluded that the lectin acts by permeabilising the epithelial cells, permitting a much stronger action of toxins.

Finally, PA-IL may also directly play a role in adhesion to tissue since it has been demonstrated that *P. aeruginosa* adheres to glycosylated fibronectin.[41] This process is enhanced by neuraminidisase treatment, resulting in the exposure of terminal galactose residues,[41] and is inhibited by the addition of several monosaccharides, which compete for the PA-IL-binding site, or of PA-IL itself, which competes for the fibronectin site.

2   PA-IIL: A High-Affinity L-Fucose-Binding Protein

2.1   Characteristics and Specificity

A mannose- and fucose-binding lectin was purified from extracts of *P. aeruginosa* by affinity chromatography on a D-mannose column.[11] From the *N*-terminus sequence, the gene was cloned as a single ORF encoding a 114 amino acid protein (excluding initiator methionine) with molecular mass of 11,732 Da.[42] When monosaccharides were tested in haemagglutination studies, the strongest ligand was L-fucose.[43] However, the lectin does not have a very limited specificity and other monosaccharides could also inhibit the haemagglutination with the following order of effectiveness: L-fucose > L-galactose > D-arabinose > D-fructose > D-mannose and no significant activity for D-galactose. When oligosaccharides were tested, PA-IIL showed a clear preference for human histo-antigens of the Lewis a (Le^a) series[44] over those of the Lewis x (Le^x) series, therefore, indicating that the αFuc1-4GlcNAc disaccharide of Le^a is bound tighter than the αFuc1-3GlcNAc disaccharide of Le^x.

Recently, the specificity study was extended to human milk oligosaccharides (HMO). Human milk contains a significant amount of structurally diverse oligosaccharides, most of them being fucosylated neutral oligosaccharides carrying lactose at their reducing end.[45] Since HMO are soluble analogues of epithelial cell surface glycoconjugates, they competitively inhibit the binding of pathogenic bacteria and viruses to epithelial ligands,[46] and it has been recently demonstrated that human milk, and not cow milk, specifically blocks haemagglutination by PA-IIL.[47] Figure 5 shows the results obtained with the enzyme-linked lectin assays (ELLA) method, which uses competition against biotinylated polymeric fucose.[48] Oligosaccharides containing a fucose residue linked to position 2 of galactose are poor inhibitors as are the Le^x series, which contain a fucose on position 3 of a GlcNAc. On the other hand, the Le^x glucose analogue series (Le^x g.a), where the GlcNAc is replaced by Glc, and the Le^a series that contain fucose linked on position 4 of a GlcNAc, are respectively six times and ten times stronger inhibitors than fucose. It could therefore be concluded that PA-IIL differentiates fucose-containing oligosaccharides with a preference in decreasing order of αFuc1-4GlcNAc > αFuc1-3Glc > Fuc > αFuc1-3GlcNAc > αFuc1-2Gal.

2.2 Affinity and Thermodynamics

The PA-IIL lectin is characterised by an unusually high affinity for carbohydrates. Early evaluations performed with equilibrium dialysis indicated a K_a of 1.5×10^6 M^{-1} for binding to fucose,[43] a value that is much higher than those classically observed for protein–carbohydrate interactions. The initial affinity studies have been enhanced with results from isothermal titration calorimetry (ITC), a method that has been demonstrated to be very well adapted to protein–carbohydrate interactions.[49] This approach not only allowed the affinity constants to be refined, but also gave direct measurements on the interaction thermodynamics, *i.e.*, values of the enthalpy and entropy contributions (Figure 6). For the binding of PA-IIL with fucose, affinity constants varied from 0.18×10^6 to 0.34×10^6 M^{-1} depending on the protein batch.[48,50] These values are lower than those previously obtained by equilibrium dialysis but confirmed the micromolar range of affinity. Oligosaccharides of the Lewis a series were confirmed to be the best ligands for the PA-IIL lectin with K_a for Le^a trisaccharide of 4.7×10^6 M^{-1} ($K_d = 2.1 \times 10^{-7}$ M). The thermodynamic calculations (Table 2) show that the interaction of PA-IIL with carbohydrate ligands is characterised by a strong enthalpy of binding and an entropy term that is either negligible or slightly favourable. This is unusual in protein–carbohydrate interactions where

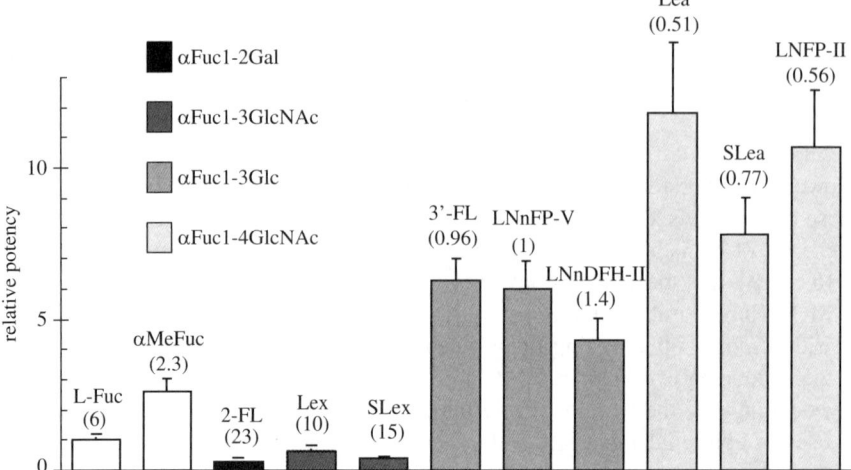

Figure 5 *Potency of different sugars when L-fucose is taken as a reference at 1.0. IC$_{50}$ values in μM (between parentheses) represent the inhibitory concentration giving 50% inhibition as determined from the inhibition curves. Abbreviations are as follows: αMeFuc, methyl-α-L-fucopyranoside; 2-FL, 2-fucosyllactose (αFuc12βGal14Glc); Le^x, Lewis x (βGal14[αFuc13]GlcNAc); SLe^x, sialyl Lewis x (αNeuAc23βGal14[αFuc13] GlcNAc); 3'-FL, 3'-fucosyllactose (βGal14[αFuc13]Glc); LNnFP-V, Lacto-N-neofucopentaose-V (βGal14βGlcNAc13Gal14[αFuc13]Glc); LNnDFH-II, Lacto-N-neodi-fucohexaose-II (βGal14[αFuc13]βGlcNAc13Gal14[αFuc13]Glc); Le^a, Lewis a (βGal13[αFuc14]GlcNAc); SiaLe^a, sialyl Lewis a (αNeuAc23βGal13 [αFuc14]GlcNAc); LNFP-II, Lacto-N-fucopentaose II (βGal13[αFuc14]βGlc NAc13βGal14 GlcNAc). Each value has been averaged from at least three independent measurements. (Reprinted from S. Perret et al., Biochem. J., 2005, **389**, 325, with permission from the Royal Biochemical Society)*

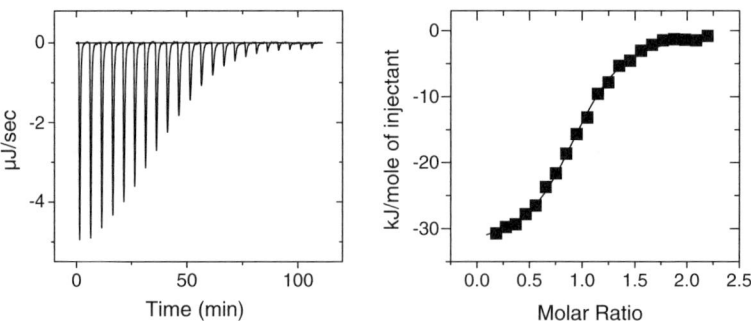

Figure 6 *Titration calorimetry results of L-fucose (0.500 mM) binding to PA-IIL (50.0 μM) in 100 mM Tris buffer containing 0.03 mM CaCl$_2$, pH 7.5 at 25 °C. Left, data obtained from 22 automatic injections, 13 μL each, of L-Fuc, into PA-IIL in the cell. Lower, plot of the total heat released as a function of total ligand concentration for the titration shown above (squares). The solid line represents the best least-square fit for the obtained data.*

Table 2 *Thermodynamics: ITC for PA-IIL with various oligosaccharides.[48] Standard deviations between several experiments are lower than 8%. Oligosaccharides are defined in legend of Figure 5*

	K_a $(10^6\ M^{-1})$	K_d $(10^{-6}\ M)$	n	$-\Delta G$ $(kJ\ mol^{-1})$	$-\Delta H$ $(kJ\ mol^{-1})$	$-T\Delta S$ $(kJ\ mol^{-1})$
Fucose	0.34	2.9	0.96	31.5	31.2	−0.3
Le^a	4.7	0.21	1.08	38.06	34.95	−3.11
3-FucLac	2.7	0.37	1.02	36.7	40.0	3.3
LNnFP-V	1.56	0.64	1.02	35.33	35.6	0.3
LNFP-II	2.01	0.50	0.94	35.96	37.8	1.8
Le^x	0.29	3.4	0.95	31.1	22.3	−8.8

several factors, usually attributed to either flexibility of the ligand[51] or to the displacement of water molecules from amphiphilic surfaces,[52] generally yield an unfavourable entropy term.

2.3 Crystal Structures of PA-IIL

The crystal structure of PA-IIL was first obtained in complex with fucose.[44] The PA-IIL crystals contain a tetramer of four independent subunits. Each monomer folds as a nine-stranded antiparallel β-sandwich that includes the so-called Greek-key structural motif. The quaternary structure is a tetrameric arrangement around a pseudo C222 axis of symmetry (Figure 7). Dimerisation is the result of a head-to-tail association of two monomers making contacts through the curved five-stranded β-sheets. The C-terminal carboxyl group (Gly114) of each monomer is closely associated to the ligand-binding site of the other monomer. Two dimers form a tetramer by antiparallel association from one of their strands with the counterpart dimer.

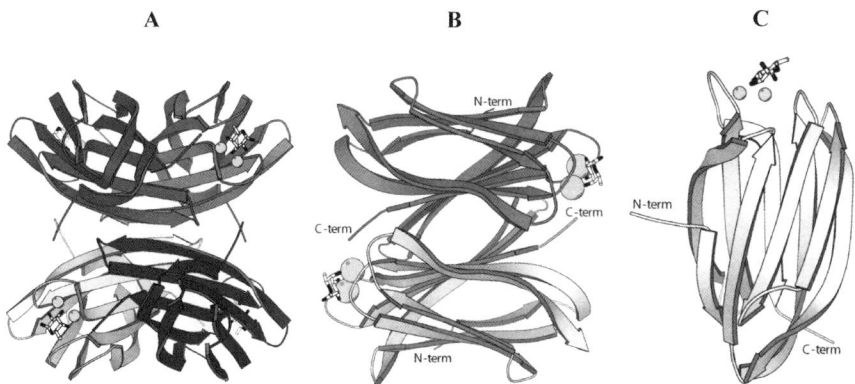

Figure 7 *(A) Ribbon representation of the PA-IIL/fucose complex tetramer with stick representation of sugar and cpk representation of the calcium ions (PDB code 1GZT).[44] (B) Representation of the tightly associated dimer. (C) Representation of one monomer*

The crystal structure of the native protein contains two calcium ions located close together (3.8 Å) in an acidic pocket at one top of each sandwich together with the C-terminus of the adjacent monomer.[53] Both calcium ions have a classical seven-ligand coordination with ten protein oxygen atoms, mainly from aspartate side chains, and three water molecules are involved. One metal-binding loop (amino acids 95–103) is mainly involved in the binding of calcium ions. This loop appears partially disordered in the crystal structure of demetallised PA-IIL.[53]

In the structure of the PA-IIL–fucose complex,[44] the monosaccharide is locked between both calcium ions, a binding mode that is unique among protein–carbohydrate interactions (Figure 8). Three of the fucose hydroxyl groups participate in the coordination of calcium ions: O2 belongs to site 1, O4 to site 2, and O3 to both sites. In addition to these electrostatic interactions, the same three hydroxyl groups participate in several hydrogen bonds with acidic groups of the calcium-binding site and the methyl group at position 6 of fucose makes favourable interactions with a small hydrophobic patch on the protein. Crystal structures were also determined for complexes with mannose and fructopyranose, yielding the molecular basis for the wide specificity of PA-IIL.[53] It appears that fucose, mannose, and fructopyranose are bound isosterically via three similarly oriented hydroxyl groups (two equatorial and one axial) that interact with the two calcium ions (Figure 8).

To understand the basis for very high affinity, the crystal structure of the tetrameric PA-IIL in complex with fucose and calcium was determined to 1.0 Å resolution.[50] From this atomic-level structure it was possible to propose locations for hydrogen atoms in the binding site and to model the hydrogen bond network. As a result, the role in stabilising the complex of a particular water molecule bridging between the fucose molecule and the protein backbone has been highlighted. Calculations of partial charges using *ab initio* computational chemistry methods demonstrated an extensive delocalisation of charges between the calcium ions, the fucose residue and aspartate and glutamate side chains. This approach rationalised the very specific role of the two calcium ions in this unusual binding mode: on the one hand, they provide charge delocalisation, associated with strong enthalpy contribution in binding, and on the other, their desolvation upon binding provide a

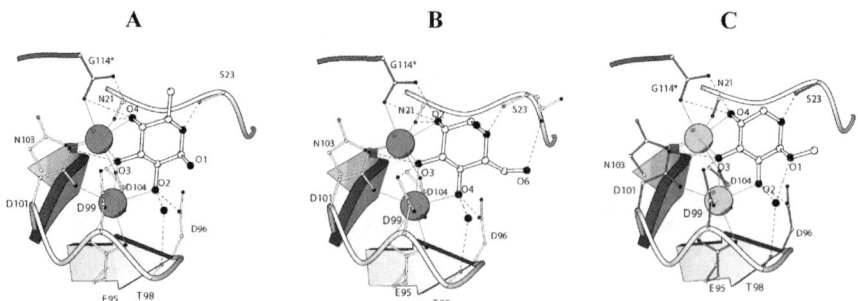

Figure 8 *Comparison of PA-IIL-binding modes for different monosaccharides. (A) Complex with α-L-fucose (PDB code 1UZV).[50] (B) Complex with α-D-mannose (PDB code 1OUR).[53] (C) Complex with β-D-fructose (PDB code 1OVP)[53]*

favourable entropy contribution that is seldom observed for protein–carbohydrate interaction.

PA-IIL has also been structurally characterised with two oligosaccharides to analyse the basis of specificity for milk oligosaccharides.[48] As predicted from the previous modelling study, Lewis a (βGal1-3[αFuc1-4]GlcNAc), which is the highest-affinity ligand known to date, establishes an additional hydrogen bond with the protein through the O6 oxygen of the GlcNAc residue (Figure 9). The galactose moiety does not play a direct role in the binding, but its presence rigidifies the trisaccharide, and therefore explains the highest observed favourable entropy contribution for this complex. A crystal structure of PA-IIL complexed with the human milk pentasaccharide lacto-*N*-neofucopentaose-V (LNnFP-V) was also refined at very high resolution (1.05 Å) and demonstrated how the Lewis x glucose analogue (βGal1-4[αFuc1-3]Glc) present in high quantity in human milk can mimic the shape of Lewis a and bind in an isosteric manner to the PA-IIL binding site. The X-ray crystal structure is in agreement with the thermodynamic data (Table 2) with a large favourable enthalpy of binding but slightly unfavourable entropy that is attributable to the flexibility of the pentasaccharide.

2.4 Comparison with other Proteins

Screening for PA-IIL sequence homologues in the genomes of other bacteria gave a small number of positive hits (Figure 10). The genomes of *Ralstonia solanacearum*, a plant pathogen, and *Chromobacterium violaceum*, an opportunistic human pathogen, contain one copy of the gene coding for a protein very similar to PA-IIL with all of the necessary amino acids for sugar and calcium binding.[54] Another gram-negative bacterium, *Burkholderia cenocepacia*, responsible for particularly dangerous

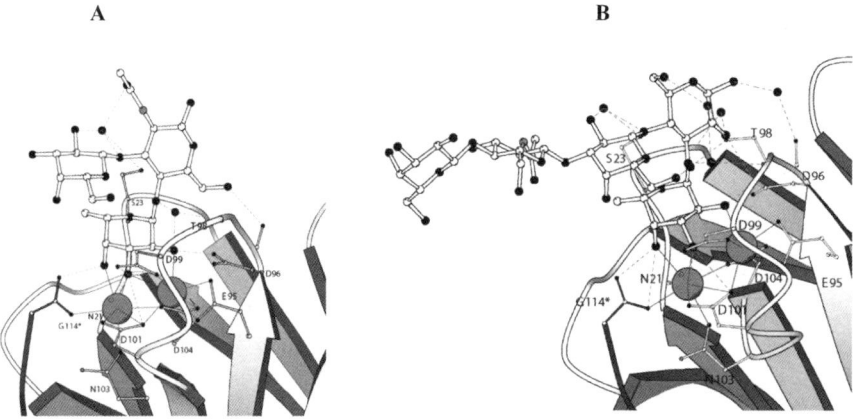

Figure 9 *(A) Representation of the binding site of PA-IIL interacting with calcium ions and Lewis a trisaccharide (PDB code 1W8H).[48] Coordination contacts are displayed as solid lines and hydrogen bonds as dashed lines. (B) Same representation for the interaction of PA-IIL with calcium ions and LNnFP-V pentasaccharide (PDB code 1W8F)[48]*

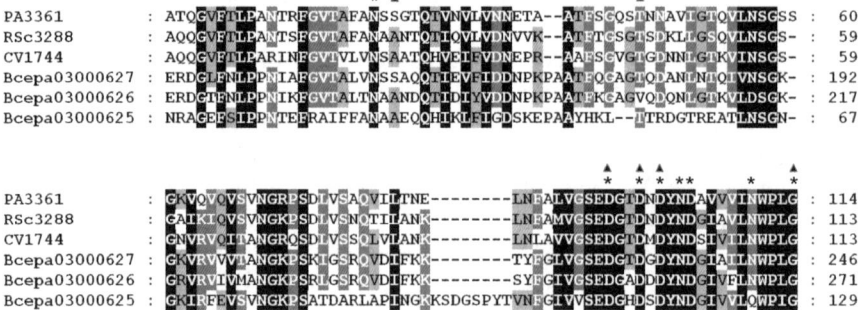

Figure 10 *Sequence alignment of PA-IIL with similar sequences identified in genomes of Ralstonia solanacearum, Chromobacterium violaceum, and Burkholderia cenocepacia. Strictly conserved amino acids are displayed with black background and similar ones with grey background. Amino acids involved in calcium binding are indicated by stars. Amino acids predicted to form hydrogen bonds with the carbohydrate ligand are indicated by triangles*

infections in cystic fibrosis patients, contains three copies of the gene coding for larger proteins, all of which contain one PA-IIL-like domain at the C-terminus. Four copies are observed in the genome of *Photorhabdus luminescens*, but they are rather different in amino acids, specially those involved in ligand binding and are not further considered here. Although these pathogens have a wide range of targets, they do share common characteristics. Except for *P. photorhabdus*, which is an insect pathogen, they are opportunistic inhabitants of soil and water and able to adapt very quickly to new environments and are highly dangerous once infection is declared. Among the different genes that have similarity to the *lec B* gene coding for PA-IIL only the *R. solanacearum* gene has been expressed and its product characterised. This gene codes for a lectin, RS-IIL, with a very high affinity for mannose and a lower affinity for fucose.[55] The RS-IIL/mannose complex has been crystallised and the structure determined[56] and the binding mode of mannose in RS-IIL can therefore be compared to the PA-IIL sugar complexes (Figure 11). A three-amino-acid motif, in the so-called monosaccharide specificity loop, plays a special role in sugar binding. The Ala-Ala-Asn motif in RS-IIL allows space for the O6 of mannose to point into the sugar-binding site and to make a strong hydrogen bond with charged side chain of Asp96, whereas the Ser-Ser-Gly in PA-IIL does not permit this optimal orientation for mannose, thereby rationalising the lower affinity. This analysis could open the route for specificity prediction for the other PA-IIL-like bacterial gene sequences.

2.5   Biological Roles

The expression of PA-IIL has been demonstrated to be under the control of the quorum sensing system, as already discussed above for PA-IL.[33,35,36] The lectin is therefore co-expressed with the virulence factors and the molecules that are involved in biofilm formation. PA-IIL is not directly cytotoxic itself, but has been shown to affect *in vitro* the ciliary beat frequency of airway epithelium cells.[57] Recent studies

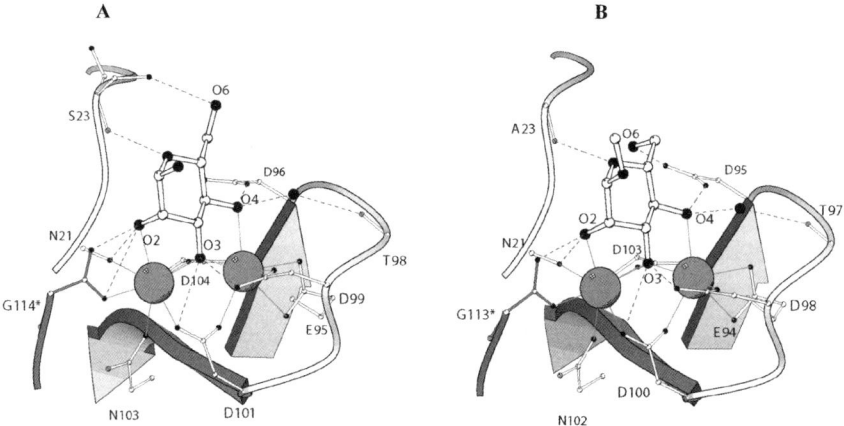

Figure 11 *Comparison of mannose-binding sites from P. aeruginosa and R. solanacearum lectins. (A) Complex between PA-IIL and α-D-mannose (PDB code 1OUR).[53] (B) Complex between RS-IIL and α-D-Me-mannoside complex (PDB code 1UQX)[56]*

have demonstrated that PA-IIL is located mainly in the outer membrane of *P. aeruginosa* where it is bound to bacterial polysaccharides.[58] At this location, and owing to its multivalent character, PA-IIL can play a role for specific recognition and binding to other molecules in the environment. The target of interest could be glycoconjugates from a host epithelium, mucus at the surface of airways, biofilms or cell walls of other bacteria (Figure 12).

At the present time, none of these hypotheses can be discarded and there is experimental evidence to support various roles for the lectin. The fact that bacterial strains having the appropriate sugars in their cell walls may be recognised and agglutinated by the lectin was proven by the effect of addition to PA-IIL to *E. coli* O128B12 cells.[59] The role in biofilm formation has been recently demonstrated since a PA-IIL-deficient bacterial strain produces a thinner biofilm than the wild strain.[58] The binding of the lectin to host fucosylated oligosaccharides has been correlated to recurrent respiratory infections of cystic fibrosis patients. Indeed, the inflammation observed in CF lungs is linked to increased fucosylation for both cell surface glycoconjugates[60] and mucin polysaccharides.[61]

3 Perspectives: Targeting *Pseudomonas* Lectins for Antibacterial Therapy?

Since the role of carbohydrate–lectin adhesion in bacterial infections of intestines, stomach, ears, and bladder was recognised, many efforts have been made to develop oligosaccharide-derived compounds with antiadhesive therapeutic activity.[62–64] The prophylactic effect of specific oligosaccharides have been proven in animal models such as the *E. coli* bladder infection in mice,[65] pneumoccocal pneumonia in rats[66] and *Helicobacter pylori* gastric infection in monkeys.[67] However, success of such treatment in human has not yet been established.

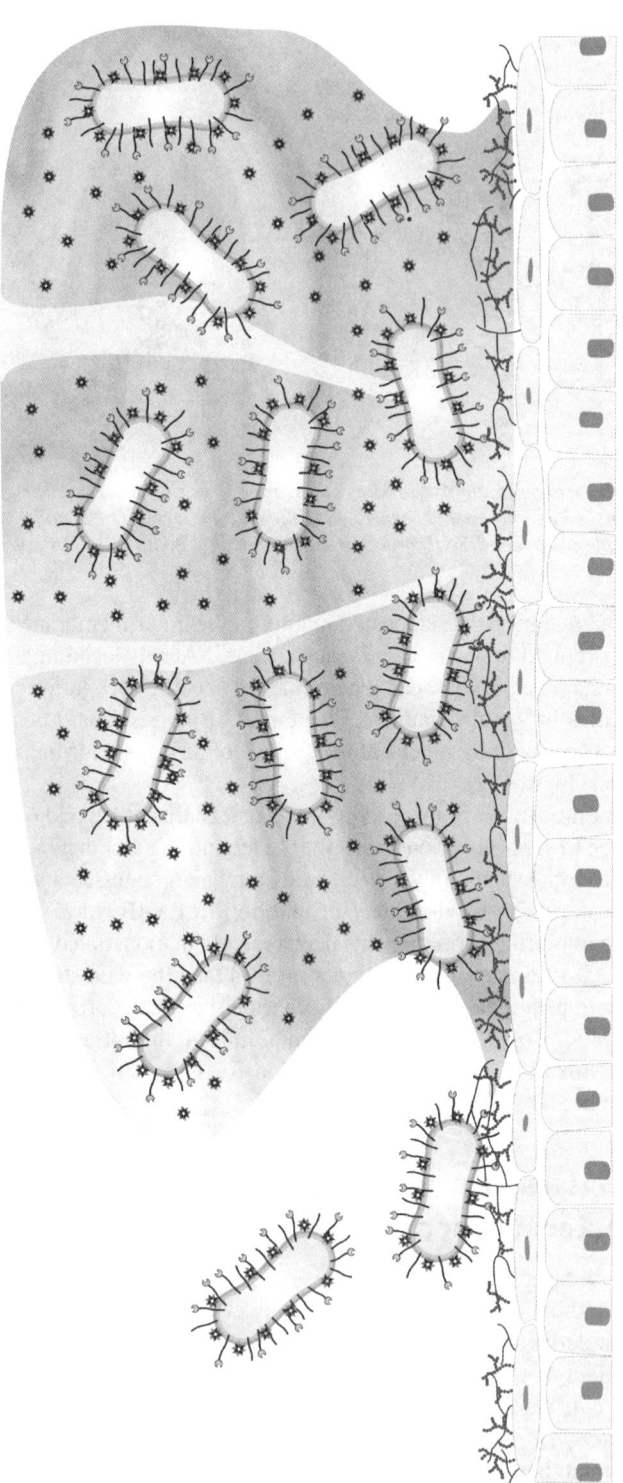

Figure 12 *Schematic representation of the possible roles of P. aeruginosa soluble lectins in host recognition and biofilm formation*

The difficulties that are encountered are multiple. In general, the carbohydrate receptors that are targeted are the fimbrial adhesins, which do not have a high affinity for oligosaccharide ligands. The use of multivalent carbohydrate ligands, to create an avidity effect by binding to several adhesins, is under development and promising results have been obtained *in vitro* for glycodendrimers,[68] self-assembling glycopeptide tectomers[69] or glyco-nanoparticles.[70] However, the main problem arises from the glyco-strategy of the bacteria, which employs several glycan-receptors with different specificities and can therefore still bind to the host, even in the presence of one competitor. For example, about ten different receptors, with different oligosaccharide specificities, have been identified for *H. pylori*.[63] If human milk is effective in protecting infants from gastrointestinal bacteria, it is because it contains astronomical numbers of oligosaccharides (more than 900 distinct fucooligosaccharides).[71]

Oligosaccharide-based prevention or treatment of *P. aeruginosa* infections should therefore take into account the different strategies that the bacteria use for recognising and binding to the host. Encouraging preliminary reports include the treatment of one case of airway infection by inhalation of fucose and galactose[72] and a relative improvement in patients with acute otitis externa by treatment with galactose, mannose and NeuAc.[73] As described in the present review, two of the lectins of *P. aeruginosa* have been fully characterised. Nevertheless, it is clear that we still need a better understanding of the other carbohydrate receptors of this bacterium to be able to design oligosaccharide ligands, or mixture of ligands, with efficient antiadhesive action to complement classical treatment by antibiotics.

Acknowledgements

The studies from our groups were supported by Ministry of Education of the Czech Republic (contract MSM 0021622413), by French Ministry of Research ACI Microbiologie programme, by Mizutani Foundation for Glycoscience and by Vaincre la Mucoviscidose. Travels and visits between NCBR and CERMAV were supported by a BARRANDE exchange programme.

References

1. A.W. Artenstein and A.S. Cross in *Pseudomonas aeruginosa as an Opportunistic Pathogen*, M. Campa, M. Beninelli and H. Friedman (eds), Plenum Press, New York, 1993, 224.
2. M. Steinert, M. Leippe and T. Roeder, *Int. J. Med. Microbiol.*, 2003, **293**, 321.
3. P.K. Singh, A.L. Schaefer, M.R. Parsek, T.O. Moninger, M.J. Welsh and E.P. Greenberg, *Nature*, 2000, **407**, 762.
4. J.W. Costerton, P.S. Stewart and E.P. Greenberg, *Science*, 1999, **284**, 1318.
5. A. Scharfman, S.K. Arora, P. Delmotte, E. Van Brussel, J. Mazurier, R. Ramphal and P. Roussel, *Infect. Immun.*, 2001, **69**, 5243.
6. H. Remaut and G. Waksman, *Curr. Opin. Struct. Biol.*, 2004, **14**, 161.
7. H.B. Sheth, K.K. Lee, W.Y. Wong, G. Srivastava, O. Hindsgaul, R.S. Hodges, W. Paranchych and R.T. Irvin, *Mol. Microbiol.*, 1994, **11**, 715.

8. H.P. Hahn, *Gene*, 1997, **192**, 99.
9. F.G. Sauer, M. Barnhart, D. Choudhury, S.D. Knight, G. Waksman and S.J. Hultgren, *Curr. Opin. Struct. Biol.*, 2000, **10**, 548.
10. I. Vallet, J.W. Olson, S. Lory, A. Lazdunski and A. Filloux, *Proc. Natl. Acad. Sci. USA*, 2001, **98**, 6911.
11. N. Gilboa-Garber, *Methods Enzymol.*, 1982, **83**, 378.
12. N. Gilboa-Garber, L. Mizrahi and N. Garber, *FEBS Lett.*, 1972, **28**, 93.
13. D. Avichezer, D.J. Katcoff, N.C. Garber and N. Gilboa-Garber, *J. Biol. Chem.*, 1992, **267**, 23023.
14. J. Glick and N.C. Garber, *J. Gen. Microbiol.*, 1983, **9**, 3085.
15. N. Gilboa-Garber, *FEBS Lett.*, 1972, **20**, 242.
16. N. Garber, U. Guempel, A. Belz, N. Gilboa-Garber and R.J. Doyle, *Biochim. Biophys. Acta*, 1992, **1116**, 331.
17. C.P. Chen, S.C. Song, N. Gilboa-Garber, K.S. Chang and A.M. Wu, *Glycobiology*, 1998, **8**, 7.
18. B. Lanne, J. Ciopraga, J. Bergstrom, C. Motas and K.A. Karlsson, *Glycoconj. J.*, 1994, **11**, 292.
19. S. Kirkeby and D. Moe, *Xenotransplantation*, 2002, **9**, 260.
20. S.R. Stoitsova, R.N. Boteva and R.J. Doyle, *Biochim. Biophys. Acta*, 2003, **1619**, 213.
21. R.N. Boteva, V.P. Bogoeva and S.R. Stoitsova, *Biochim. Biophys. Acta*, 2005, **1747**, 143.
22. C.V. Gegg, D.D. Roberts, I.H. Segel and M.E. Etzler, *Biochemistry*, 1992, **31**, 6938.
23. K. Karaveg, Z.J. Liu, W. Tempel, R.J. Doyle, J.P. Rose and B.C. Wang, *Acta Crystallogr. D Biol. Crystallogr.*, 2003, **59**, 1241.
24. H.M. Berman, J. Westbrook, Z. Feng, G. Gilliland, T.N. Bhat, H. Weissig, I.N. Shindyalov and P.E. Bourne, *Nucl. Acids Res.*, 2000, **28**, 235.
25. G. Cioci, E.P. Mitchell, C. Gautier, M. Wimmerova, D. Sudakevitz, S. Pérez, N. Gilboa-Garber and A. Imberty, *FEBS Lett.*, 2003, **555**, 297.
26. A.G. Murzin, S.E. Brenner, T. Hubbard and C. Chothia, *J. Mol. Biol.*, 1995, **247**, 536.
27. S. Jamal-Talabani, A.B. Boraston, J.P. Turkenburg, N. Tarbouriech, V.M. Ducros and G.J. Davies, *Structure*, 2004, **12**, 1177.
28. K. Drickamer and A.J. Fadden, *Biochem. Soc. Symp.*, 2002, **69**, 59.
29. A.R. Kolatkar and W.I. Weis, *J. Biol. Chem.*, 1996, **271**, 6679.
30. S.F. Poget, G.B. Legge, M.R. Proctor, P.J. Butler, M. Bycroft and R.L. Williams, *J. Mol. Biol.*, 1999, **290**, 867.
31. N.A. Whitehead, A.M. Barnard, H. Slater, N.J. Simpson and G.P. Salmond, *FEMS Microbiol. Rev.*, 2001, **25**, 365.
32. D.G. Davies, M.R. Parsek, J.P. Pearson, B.H. Iglewski, J.W. Costerton and E.P. Greenberg, *Science*, 1998, **280**, 295.
33. K. Winzer, C. Falconer, N.C. Garber, S.P. Diggle, M. Camara and P. Williams, *J. Bacteriol.*, 2000, **182**, 6401.
34. C.K. Stover, X.Q. Pham, A.L. Erwin, S.D. Mizoguchi, P. Warrener, M.J. Hickey, F.S. Brinkman, W.O. Hufnagle, D.J. Kowalik, M. Lagrou, R.L. Garber, L.

Goltry, E. Tolentino, S. Westbrock-Wadman, Y. Yuan, L.L. Brody, S.N. Coulter, K.R. Folger, A. Kas, K. Larbig, R. Lim, K. Smith, D. Spencer, G.K. Wong, Z. Wu, I.T. Paulsen, J. Reizer, M.H. Saier, R.E. Hancock, S. Lory and M.V. Olson, *Nature*, 2000, **406**, 959.

35. V.E. Wagner, D. Bushnell, L. Passador, A.I. Brooks and B.H. Iglewski, *J. Bacteriol.*, 2003, **185**, 2080.

36. M. Schuster, C.P. Lostroh, T. Ogi and E.P. Greenberg, *J. Bacteriol.*, 2003, **185**, 2066.

37. O. Bajolet-Laudinat, S. Girod-de Bentzmann, J.M. Tournier, C. Madoulet, M.C. Plotkowski, C. Chippaux and E. Puchelle, *Infect. Immun.*, 1994, **62**, 4481.

38. J. Alverdy, C. Holbrook, F. Rocha, L. Seiden, R.L. Wu, M. Musch, E. Chang, D. Ohman and S. Suh, *Ann. Surg.*, 2000, **232**, 480.

39. R.S. Laughlin, M.W. Musch, C.J. Hollbrook, F.M. Rocha, E.B. Chang and J.C. Alverdy, *Ann. Surg.*, 2000, **232**, 133.

40. L. Wu, C. Holbrook, O. Zaborina, E. Ploplys, F. Rocha, D. Pelham, E. Chang, M. Musch and J. Alverdy, *Ann. Surg.*, 2003, **238**, 754.

41. J. Rebiere-Huet, P. Di Martino and C. Hulen, *Can. J. Microbiol.*, 2004, **50**, 303.

42. N. Gilboa-Garber, D.J. Katcoff and N.C. Garber, *FEMS Immunol. Med. Microbiol.*, 2000, **29**, 53.

43. N. Garber, U. Guempel, N. Gilboa-Garber and R.J. Doyle, *FEMS Microbiol. Lett.*, 1987, **48**, 331.

44. E. Mitchell, C. Houles, D. Sudakevitz, M. Wimmerova, C. Gautier, S. Pérez, A.M. Wu, N. Gilboa-Garber and A. Imberty, *Nat. Struct. Biol.*, 2002, **9**, 918.

45. P. Chaturvedi, C.D. Warren, M. Altaye, A.L. Morrow, G. Ruiz-Palacios, L.K. Pickering and D.S. Newburg, *Glycobiology*, 2001, **11**, 365.

46. C. Kunz, S. Rudloff, W. Baier, N. Klein and S. Strobel, *Ann. Rev. Nutr.*, 2000, **20**, 699.

47. E. Lesman-Movshovich, B. Lerrer and N. Gilboa-Garber, *Can. J. Microbiol.*, 2003, **49**, 230.

48. S. Perret, C. Sabin, C. Dumon, M. Pokorná, C. Gautier, O. Galanina, S. Ilia, N. Bovin, M. Nicaise, M. Desmadril, N. Gilboa-Garber, M. Wimmerova, E.P. Mitchell and A. Imberty, *Biochem. J.*, 2005, **389**, 325.

49. T.K. Dam and C.F. Brewer, *Chem. Rev.*, 2002, **102**, 387.

50. E.P. Mitchell, S. Sabin, L. Šnajdrová, M. Budová, S. Perret, C. Gautier, C. Hofr, N. Gilboa-Garber, J. Koca, M. Wimmerová and A. Imberty, *Proteins: Struct. Funct. Bioinfo.*, 2005, **58**, 735.

51. J.P. Carver, *Pure Appl. Chem.*, 1993, **65**, 763.

52. R.U. Lemieux, L.T. Delbaere, H. Beierbeck and U. Spohr, *Ciba Found. Symp.*, 1991, **158**, 231.

53. R. Loris, D. Tielker, K.-E. Jaeger and L. Wyns, *J. Mol. Biol.*, 2003, **331**, 861.

54. A. Imberty, M. Wimmerova, E.P. Mitchell and N. Gilboa-Garber, *Microb. Infect.*, 2004, **6**, 222.

55. D. Sudakevitz, A. Imberty and N. Gilboa-Garber, *J. Biochem.*, 2002, **132**, 353.

56. D. Sudakevitz, N. Kostlanova, G. Blatman-Jan, E.P. Mitchell, B. Lerrer, M. Wimmerova, f.D.J. Katcof, A. Imberty and N. Gilboa-Garber, *Mol. Microbiol.*, 2004, **52**, 691.

57. E.C. Adam, B.S. Mitchell, D.U. Schumacher, G. Grant and U. Schumacher, *Am. J. Respir. Crit. Care Med.*, 1997, **155**, 2102.
58. D. Tielker, S. Hacker, R. Loris, M. Strathmann, J. Wingender, S. Wilhelm, F. Rosenau and K.-E. Jaeger, *Microbiology*, 2005, **151**, 1313.
59. D. Sudakevitz and N. Gilboa-Garber, *Microbios*, 1982, **34**, 159.
60. A.D. Rhim, L.I. Stoykova, A.J. Trindade, M.C. Glick and T.F. Scanlin, *J. Cyst. Fibros.*, 2004, **3**, 95.
61. P. Roussel and G. Lamblin, *Adv. Exp. Med. Biol.*, 2003, **535**, 17.
62. D. Zopf and S. Roth, *Lancet*, 1996, **347**, 1017.
63. C.A. Lingwood, *Curr. Opin. Chem. Biol.*, 1998, **2**, 695.
64. I. Ofek, D.L. Hasty and N. Sharon, *FEMS Immunol. Med. Microbiol.*, 2003, **38**, 181.
65. M. Aronson, O. Medalia, L. Schori, D. Mirelman, N. Sharon and I. Ofek, *J. Infect. Dis.*, 1979, **139**, 329.
66. I. Idanpaan-Heikkila, P.M. Simon, D. Zopf, T. Vullo, P. Cahill, K. Sokol and E. Tuomanen, *J. Infect. Dis.*, 1997, **176**, 704.
67. J.V. Mysore, T. Wigginton, P.M. Simon, D. Zopf, L.M. Heman-Ackah and A. Dubois, *Gastroenterology*, 1999, **117**, 1316.
68. R. Autar, A.S. Khan, M. Schad, J. Hacker, R.M. Liskamp and R.J. Pieters, *Chembiochem*, 2003, **4**, 1317.
69. N.V. Bovin, A.B. Tuzikov, A.A. Chinarev and A.S. Gambaryan, *Glycoconj. J.*, 2004, **21**, 471.
70. L. Qu, P.G. Luo, S. Taylor, Y. Lin, W. Huang, N. Anyadike, T.R. Tzeng, F. Stutzenberger, R.A. Latour and Y.P. Sun, *J. Nanosci. Nanotechnol.*, 2005, **5**, 319.
71. D.S. Newburg, *J. Nutr.*, 1997, **127**, 980S.
72. P. von Bismarck, R. Schneppenheim and U. Schumacher, *Klin. Padiatr.*, 2001, **213**, 285.
73. M.K. Steuer, H. Herbst, J. Beuth, M. Steuer, G. Pulverer and R. Mathhias, *Otorhinolaryngol. Nova*, 1993, **3**, 19.
74. P. Kraulis, *J. Appl. Crystallogr.*, 1991, **24**, 946.

CHAPTER 4

Protein–Carbohydrate Interactions in Enterobacterial Infections

NATHAN SHARON[1] AND ITZHAK OFEK[2]

[1]Department of Biological Chemistry, The Weizmann Institue of Science, Rehovot 76110, Israel
[2]Department of Human Microbiology, Sackler School of Medicine, Tel Aviv 69978, Israel

1   Introduction

For many organisms, to stick somewhere is key to survival. Nature's varied adhesive structures and substances enable organisms to stick to inert substrates, to each other, and even to parts of themselves. Microorganisms are no exception to this rule; animals provide a wide variety of surfaces microorganisms need to stick to and survive. In most cases, these microbes cause no harm and often their existence is beneficial. However, when the host's antimicrobial defenses are compromised, even the most benign of microbes can cause life-threatening infections. On the other hand, pathogenic microbes cause infections in otherwise healthy individuals and, by and large, these infections are initiated by microbial colonization of mucosal epithelial surfaces of the host. The majority of the microorganisms, whether benign or pathogenic, latch onto host cells or tissues by proteinaceous surface structures, called adhesins.[1] In the case of pathogenic bacteria, the adhesins are important virulence determinants of the organisms. Many of the adhesins mediate the binding of the microorganisms to host cell complex carbohydrates.[2,3] Such adhesins belong to the class of proteins called lectins, sugar-specific proteins that function in cell recognition.[4,5] Certain microbial toxins have also adopted the protein–carbohydrate strategy for attachment to host tissue (Bundle, this volume), and are therefore considered as lectins, too. Many concepts concerning the structure–function relationships of bacterial lectins evolved from studies of surface lectins expressed by members of enterobacterial species.

In this chapter, we review the common themes underlying the specificity, structure, and functions of enterobacterial lectins. We then deal with strategies that have been employed to inhibit the protein–carbohydrate interactions, in order to prevent or treat enteric infectious diseases.

2 Enterobacterial Surface Lectins

2.1 Specificity and Structure

A variety of enterobacterial species and genera express surface lectins, frequently of more than one type and with distinct specificities[1,2,5] (Table 1). The diverse specificities of these lectins are among the factors determining the organ and animal tropism of the enterobacteria[6,7] as well as those of other bacteria. It is not known whether individual cells co-express multiple lectins or if each lectin is confined to a distinct cell subpopulation. We know, however, that the expression of such proteinaceous adhesins is commonly regulated by a mechanism known as phase variation, which controls the back-and-forth conversion of the adhesin-expressing cells in a bacterial population to non-expressing ones.[8] In *Escherichia coli*, *Klebsiella pneumoniae*, and *Salmonellae* spp., the lectins often are in the form of submicroscopic hair-like appendages, known as fimbriae or pili, which protrude from the surface of the cells[9,10] (Figure 1). During the fimbriated phase, a typical Gram-negative bacterium carries 200–500 peritrichously arranged fimbriae.

Knowledge of the specificity of bacterial lectins is based primarily on inhibition experiments, in which the effect of different carbohydrates is examined on adhesion of the bacteria to relevant animal cells or on the agglutination by the bacteria of erythrocytes[17] or other kind of cell, such as yeasts for mannose-specific bacteria.[18,19] In only few cases have the association constants of carbohydrates to the isolated lectins been measured by physicochemical techniques, such as surface plasmon

Table 1 *Enterobacterial surface lectins**

Organism	Carbohydrate specificity	Form†
Campylobacter jejuni[14]	Fucα2Galβ4GlcNAc	GP
Escherichia coli type 1	Manα3(Manα6)Man	GP
F1C[15]	GalNAcβ4Gal	GSL
P	Galα4Gal	GSL
S	Neu5Acα2,3Galβ3GalNAc	GSL
CFA/1	Neu5Acα2,8-	GP
K1	GlcNAcβ4GlcNAc	GP
K99	Neu5Acα2,3Galβ4Glc	GSL
F17[16]	GlcNAc	GP
Klebsiella pneumoniae type 1	Man	GP
Salmonella typhimurium Type 1	Man	GP

*Unless otherwise noted, see Ref. 13.
†Predominant form of ligand on cells: GP, glycoproteins; GSL, glycosphingolipids.

resonance (SPR).[20,21] Specificity for the carbohydrate of glycolipids is usually assessed by binding of the bacterial lectins to these glycoconjugates separated on thin layer chromatograms.[22,23] Recently, the application of carbohydrate microarrays for specificity studies of bacterial surface lectins has been described[24] (Figure 2).

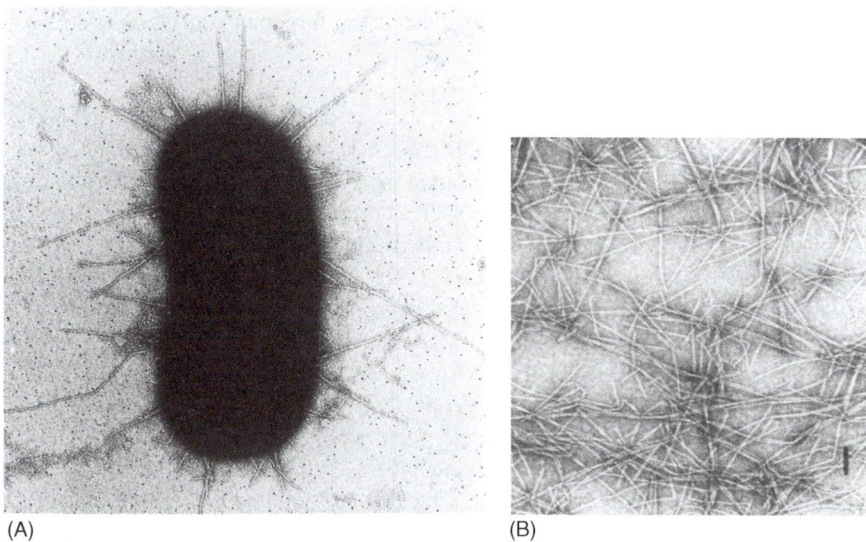

(A) (B)

Figure 1 *(A) Type 1 fimbriated E. coli;[11] (B) the isolated fimbriae; bar is 0.1 μm in size[12]*

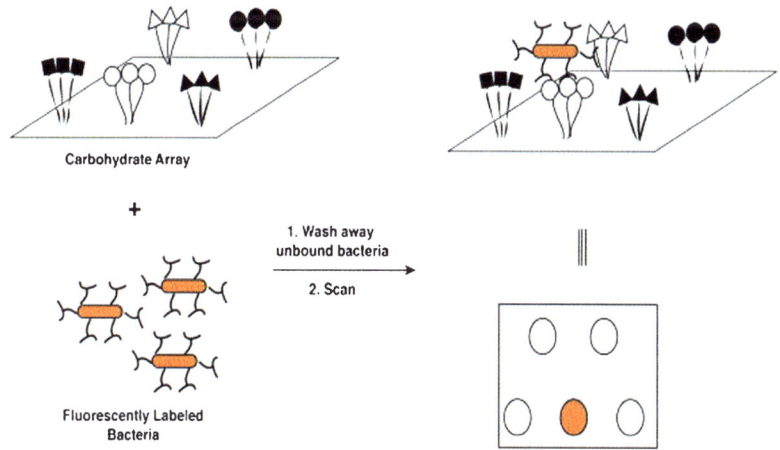

Figure 2 *Examination of the specificity of fluorescent-labeled E. coli, by the use of a carbohydrate array. The bacteria are deposited in the wells of the array, each of which carries a different carbohydrate. Unbound bacteria are washed away and the array is scanned, resulting in a characteristic binding pattern based on sugar-binding affinities. The multivalent nature of the carbohydrates on the array is indicated[25]*

Although the sugar specificity of a considerable number of enterobacterial surface lectins has been established (Table 1), only very few of them have been well studied with respect to their structure and biosynthesis. By far best characterized of these lectins, as well as the most prevalent ones among enterobacterial species, are the mannose-specific type 1 fimbriae of *E. coli*. Another well-characterized group is the galabiose (Galα4Gal)-specific P fimbriae of uropathogenic *E. coli* (UPEC). More recently the F17 fimbriae of *E. coli* have also been studied in considerable detail.

2.1.1 Type 1 fimbriae

A large number of enterobacterial strains express mannose-specific fimbriae, classified as type 1. In particular, they are found in more than 95% of *E. coli* isolates from intestinal and extraintestinal infections such as urinary and enteric ones.[1,10] Many of the type 1-fimbriated *E. coli* have been studied extensively. In a few cases type 1 fimbriae of other enterobacterial species, such as *K. pneumoniae, Salmonella typhimurium*, and *S. enteritidis*, have also been investigated, although usually to a limited extent. Most phenotypes have a considerably higher affinity for oligosaccharides, such as Manα3Manβ4GlcNAc or Manα6(Manα3)Manα6(Manα3)Man that are constituents of cell surface glycoproteins, than they do for mannose (up to 40 times more).[26] The affinity of different phenotypes of *E. coli* type 1 or of the isolated fimbriae to mannose or methyl α-mannoside is generally low, in the millimolar range; it may however differ within a factor of 15 and the bacteria can therefore be functionally subdivided into either low mannose-binding (M1L) or high mannose-binding (M1H) phenotypes.[27] *E. coli* exhibiting these two basic phenotypes have been found to predominate in different niches. Most isolates from the large intestine of healthy humans (around 80%) express a distinct M1L phenotype, whereas most isolates from UTIs (more than 70%) express M1H variants. These naturally occurring variations dramatically change the tissue tropism of *E. coli* and can be a major factor in shifting the bacterial adaptation from commensal to pathologic habitats. Thus, the transition from commensal to virulent phenotype may be mediated not only by acquisition of "virulence genes" but also by selection of genetic variations in a commensal trait that are adaptive to a pathologic environment.

In the urinary tract, the fimbriae mediate binding of the bacteria to uroplakins Ia and Ib, two major glycoproteins of urothelial apical plaques.[28] Anchorage of *E. coli* to the urothelial surface via type 1 fimbriae–uroplakin I interactions may play a role in bladder colonization by the bacteria and their eventual ascent through the ureters, against urine flow, to invade the kidneys.

Structurally, the fimbriae are 1–2 μm long and 7 nm thick, rod-like fibers, which are made up largely of repeating immunoglobulin-like FimA subunits (MW 17 kDa) that are helically arranged in a structure referred to as the "shaft."[29,30] The shaft is joined to a short 3 nm thick distal tip fimbrillum structure that consists of two adaptor proteins, FimF and FimG, and a third protein of a different kind, FimH (MW 29–31 kDa) (Figure 3). The latter is the only subunit that possesses a carbohydrate-binding site and is thus responsible for the sugar-binding activity of the fimbriae. FimH may also be present in small numbers at intervals along the fimbrial filament, but only the subunit at the tip appears to be able to mediate mannose-specific adhesive interactions,

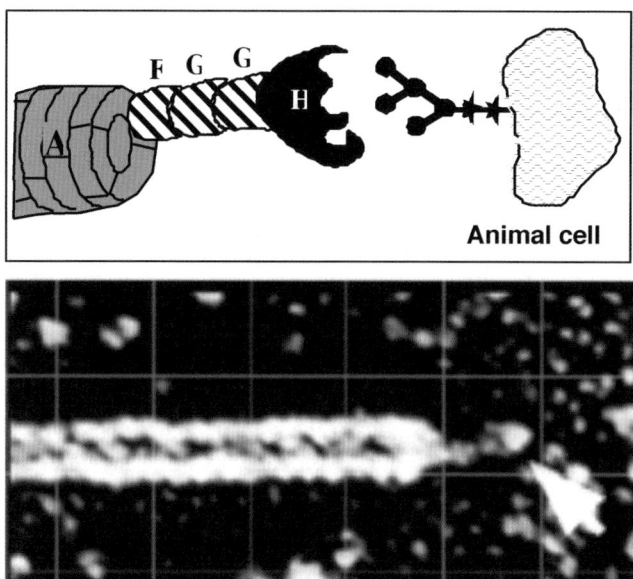

Figure 3 *Composite structure of type 1 fimbriae. Top, schematic representation, depicting the distal end of the shaft as well as the tip of a fimbrium (or pilus) and a typical attachment site on host cells; A, F, G, and H denote the fimbrial subunits. Bottom, electron micrograph, in which arrowhead indicates the fimbrial tip containing the FimH subunit. (Reproduced with permission from Ref. 31. Copyright University of Chicago Press)*

whereas the subunits at the other positions are inaccessible to the carbohydrate ligand. Marked differences among different enterobacterial strains and species are found in the size and antigenic properties of FimA. Minor sequence variations in FimH alleles from different clinical isolates have been shown to correlate with altered carbohydrate-binding profiles of the fimbriae.[27]

The FimH subunits of *E. coli* and *K. pneumoniae* are 88% homologous. Still the two organisms differ in their fine specificity, for example, in their relative affinity for Manα3Manβ4GlcNAc and *p*-nitrophenyl α-mannoside.[32,33] Other aromatic α-mannosides are also high-affinity ligands for type 1 fimbriated *E. coli*, up to 3 orders of magnitude stronger than mannose, suggesting the presence of a hydrophobic binding region next to the monosaccharide-combining site of FimH.[34] With several *Salmonella* species examined, aromatic α-mannosides as well as the trisaccharide Manα3Manβ4GlcNAc were weaker inhibitors than methyl α-mannoside (Table 2). The combining site of Salmonella species appears thus to be smaller than that of *E. coli* and *K. pneumoniae*, and to be devoid of an adjoining hydrophobic region. Different combining sites were found in other mannose-specific bacterial lectins. Therefore, although classified together on the basis of their monosaccharide (or primary) specificity, these lectins differ in their fine specificity.

Genetically engineered hybrid fimbriae, in which the FimH of one species (*e.g.*, *E. coli*) was presented on the shaft of another species (*e.g.*, *K. pneumoniae*), have

Table 2 *Inhibitory activity of mannose derivatives on yeast agglutination by type 1 fimbriated enterobacteria*

Inhibitor	Organism	
	E. coli	*S. enteritidis*
MeαMan	1.0 (0.5–2.4 mM)	1.0 (0.4 mM)[1]
*p*NPheαMan	30–72	0.3
4MeUmbαMan	600	
Manα3Manβ4GlcNAc	21–35	0.1–0.4
Manα6(Manα3)Manα6-(Manα3)ManαOMe	30–45	

Data from Refs. 32, 33. *p*NPheαMan, p-nitrophenyl.
[1]Concentration of MeαMan required for 50% inhibition of yeast aggregation by the bacteria.

shown that the shaft (*i.e.*, one or more of the subunits other than FimH) plays a role in modulating the specificity of fimbriae, probably by imposing conformational constraints on the carbohydrate-binding subunit.[35] That the shaft influences the carbohydrate specificity of FimH is supported by recent studies in which the specificity of soluble hybrids of the lectin domain of FimH and maltose-binding protein was compared to the native type 1 fimbriae of *S. typhimurium* and *E. coli*.[36]

The first three-dimensional structure of a fimbrial carbohydrate-binding subunit to be solved was that of a FimC–FimH complex cocrystallized with the mannose analog cyclohexylbutanoyl-*N*-hydroxyethyl-D-glucamide.[37] As will be explained below, FimH alone is unstable and requires the chaperone FimC for stability. In the FimC–FimH complex, FimH is seen folded into two all-β class domains connected by a short extended linker (Figure 4).[31] One of these, located in the N-terminal half of the protein subunit (residues 1–156), is the lectin domain, with the mannose-binding site at its tip. The C-terminal half (residues 160–279), known as the pilin domain, serves to anchor the subunit to the fimbriae. It binds in the cleft of FimC, although there is only limited contact between FimH and the C-terminal domain of the chaperone.

The carbohydrate recognition domain of FimH is an 11-stranded elongated β-barrel with a jelly roll-like topology, while the pilin domain has an immunoglobulin fold that lacks the seventh (C-terminal) β strand present in the canonical immunoglobulin fold. In the FimC–FimH complex the missing β strand is donated by the seventh strand of the N-terminal domain of the chaperone to complete the immunoglobulin-like fold of FimH. This kind of "donor-strand complementation" (see later) is thought to initiate folding of FimH directly on the chaperone, thus accounting for the function of the latter in biogenesis of the fimbriae.

The structure of the FimC–FimH in complex with bound mannose has now also been elucidated by X-ray crystallography (Figure 5).[38] Although mannose exists in solution as a mixture of α- and β-anomers, only the former was found in the crystal. It was buried at a deep and negatively charged site at the edge of FimH, opposite to the region through which the latter combines with the chaperone. All the mannose hydroxyls, except the anomeric one, interacted extensively with combining site residues, almost all of which are situated at the ends of β strands or in the loops extending from them. Part of the hydrogen-bonding network is identical to that

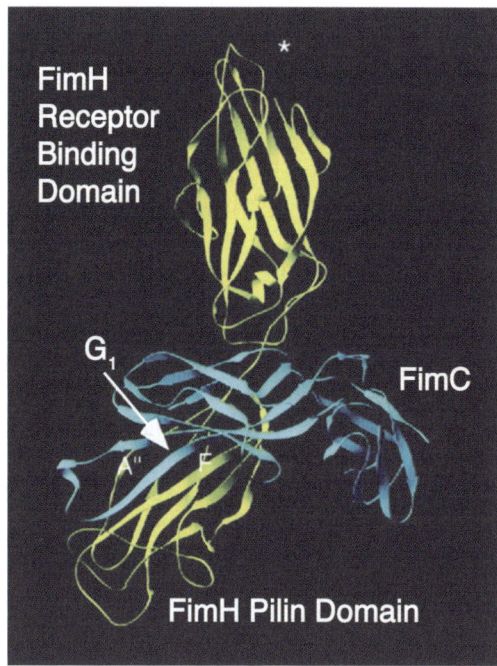

Figure 4 *Crystal structure of FimH, the carbohydrate-binding subunit of type 1 fimbriae, in complex with the FimC chaperone. The * indicates the mannose-binding pocket of FimH. Also shown is the insertion of the G1 strand of the FimC chaperone into the hydrophobic groove formed between the A and F strands of the FimH pilin domain, an interaction known as donor-strand complementation. (Reproduced with permission from Ref. 31. Copyright University of Chicago Press)*

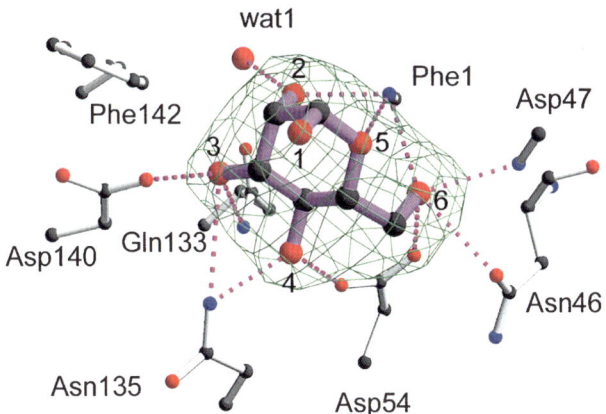

Figure 5 *The α-anomer of mannose in the combining site of FimH. The eight marked residues combine with the mannose by hydrogen bonds and hydrophobic interactions. Contact residues are shown as ball-and-stick model. Wat stands for water. (From C. S. Hung et al.[38])*

found in mannose complexes of other lectins. Thus, Asp54 makes cooperative hydrogen bonds with the 4- and 6-OH, and Asn140 with the 3- and 4-OH, of the ligand, similar to those made by equivalent residues with the same sugar in the combining sites of legume lectins with the same primary specificity.[39] In addition, the N-terminal amine of the FimH polypeptide is H-bonded to the 2-OH, 6-OH, and the ring oxygen of the mannose.[38] This is one of the rare cases in which an N-terminal amine of a protein participates in ligand binding. Phe142 of FimH interacts hydrophobically with the C2–C3 bond of the mannose. The same residue, together with Ile13, Tyr 48, and Ile52 forms part of a hydrophobic ridge that surrounds the site and which may help to direct the ligand into it. Site-directed mutagenesis showed that combining site residues Asp54, Gln133, Asn135, and Asp140 are essential for carbohydrate binding by FimH, since their replacement by alanine, asparagine, or aspartic acid resulted in complete loss of this activity.

Elucidation of the binding site region of FimH has provided confirmation of the suggestion, made some time ago, that the combining site of this lectin is extended, and fits best mannose-containing trisaccharides such as Manα3Manβ4GlcNAc or mannotriose (Manα6(Manα3)Man). Recent X-ray crystallography and modeling studies of the complexes of FimH with hydrophobic mannosides[21] have provided a molecular explanation for the high affinity of type 1 fimbriae to such compounds (Table 3 and Figure 6). It is likely that the hydrophobic character of the ridge of the mannose-binding site of FimH distinguishes *E. coli* type 1 fimbriae from those of Salmonella species that do not exhibit an increased affinity for mannose with hydrophobic substituents or for oligomannosides.[32,33]

Almost all mutations in the combining site of FimH abolished or reduced its binding not only to mannose, but also to urinary epithelial cells, indicating that the site may be highly conserved.[40] Strong support for this conclusion comes from the finding that

Table 3 *Relative affinity of mannose glycosides for type 1 E. coli, the isolated fimbriae and FimH*

Glycoside	Relative affinity			
	E. coli 346		*E. coli K12*	
	Cells(a)	*Fimbriae(b)*	*Cells(c)*	*FimH(d)*
MeαMan	1.00	1.00	1.00	1.00
PheαMan	40	48		
*p*NPheαMan(1)	30–72	48	70	92
*p*NoClPheαMan(2)	720		470	
MeUmbαMan(3)		600	1015	200
ButαMan(4)				16
HepαMan(5)			500	75

(1): *p*NPhe, *p*-nitrophenyl; (2): *p*NoCl, *p*-nitro-*o*-chlorophenyl; (3): MeUmb, methylumbeiferyl; (4): But, butyl; (5): Hep, heptyl.
Assayed by (a): Inhibition of yeast aggregation [Firon *et al., Infect. Immun.*, 1984, **43**, 1088]; (b): Same [Firon *et al., Carbohyd. Res.*, 1983, **120**, 235]; (c): Inhibition of binding to guinea pig intestinal cells [Firon *et al., Infect. Immun.*, 1987, **55**, 472]; (d): Surface plasmon resonance [Bouckaert *et al., Mol. Microbiol.*, 2005, **55**, 441].

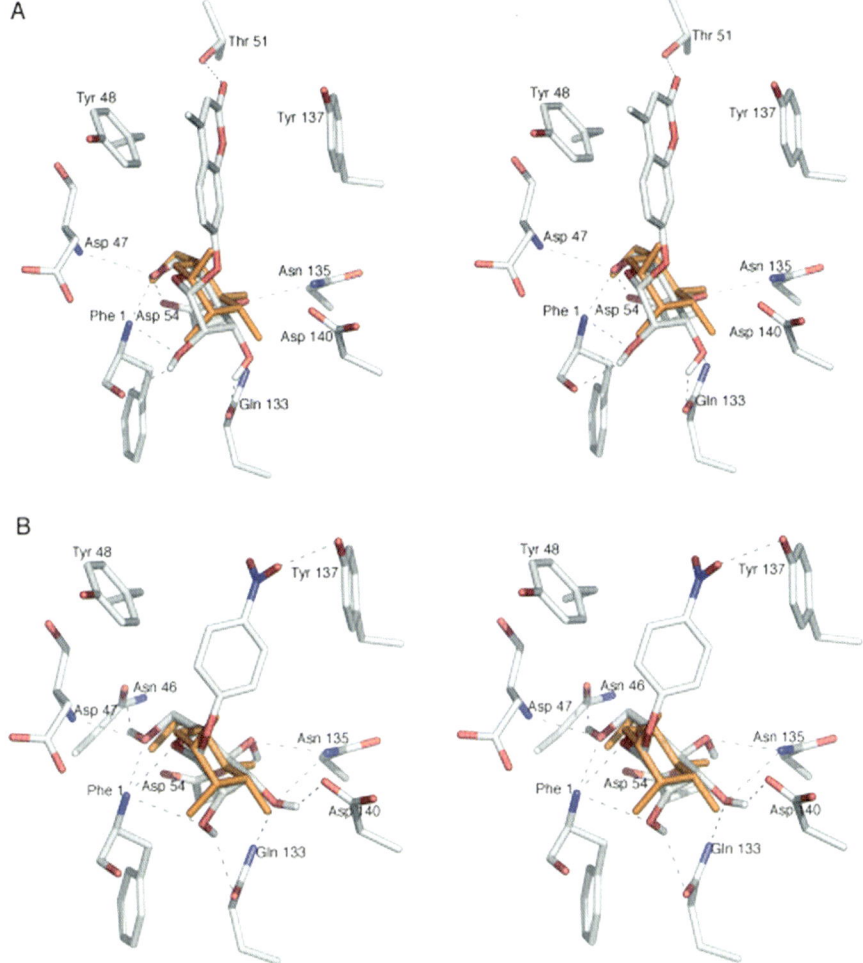

Figure 6 *Stereodiagrams of MeUmbαMan (A) and pNPheαMan (B) docked in the FimH-binding site. The crystallographically determined position of the α-anomer of mannose is shown in orange for comparison[21]*

there are very few variations in the sequences of the mannose-binding site of over 200 uropathogenic strains of *E. coli* examined, in contrast to enterohemorrhagic strains of *E. coli*, in which there are sequence variations at this site.[38]

On the other hand, replacement of residues 185–279 within the FimH pilin domain with a corresponding segment of the type 1C fimbrial adhesin FocH has led to the loss of the multivalent mannotriose-specific binding property, accompanied by the acquisition of a distinct mannose-specific (*i.e.*, monovalent) binding capability.[40] Bacteria expressing the monovalent hybrid FimH were capable of attaching strongly to uroepithelial tissue culture cells and guinea pig erythrocytes. They could not, however, agglutinate yeasts or bind human buccal cells, functions readily accomplished by the

E. coli-expressing mannotriose-specific FimH variants. Based on the relative potency of inhibiting compounds of different structures, the receptor-binding site within the monovalent FimH–FocH subunit has an extended structure with an overall configuration similar to that within the multivalent FimH of natural origin. The monovalent specific phenotype could also be invoked by a single point mutation, E89K, located within the lectin domain of FimH, but distant from the receptor-binding site. The structural alterations influence the receptor-binding valency of the FimH via distal effects on the combining pocket, obviously by affecting the FimH quaternary structure.

Another class of high-affinity ligand for type 1 fimbriated *E. coli* are mannose-derived neoglycoproteins and dendrimers (Figure 7). The former are proteins to which varying numbers of α-mannose residues are covalently attached, while the latter are multifunctional spherical branched polymers of well-defined molecular size that carry large numbers of such residues on their surface.[41] When the mannose in the neoglycoproteins or dendrimers was in the form of α-glycosides of long aliphatic chain fatty acids or of aromatic ring compounds, they were found to be up to 1600 times more powerful inhibitors of the agglutination by type I fimbriated *E. coli* of horse red blood cells than the monomeric counterpart.[42]

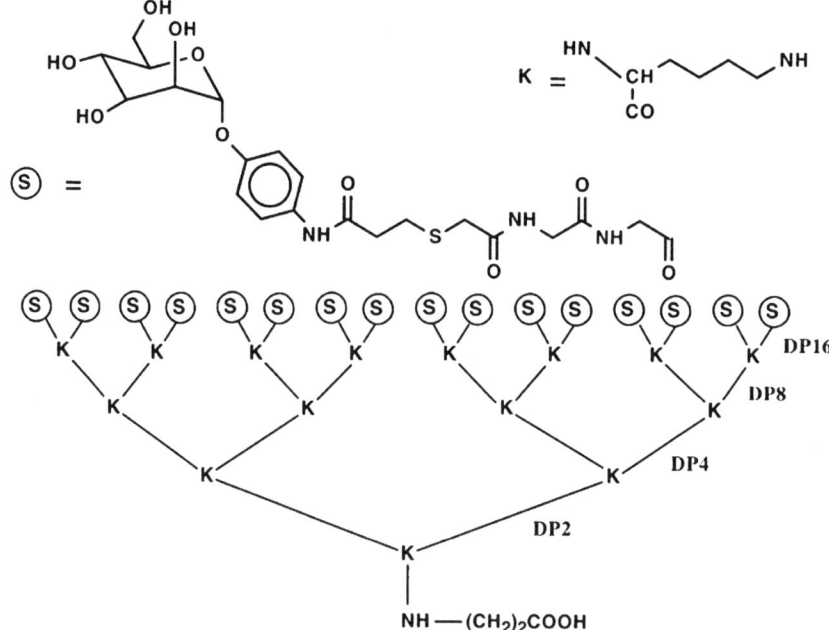

Figure 7 *Structure of dendrimer DP=16 which is a powerful inhibitor of type 1 fimbriated E. coli. On a molar basis, this compound is 11,000 times more potent inhibitor of the bacterial fimbriae than methyl α-mannoside, whereas on the basis of mannose content it is 690-fold better than the mannose glycoside. Enhancement of potency is both due to the presence of multiple the mannose residues and of an aromatic aglycone*[42]

2.1.2   P fimbriae

In contrast to type 1 *E. coli*, which recognize only structures that are present on glyco-proteins, P fimbriated *E. coli* are specific for glycosphingolipids of the globoseries, and recognize specifically galabiose, Galα4Gal.[20,43] These bacteria bind the disaccharide when it is present on the glycolipids. They adhere mainly to the upper part of the kidney, where galabiose is more abundant. P fimbriae of uropathogenic *E. coli* were subdivided into several classes based on the particular location of their galabiose ligand within different glycolipids (Figure 8).

Similar to type 1 fimbriae, P fimbriae are composite structures consisting of a long, rigid rod and a short, flexible, open helical part (Figure 9).[44] The rigid section is about 7 nm in diameter and is composed mainly of about 1200 to 2400 copies of the PapA subunit (16–22 kDa) arranged in a tightly packed right-handed helix. A short flexible tip, consisting of PapE monomers, is 2 nm in diameter and is joined to the rigid rod by PapK adapter subunit. The carbohydrate-binding subunit, PapG, is located exclusively at the N-terminal domain of the flexible tip and appears to be the sole determinant of binding specificity. It has a mostly β sheet structure that can be subdivided into two regions (Figure 10). One is in the form of a β barrel similar to the corresponding region of FimH; the other, with a structure that has not been encountered elsewhere, contains the carbohydrate-binding site.[45] Like in type 1 fimbriae, donor-strand complementation

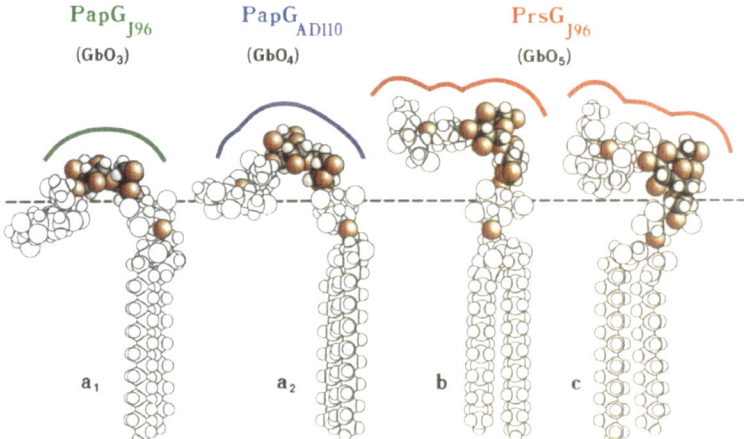

Figure 8 *Preferred conformations and epitope presentations of globo class glycosphin-golipids (Globo-GSLs). The conformations calculated for globo-GSLs and illus-trated by GbO$_5$ (GalNAcα3GalNAcβ3Galα4Galβ4GlcβCer) shown with the Galα4Gal unit and glycosidic oxygens are shaded; GbO$_4$ is GalNAcβ3Galα4Galβ4GlcβCer and GbO$_3$ is Galα4Galβ4GlcβCer. The internal conformation of the saccharide chain is practically identical for all four conform-ers and is independent of the sugar residues distal to the Galα4Gal unit. The four conformers differ in the conformation of the saccharide–ceramide linkage, result-ing in different saccharide orientations with respect to the membrane surface (dashed line). The preferred isoreceptor for the various lectins is indicated in parentheses and the possible binding surfaces by the lines above the models. (Reproduced with permission from Ref. 44. Copyright 2000).*

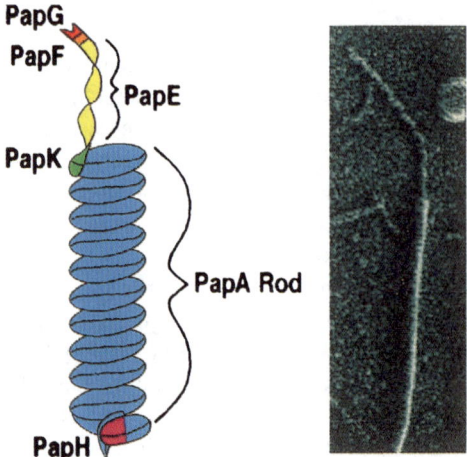

Figure 9 *Structure of P fimbriae (pili). On the left, a schematic diagram of a single P fimbrium, showing the location of each subunit within the structure; on the right, electron micrograph showing the two sub-assemblies of a P fimbrium[44]*

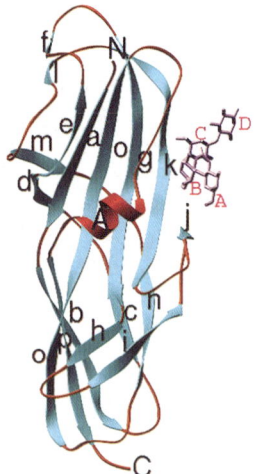

Figure 10 *Ribbon presentation of the N-terminal region of PapG. The upper part, made up
of seven β strands, forms a β barrel. The lower part is composed of a six-stranded
central antiparallel β sheet, flanked on one side by two double-stranded β sheets
and, on the other side, by an α helix and a large loop connecting this helix to the
central β sheet. The upper part contains the carbohydrate-binding site, showing
also a stick model of the tetrasaccharide ligand, GalNAcβ3Galα4Galβ4Glc, starting with A at the nonreducing end. N- and C- mark the respective ends of the
polypeptide. (Reproduced with permission from Ref. 45. Copyright 2001,
American Society for Microbiology)*

between PapD and PapK has been observed in the crystal structure of the chaperone-subunit complex of type P fimbriae.

2.1.3   F17 fimbriae

These fimbriae, produced by enterotoxigenic *E. coli*, are 3 nm wide, flexible, and wire-like organelles, built up of the major pilin subunit F17-A and exposing the F17-G subunit at their tip.[16] They mediate binding of the bacteria to *N*-acetylglucosamine- presenting receptors on the microvilli of the intestinal epithelium of ruminants, leading to diarrhea or septicemia. Binding of F17 *E. coli* to the microvilli could be inhibited by *N*-acetylglucosamine as well as its β4-linked oligomers. The appearance of F17 fimbriae is similar to that of type 1 and P fimbriae, being composed of a flexible tip fibrillum with an open helical structure connected to the end of a tightly wound helical rod made up of their major fimbrial subunit. F17-G, the carbohydrate-binding subunit of F17 is also similar to FimH and PapG: it is a two-domain protein linking a C-terminal pilin domain with an N-terminal carbohydrate-specific lectin domain.

The high-resolution crystal structure of the lectin domain of F17-G, in complex with *N*-acetylglucosamine revealed that the monosaccharide is bound on the side of the ellipsoid-shaped protein in a conserved site around which all natural variations of F17-G are clustered (Figure 11). Unexpectedly, the F17-G structure showed that the lectin domains of the F17-G, PapGII, and FimH fimbrial subunits, all share the immunoglobulin-like fold of the structural components (pilin domains) of their fimbriae, despite the lack of any sequence identity.

The binding site of the lectin domain of F17a-G, one of the several cloned variants of F17-G examined in complex with *N*-acetylglucosamine, is formed by the carbonyl group of Ala43, the side chains of Asp88, Thr89, Trp109, Ser117, Thr118, Gln119, and the nitrogen of Gly120 (Figure 12).[16] Interactions between the carbohydrate and the protein include 11 possible hydrogen bonds, of which four are mediated by water molecules, as well as the hydrophobic stacking of the Trp109 side-chain against the C5 and C6 atoms of the sugar. The *N*-acetyl group of *N*-acetylglucosamine contributes significantly to the affinity of the ligand Fim17a, due to a good complementarity of van der Waals surfaces between this group and the side chains of Thr118 and Asn44, as well as the carbonyl group of Ala43.

2.2   Genetics and Biosynthesis

Fimbrial genes are organized in operons that encode the major and minor fimbrial subunits, as well as the chaperones and other proteins required for the biosynthesis, secretion, and proper assembly of the fimbriae on the bacterial cell surface. Each fimbrial gene cluster also encodes one or more proteins that regulate gene expression.[46,47] Type 1 fimbriae are encoded by a 9.5 kb chromosomal region encompassing the *fim* gene cluster (Figure 13). The cluster contains at least nine open reading frames. *FimA* encodes the major subunit of the fimbrial shaft. *FimF, FimG,* and *FimH* encode minor structural proteins whereby *FimH* encodes the

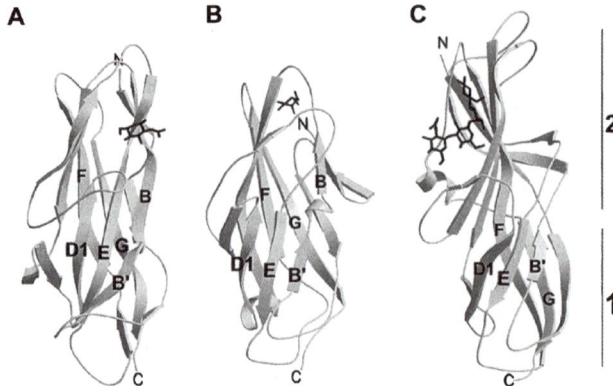

Figure 11 *Localization of the sugar-binding sites of F17a-G (A), FimH (B), and PapGII (C).
The proteins (represented in grey) were superimposed, based on the structural
core of the immunoglobulin fold which was identified in the three domains.
Representative carbohydrate ligands are shown in black. The C-termini of the
lectin domains, which precede the linker to the pilin domain, coincide approximately. Structurally equivalent strands are labelled with their names as defined for
F17-G. The two parts of the PapGII domain are indicated: part 1 has the
immunoglobulin-like core, whereas part 2 holds the sugar-binding site[16]*

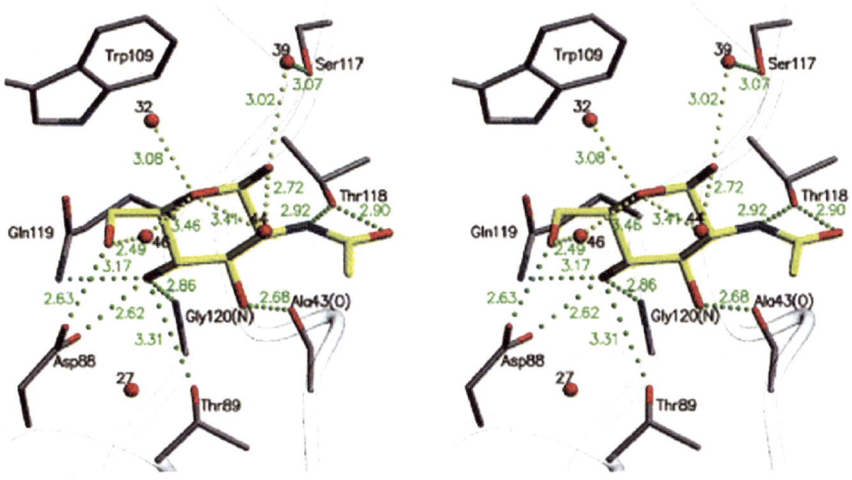

Figure 12 *Stereo view of N-acetylglucosamine in the binding site of F17a-G. Hydrogen
bonds are indicated by the dotted lines with heavy-atom distances in Å. Water molecules are represented as small spheres[16]*

carbohydrate-binding subunit of the fimbriae. *FimD* encodes the large outer membrane protein that anchors the fimbrial lectin to the the membrane and *FimC* encodes a chaperone protein that initiates proper folding of the subunits and protects them from proteolytic digestion. In fact, none of the structural fimbrial proteins is stable in soluble form; such forms of these proteins can be obtained only when they are complexed with their chaperone protein. *FimB* and *FimE* are genes involved in regulation of the expression of the fimbrial lectin.

As mentioned earlier, the expression of the fimbriae is phase-variable, *i.e.*, the bacteria shift periodically back and forth between a fimbriated and non-fimbriated state. As a result, a given bacterial population always contains cells of both phenotypes. The on and off phase variation is controlled at the transcriptional level. It involves the inversion of a 314-base-pair DNA sequence harboring the promoter of the *fimA* gene in the case of type 1 fimbriae and the methylation state of two particular sites of the DNA of the regulatory region of the *pap* operon in that of P fimbriae.

The FimH (or PapG) subunit is produced as a precursor with an N-terminal signal sequence that is removed during transport from the cell interior through the cytoplasmic membrane (Figure 14).[46,47] Further movement to the cell exterior relies on a fimbriae-specific export and multisubunit assembly system, a key component of which is the FimC chaperone. The chaperone stabilizes the fimbrial subunits in the periplasm through the formation of distinct complexes; in its absence the subunits are degraded. Stabilization is achieved by a donor complementation mechanism, in which the G1 β strand of the chaperone occupies the groove of FimH, thus completing its Ig fold (Figure 15).[48] The subunit-chaperone complexes are targeted to FimD, an outer membrane protein, that controls their ordered secretion and formation of an extracellular polymer. In the biosynthesis of P fimbriae, PapD acts as the chaperone, while PapC is the outer membrane protein, also designated as an "usher," due to its ability to allow the ordered passage of different fimbrial subunits through the cell membrane. Two additional proteins, PapH and PapK, appear to regulate the length of the fimbrial rod and the flexible tip, respectively. Similar chaperones are also required for the assembly of other types of fimbriae, and of non-fimbrial lectins.

During fimbrial assembly, the last strand of the N-terminal domain of an incoming subunit displaces the chaperone G1 strand, and occupies the cleft of the most recently incorporated subunit, a process referred to as "donor-strand exchange." This leads, simultaneously, to the release of the chaperone (Figure 15).[49] Therefore, in the mature fimbriae, every subunit completes the Ig fold of its neighbor. This mechanism of assembly deviates from the Anfinsen principle, according to which the primary structure of a protein is sufficient to define how it folds into a native, fully active tertiary structure.

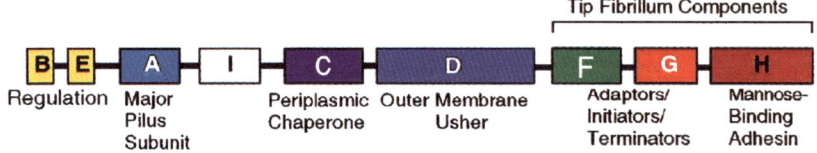

Figure 13 *Organization of the type 1 fimbrial gene cluster. (Reproduced with permission from Ref. 31. Copyright University of Chicago Press)*

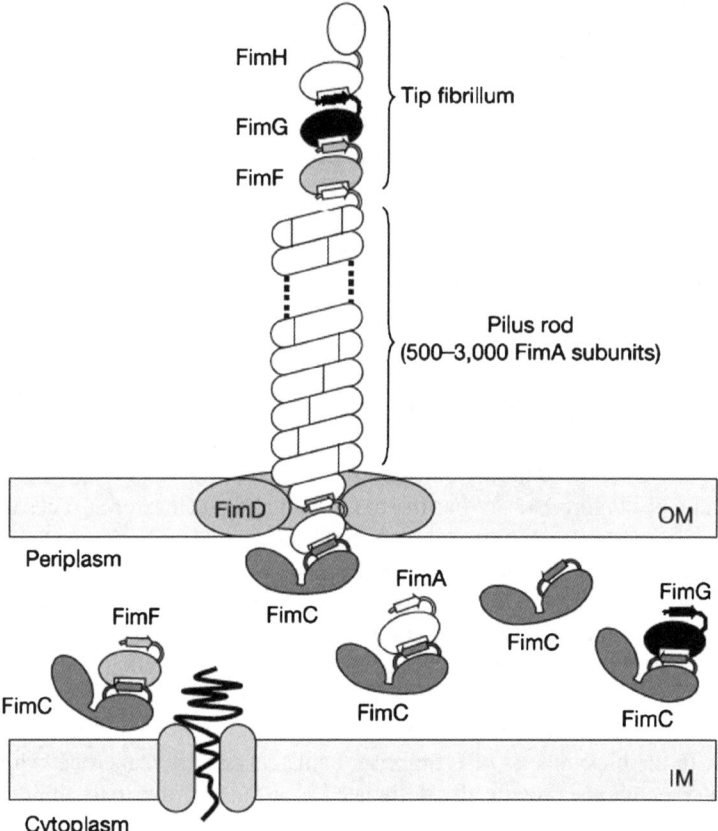

Figure 14 Schematic representation of type 1 pilus assembly according to the chaperone–usher pathway. Pilus subunits (FimA, FimF, FimG, and FimH) enter the periplasm in an unfolded conformation. They subsequently form 1:1 complexes with the pilus chaperone FimC. Upon interaction with the usher FimD, which forms a pore in the outer membrane, the chaperone–subunit complexes dissociate. While the chaperone remains in the periplasm, the subunits cross the outer membrane and become incorporated into the growing pilus. IM, inner membrane; OM, outer membrane (Reproduced with permission from Ref. 48)

Proof that donor-strand exchange from the neighboring subunit completes the folding of the pilin domain of FimH was obtained in experiments with a strand-complemented construct of FimH, consisting of the latter subunit extended at its C-terminus by a peptide corresponding to the G1 β strand of the N-terminal domain of FimG.[50] Unlike wild-type FimH, the strand-complemented variant could be expressed in *E. coli* in the absence of the FimC chaperone in the form of a proteolytically stable protein with wild-type mannose-binding activity.

Recent work has shown that FimC binds non-native fimbrial subunits and accelerates the folding of FimG by 100-fold.[48] These, and other results, have led to the conclusion that FimC represents a previously unknown type of protein-folding catalyst and simultaneously acts as a kinetic trap preventing spontaneous association in the cytoplasm.

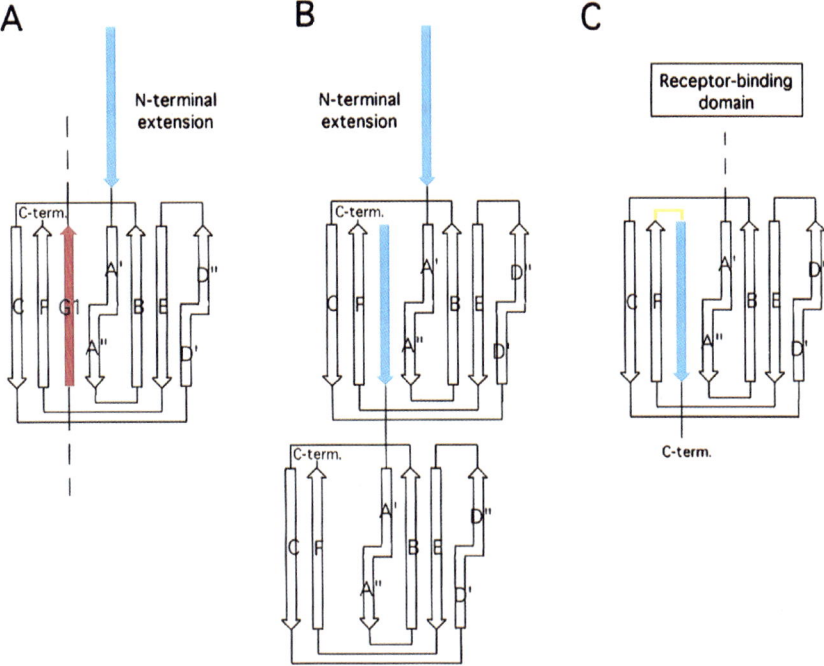

Figure 15 *P fimbrial domain topology diagrams. Dashes indicate an additional polypeptide not shown. (A) In donor-strand complementation, the chaperone contributes its G1 strand (black) to complete the immunoglobulin-like fold of the subunit (white). The completed fold is noncanonical because the G1 strand runs parallel to the subunit C-terminal F strand. The N-terminal extension is shown as a gray strand. (B) After donor-strand exchange, the N-terminal extension of one subunit completes the Ig fold of its neighbor in a canonical manner, as the N-terminal extension runs anti-parallel to the F strand. (C) Donor-strand-complemented FimH was constructed by fusing the N-terminal extension of FimG (grey), which is predicted to complete the fold of FimH in the pilus, till the C-terminus of FimH with a tetrapeptide linker (light gray). The topology of the receptor-binding domain is not shown, but its position relative to the FimH pilin domain is indicated by the framed box. (Reprinted with permission from Ref. 49. Copyright 2000, The National Academy of Sciences)*

2.3 Functions

As mentioned in the introduction, the major function of the enterobacterial surface lectins, similar to that of such lectins of other microorganisms, is to mediate the adhesion of the organisms to host cells, at the initial stage of infection. This has been extensively demonstrated both *in vitro*, in studies with isolated cells and cell cultures, and *in vivo* in experimental animals, and is supported in some cases also by clinical data. It is best documented for *E. coli* type 1 (for reviews of the earlier literature, see Refs. 2 and 51) and is summarized in Table 4. As for P fimbriae, they have been shown to enhance the early establishment of *E. coli* in the human urinary tract,[52] and a strong association has been found between the presence of P fimbriae

Table 4 *Evidence for the role of type 1 fimbriae of E. coli in infection**

---

1. Mutants of *E. coli* deficient in FimH, the carbohydrate-binding subunit of the fimbriae, are unable to cause cystitis in monkeys
2. Vaccination with a FimH–FimC complex protects mice and monkeys from colonization and infection by uropathogenic *E. coli*
3. Knockout mice lacking Tamm-Horsfall urinary glycoprotein, an inhibitor of type 1 fimbriated *E. coli*, are more susceptible than wild type mice to bladder colonization by the bacteria, but are equally susceptible to P fimbriated *E. coli*
4. Polystyrene latex beads coated with FimH readily associate and are internalized by human bladder cells *in vitro*
5. Type 1 fimbrial expression is associated with severity of urinary tract infection in children[54]
6. Mannose and methyl α-mannoside prevent urinary tract infection by type 1 fimbriated *E. coli* in experimental animals

---

*For references, see text, unless otherwise noted.

and disease severity, suggesting that adherence mediated by these organelles has also a direct effect on mucosal inflammation *in vivo*.[53]

Infection was also shown to depend on the presence in the target tissues of sugars recognized by the lectins, as illustrated by *E. coli* K99.[21] This organism binds to glycolipids containing *N*-glycolylneuraminic acid but not to those containing *N*-acetylneuraminic acid (Table 5). The former sialic acid is found on intestinal cells of newborn piglets, but is replaced by *N*-acetylneuraminic acid when the animals develop and grow. It is also not normally formed by humans, explaining why *E. coli* K99 can cause lethal diarrhea in piglets, but not in adult pigs or in humans.

Attachment of a pathogen to a tissue does not by itself initiate the disease. It must be coupled to specific responses that lead to infection. Adherence of P fimbriated *E. coli* or of the purified P-type fimbriae to galabiose of uroepithelial cells induces a two-way flow of biological cross-talk via the lectin bridge, affecting both partners.[55] Following adherence, the target cells are activated, with resultant production of cytokines that engender acute inflammation and other symptoms of disease, while in the bacteria the interaction leads to up-regulation of signal transduction systems that allow responses to the changing environment.

The FimH subunit of both *E. coli* and *K. pneumoniae* mediates not only bacterial adhesion, but also invasion of human bladder and intestinal epithelial cells, respectively.[56,57] In contrast, adhesion mediated by another pilus adhesin, PapG, did not initiate bacterial internalization. FimH-mediated invasion required localized host actin reorganization, phosphoinositide 3-kinase (PI 3-kinase) activation, and host protein tyrosine phosphorylation, but not activation of Src-family tyrosine kinases.[57] Phosphorylation of focal adhesin kinase (FAK) at Tyr397 and the formation of complexes between FAK and PI 3-kinase and between alpha-actinin and vinculin were found to correlate with type 1 fimbriae-mediated bacterial invasion. Inhibitors that prevented bacterial invasion also blocked the formation of these complexes. Thus, *E. coli* strains that cause urinary tract infections may not be strictly extracellular pathogens, and FimH can directly trigger host cell signaling cascades that lead to bacterial internalization.

Type 1 fimbriae mediate the attachment of *E. coli* not only to epithelial cells, but also to human polymorphonuclear cells and human and mouse macrophages, in the

Table 5 *Presence of N-glycolylneuraminic acid (NeuGc) in cell surface glycoconjugates and susceptibility to infection by enterotoxigenic E.coli K99**[*]

Animal	NeuGc	K99 infection
Piglets	+	+
Pigs	−	−
Humans	−	−

[*]Data from Ref. 22.

absence of opsonins. This is often followed by the ingestion and killing of the bacteria, a phenomenon we named "lectinophagocytosis."[58] Lectinophagocytosis is an early example of innate immunity; it may function *in vivo*, for example in sites poor in opsonins, such as the peritoneal cavity during peritoneal dialysis. Injection of type 1 fimbriated *E. coli* into the peritoneal cavity of mice led to the activation of the peritoneal macrophages; no activation was observed in the presence of methyl α-mannoside or when non-fimbriated bacteria were used.[59]

Enterobacteria can attach by their surface lectins to mast cells as well, with resultant activation of the target cells and production of high levels of certain cytokines, in particular TNF-α.[60] Activation of mast cells can also be induced by purified type 1 fimbriae, and by FimH. The cytokines released by the activated mast cells cause rapid recruitment of neutrophils into the site of infection, resulting in early clearance of the bacteria. As expected, mice, lacking mast cells, were significantly less efficient in clearing intranasal or intraperitoneal infection caused by *K. pneumoniae*.

Evidence for the role of type 1 fimbriae in infection comes also from the study of endogenous inhibitors of adhesion, which may be considered components of the innate immune system of the body. A case in point are the two major mucus-associated glycoproteins that carry oligomannosides, namely IgA[61,62] and Tamm-Horsfall glycoprotein[63,64] that interact with the type 1 fimbrial lectin in a mannose-specific fashion. Significantly, Tamm-Horsfall knockout mice were very recently shown to be considerably more susceptible to bladder colonization by type 1 fimbriated *E. coli* whereas they were equally susceptible to P fimbriated *E. coli*.[65,66] These results provide the first *in vivo* evidence indicating that under physiological conditions, Tamm-Horsfall glycoprotein can serve as an effective soluble receptor for type 1-fimbriated *E. coli* competitively inhibiting them from adhering to the uroplakin Ia receptors present on the urothelial surface.

Many of the different oligosaccharides present in human milk in high concentrations (a few mg to a few hundred mg per liter) are inhibitors of bacterial surface lectins of various bacteria. These oligosaccharides can be considered as contributing to the protection of breast-fed babies against infectious bacteria that carry lectins specific for such oligosaccharides.[67,68] Of particular interest in this respect are the fucosylated oligosaccharides (*e.g.*, Fucα2Galβ4GlcNAc) that act effectively as inhibitors of adhesion of the enteropathogen *Campylobacter jejuni*. Infants breast-fed on milk containing high levels of these oligosaccharides suffer from a considerably lower incidence of diarrhea than infants fed on milk with low levels of such oligosaccharides.

3   Anti-Adhesion Therapy

Because adhesion is an important aspect of the life style of microorganisms on surfaces in general and, in particular, on host tissues, attempts to employ various agents that block adhesion as therapeutics in animal models have started already in the late 1970s.[69,70] An attractive feature of anti-adhesive drugs is that they do not act by killing or arresting the growth of the pathogen, as antibiotics do. It is therefore highly likely that the spread of strains resistant to such agents would be significantly delayed as compared to that of strains resistant to antibiotics.

The first proof for the validity of the concept of anti-adhesive therapy was presented nearly three decades ago when we demonstrated that urinary tract infection by type 1 fimbriated *E. coli* could be effectively prevented by methyl α-mannoside, an inhibitor of adhesion of these bacteria to epithelial cells, but not by methyl α-glucoside, a non-inhibitory sugar.[71] In the following years, it has been shown that suitable carbohydrates can specifically block infection by different enterobacteria, as well as by other microorganisms, in a variety of experimental animals (Table 6).

Lectin-mediated adhesion and infection can be inhibited by suitable antibodies as well. Hybridoma antibodies directed against quaternary structural epitopes of type 1 fimbriae or against mannose, the sugar determinant in the complementary host cell receptor, prevented the attachment of the parent bacteria to various eukaryotic cells.[72] Intraperitoneal administration of the fimbriae-specific or mannose-specific antibodies protected mice against retrograde colonization with mannose-specific *E. coli* instilled into their urinary bladders. Monoclonal antibodies directed against *N*-acetylgalactosamine rather than mannose residues lacked protective activity. Antibodies against the putative carbohydrate-binding domain of FimH (amino acid residues 1–25) also protected mice against infection by *E. coli* type 1.[73]

Immunization of mice and non-human primates with the FimC–FimH complex protected them against urinary tract infection caused by uropathogenic *E. coli*.[74] Vaccination of cynomolgus monkeys with purified PagG–PapD complex of P fimbriae resulted in the development of specific antibody response in the animals. It also protected them from subsequent urethral challenge with a pyelonephritogenic P fimbriated *E. coli*.[75]

The evidence summarized above provides unequivocal proof that sugars can prevent infection of experimental animals by lectin-carrying bacteria. Only two clinical studies of this approach have been carried out to date, in which attempts were made to use specific oligosaccharides to prevent or clear infections caused by bacteria other than enteric ones.

In one of these, children 10–24 months old were treated prophylactically for 3 months with a nasal spray containing the antiadhesive pentasaccharide Sia3'LnnT.[76] The treatment failed to reduce the incidence of nasopharyngal colonization with *Streptococcus pneumoniae* and *Hemophilus influenzae*, and of acute otitis media. In another trial, orally administered NeuAcα2,3Galβ4GlcNAc failed to clear gastric colonization by *Helicobacter pylori* in human patients.[77] These failures could be either (a) because the inhibitors used were at insufficient levels or not powerful enough or (b) since during natural infection, bacteria express multiple lectins with diverse specificities, a cocktail of saccharides may be required for efficient action.

Table 6 *Inhibitors of carbohydrate-specific adhesion prevent enterobacterial infections in experimental animals**

Bacterial pathogen	Animal (site)	Inhibitor
Campylobacter jejuni	Mouse (GIT)†	Fucosyl oligosaccharides of human milk
Escherichia coli type 1	Mouse (UT)‡	MeαMan
	Mouse (GIT)	Mannose
	Mouse (UT)	Anti-Man-antibody
E. coli P	Mouse (UT)	Globoside
	Mouse (UT)	Galα4GalβOMe
E. coli K99	Calf (GIT)	Bovine serum glycopeptides
Klebsiella pneumoniae type 1	Rat (UT)	MeαMan
Shigella flexneri type 1	Guinea pig (eye)	Mannose
Streptococcus pnemoniae	Rabbit (lungs)	Lacto-*N*-tetraose

*For references, see Ref. 70.
†GIT, gastrointestinal tract
‡UT, urinary tract

The future of anti-adhesion therapy will depend on the development of powerful, non-toxic inhibitory agents, primarily based on carbohydrates or glycomimetics, which are targeted at one or several bacterial surface lectins, and that would also be capable of displacing bacteria that are attached to host tissues. Once such drugs become available, anti-adhesion therapy will turn out to be an important means for the fight against infectious diseases.

References

1. I. Ofek, D.L. Hasty and R.J. Doyle, *Bacterial Adhesion to Animal Cells and Tissues*, ASM Press, Washington, DC, 2003, 416pp.
2. I. Ofek and N. Sharon, *Curr. Topics Microbiol. Immunol.*, 1990, **151**, 91.
3. K.A. Karlsson, *Mol. Microbiol.*, 1998, **29**, 1.
4. N. Sharon and H. Lis, *Science*, 1989, **246**, 227.
5. N. Sharon and H. Lis, *Lectins*, 2nd edn, Kluwer Academic Publishers, Dordrecht, The Netherlands, 2003, 452pp.
6. N. Sharon and H. Lis, in *Glycoproteins II*, J. Montreuil, H. Schachter and J.F.G. Vliegenthart (eds), Elsevier, Amsterdam, 1997, 473.
7. S.N. Abraham, N. Sharon and I. Ofek, Adhesion and colonization. In *Molecular Medical Microbiology* (Sussman M., ed). Academic Press, New York, 2002, pp 629–644.
8. B.I. Eisenstein, *Science*, 1981, **214**, 337.
9. P. Klemm, *EMBO J.*, 1986, **5**, 1389.
10. G.E. Soto and S.J. Hultgren, *J. Bacteriol.*, 1999, **181**, 1059.
11. T. Iwahi, Y. Abe and K. Tsuchya, *J. Med. Microbiol.*, 1982, **15**, 303.
12. Y. Eshdat, F.J. Silverblatt and N. Sharon, *J. Bacteriol.*, 1981, **148**, 308.
13. N. Sharon and I. Ofek, *Glycoconjugate J.*, 2000, **17**, 669.

14. G.M. Ruiz-Palacios, L.E. Cervantes, P. Ramos, B. Chavez-Munguia and D.S. Newburg, *J. Biol. Chem.*, 2003, **278**, 14112.

15. A.S. Khan, B. Kniep, T.A. Geldschlaeger, I. Van Die, T. Korhonen and J. Hacker, *Infect. Immun.*, 2000, **68**, 3541.

16. L. Buts, J. Bouckaert, E. De Genst, R. Loris *et al.*, *Mol. Microbiol.*, 2003, **49**, 705.

17. J. Goldhar, *Methods Enzymol.*, 1999, **253**, 43.

18. I. Ofek, D. Mirelman and N. Sharon, *Nature*, 1977, **265**, 623.

19. I. Ofek and E.H. Beachey, *Infect. Immun.*, 1978, **22**, 47.

20. A. Larsson, J. Ohlsson, K.W. Dodson, S.J. Hultgren, U.L. Nillson and J. Kihlberg, *Bioorg. Medicinal Chem.*, 2003, **11**, 2255.

21. J. Bouckaert, J. Berglund, M. Schembri, E. De Genst, L. Cools, M. Wuhrer *et al.*, *Mol. Microbiol.*, 2005, **55**, 441.

22. M. Kyogashima, V. Ginsburg and H.C. Krivan, *Arch. Biochem. Biophys.*, 1989, **270**, 391.

23. S. Teneberg, P.T. Willemsen, F.K de Graaf, G. Stenhagen, W. Pimlott *et al.*, *J. Biochem.*, 1994, **116**, 560.

24. M.D. Disney and P.H. Seeberger, *Chem. Biol.*, 2004, **11**, 1701.

25. L.K. Mahal, *Chem. Biol.*, 2004, **11**, 1603.

26. N. Sharon, *FEBS Lett.*, 1987, **217**, 145.

27. E.V. Sokurenko, V. Chesnokova, D.E. Dukhuizen, I Ofek *et al.*, *Proc. Natl. Acad. Sci. USA*, 1998, **95**, 8922.

28. X.R. Wu, T.T. Sun and J.J. Medina, *Proc. Natl. Acad. Sci. USA*, 1996, **93**, 9360.

29. B. Madison, I. Ofek, S. Clegg and S.N. Abraham, *Infect Immun.*, 1994, **62**, 843.

30. I. Ofek, D.L. Hasty, S. Abraham and N. Sharon, *Adv. Exp. Med. Biol.*, 2000, **485**, 183.

31. J.D. Schilling, M.A. Mulvey and S.J. Hultgren, *J. Infect. Dis.*, 2001, **183**, S36.

32. N. Firon , I. Ofek and N. Sharon, *Carbohydr. Res.*, 1983, **120**, 235.

33. N. Firon, I. Ofek and N. Sharon, *Infect. Immun.*, 1984, **43**, 1088.

34. N. Firon, S. Ashkenazi, D. Mirelman, I. Ofek and N. Sharon, *Infect. Immun.*, 1987, **55**, 472.

35. B. Madison, I. Ofek, S. Clegg and S.N. Abraham, *Infect. Immun.*, 1994, **62**, 843.

36. M.J. Duncan, E.L. Mann, M.S. Cohen, I. Ofek, N. Sharon and S.N. Abraham, *J. Biol. Chem.*, 2005, **280**, 37707–37716.

37. D. Choudhury, A. Thompson, V. Stojanoff, S. Langermann *et al.*, *Science*, 1999, **285**, 1061.

38. C.S. Hung, J. Bouckaert, D. Hung, J. Pinkner, C. Widberg *et al.*, *Mol. Microbiol.*, 2002, **44**, 903.

39. N. Sharon and H. Lis, *Adv. Exp. Med. Biol.*, 2001, **491**, 1.

40. E.V. Sokurenko, M.A. Schembri, E. Trinchina, K. Kiaegaard *et al.*, *Mol. Microbiol.*, 2001, **41**, 675.

41. R.J. Pieters, *Trends Glycosc. Glycotech. TIGG*, 2004, **16**, 243.

42. N. Nagahori, R.T. Lee, S.-I. Nishimura, D. Pagé, R. Roy and Y.C. Lee, *Chem. Biochem.*, 2002, **3**, 836–844.

43. N. Strömberg, P.-G. Nyholm, I. Pascher and S. Normark, *Proc. Natl. Acad. Sci. USA*, 1991, **88**, 9340.

44. K.W. Dodson, J.S. Pinkner, T. Rose, G. Magnuson, S.J. Hultgren and G. Waksman, *Cell*, 2001, **105**, 733.

45. F.G. Sauer, M. Barnhart, D. Choudry, S.D. Knight *et al.*, *Curr. Opin. Struct. Biol.*, 2000, **10**, 548.
46. D.L. Hung and S.J. Hultgren, *J. Struct. Biol.*, 1998, **124**, 201.
47 S.D. Knight, J. Berglund and D. Choudhury, *Curr. Opin. Chem. Biol.*, 2000, **4**, 653.
48. M. Vetsch, C. Puorger, T. Spirig, U. Grauschopf, E.U. Weber-Ban, R. Glockshuber, *Nature*, 2004, **431**, 329.
49. Normark, *Proc. Natl. Acad. Sci. USA*, 2000, **97**, 7670.
50. M.M. Barnhart, J.S. Pinkner, G.E. Soto, F.G. Sauer *et al.*, *Proc. Natl Acad. Sci. USA*, 2000, **97**, 7709.
50. P.E. Orndorff and C.A. Black, *Microbiol. Pathog.*, 1990, **9**, 75.
52. B. Wullt, G. Bergsten, H. Connell, P. Rollano, N. Gebratisedik *et al.*, *Mol. Microbiol.*, 2000, **38**, 456.
53. B. Wullt, G. Bergsten, H. Connell, P. Rollano, N. Gebratisedik *et al.*, *Cell Microbiol.*, 2001, **3**, 255.
54. H. Connell, W. Agace, P. Klemm, M. Schembri, S. Märild and C. Svanborg, *Proc. Natl. Acad. Sci. USA*, 1996, **93**, 9827.
55. S.N. Abraham, A.B. Jonsson and S. Normark, *Curr. Opin. Microbiol.*, 1998, **1**, 75.
56. J. Schilling, M.A. Mulvey and S.J. Hultgren, *Urology*, 2001, **57**(6 Suppl. 1), 56.
57. J.J. Martinez, M.A. Mulvey, J.D. Schilling, J.S. Pinkner and S.J. Hultgren, *EMBO J.*, 2000, **19**, 2803.
58. I. Ofek and N. Sharon, *Infect. Immun.*, 1988, **56**, 539.
59. W. Bernhard, A. Gbarah and N. Sharon, *J. Leuk. Biol.*, 1993, **52**, 343.
60. R. Malaviya and S.N. Abraham, *Immunol. Rev.*, 2001, **176**, 16.
61. R.C. Williams and R.J. Gibbons, *Science*, 1972, **177**, 697.
62. A.E. Wold, J. Mestecky, M. Tomana, A. Kobata *et al.*, *Infect. Immun.*, 1990, **58**, 3073.
63. S.M. Kuriyama and F.J. Silverblatt, *Infect. Immun.*, 1986, **51**, 193.
64. F. Serafini-Cessi, A. Monti and D. Cavallone, *Glycoconjugate J.*, 2005, **22**, 383.
65. L. Mo, X.H. Zhu, H.Y. Huang, E. Shapiro, D.L. Hasty and X.R. Wu, *Am. J. Physiol. Renal Physiol.*, 2004, **286**, F795.
66. J.M. Bates, H.M. Raffi, K. Prasadan, R. Mascarenhas *et al.*, *Kidney Int.*, 2004, **65**, 791.
67. D.S. Newburg, G.M. Ruiz-Palacios and A.L. Morrow, *Annu. Rev. Nutr.*, 2005, **25,** 37.
68. A.L. Morrow, G.M. Ruiz-Palacios, X. Jiang and D.S. Newburg, *J. Nutr.*, 2005, **135**(Suppl.), 1306.
69. I. Kahane and I. Ofek (eds), *Toward Anti-adhesion Therapy of Microbial Diseases*, Plenum Press, New York, 1996, 288pp.
70. I. Ofek, D.L. Hasty and N. Sharon, *FEMS Immun. Med. Microbiol.*, 2003, **38**, 181.
71. M. Aronson, O. Medalia, D. Schori, D. Mirelman, N. Sharon and I. Ofek, *I. Infect. Dis.*, 1979, **139**, 329.
72. S.N. Abraham, J.P. Babu, C.S. Giampapa, D.L. Hasty, W.A. Simpson and E.H. Beachey, *Infect Immun.*, 1985, **48**, 625–628.
73. K. Thankavel, B. Madison, T. Ikeda, R. Malaviya *et al.*, *J. Clin. Invest.*, 1997, **100**, 1123.
74. S. Langermann and R. Ballou Jr., *J. Infect. Dis.*, 2001, **183**, S84.

75. J.A. Roberts, M.B. Kaack, G. Baskin, M.R. Chapman *et al.*, *J. Urol.*, 2004, **171**, 1686.
76. F. Ukkonen, K. Varis, M. Jernfors and E. Herva, *Lancet*, 2000, **356**, 1390.
77. F. Parente, C. Cucino, A. Anderloni, G. Grandinetti and G.B. Porro, *Helicobacter*, 2003, **8**, 252.

CHAPTER 5

GM1 Glycomimetics and Bacterial Enterotoxins

ANNA BERNARDI[1], ČRTOMIR PODLIPNIK[1], AND JESÚS JIMÉNEZ-BARBERO[2]

[1]Dipartimento di Chimica Organica e Industriale, Universita' di Milano, via Venezian 21, 20133 Milano, Italy
[2]Centro de Investigaciones Biológicas, CSIC, Ramiro de Maeztu 9, 28040 Madrid, Spain

1 Introduction

Molecular recognition processes between carbohydrates and proteins play crucial roles in numerous biological mechanisms, including intracellular and intercellular routing processes, endocytosis of deleterious glycoconjugates, and inflammation.[1–3] Recognition of carbohydrate ligands by bacterial and mammalian lectins are key examples of these types of phenomena, and specific inhibition of recognition events of this type have been proposed as therapeutic approaches for prevention of viral and bacterial infections.[4,5] Owing to the relatively weak affinity of the interactions between single sugar and protein molecules, multivalency is a frequent strategy in carbohydrate–protein systems to achieve tight binding.[6] Among the processes of biomedical interest in which glycoconjugates are implicated, those involving the AB5 toxins are the focus of this chapter. The AB5 bacterial toxins[7] are dubbed after their characteristic architecture comprising a single catalytically active component, A, and a nontoxic receptor-binding component, a pentamer of B subunits.[8] The B pentamer is responsible for binding to gangliosides at the cell surface.[9] This recognition function is retained even in the absence of the A subunit. In contrast, the complete AB5 holotoxin is required for the toxic effects. There are several families of AB5 toxins, classified according to their sequence homology and catalytic activity. Here, we will focus on the cholera toxin (CT) family, which includes CT itself, produced by *Vibrio cholerae*, and the *Escherichia coli* heat-labile toxins (LTs) LT-I and LT-II,[10] among others. They are enterotoxins responsible for different effects on human populations, from the relatively mild travelers' diarrhea caused by LT-I to the much more serious and sometimes life-threatening cholera caused by *Vibrio cholerae*. Cholera toxin and

LT-I share 80% sequence homology in both the A and the B subunits, whereas they have lower sequence homology to LT-II. In terms of size, both CT and LT consist of a 27-kDa catalytic A domain anchored in the ring formed by the five identical 11.7-kDa B subunits.[7] However, there are unique fine details in the molecular recognition process of either toxin. The B pentamer of CT (CTB) binds exclusively to the GM1 ganglioside, while that of LT (LTB) additionally binds to other gangliosides.[11]

Other families include the Shiga toxin (SHT) family, members of which are the causative agents of the hemolytic uremic syndrome ("hamburger disease"), and the Shiga-like toxins (SLTs), also known as verotoxins.[12–19]

The AB5 toxins have a wide range of toxic effects on humans. It has been estimated[9] that they are probably responsible for over a million deaths annually worldwide and therefore remain a severe medical and social problem.

In terms of the mechanism of action, the first steps of the infection process have been elucidated at the molecular level. The threatening action of CT is initiated by binding of the B subunits to the GM1 ganglioside on intestinal epithelial cell membranes. This binding event is followed by nicking of the A chain, and disulfide bond reduction, which yield the two fragments A1 and A2. The A1 subunit is then translocated into the cell where it mono-ADP-ribosylates Gsα, a guanine nucleotide-binding regulatory protein, resulting in persistent adenylyl cyclase activation.[20,21]

2 Structure and Interaction Features of AB5 Toxins: An Overview

The structure and function of the AB5 toxins have been reviewed in detail on several occasions.[9] Several high-resolution structures of AB5 toxins with or without bound ligands are available,[22–29] as well as binding data obtained through a variety of biophysical techniques, including NMR, X-ray, solid-phase and TLC overlay assays,[30,31] surface plasmon resonance (SPR),[32,30] fluorescence spectroscopy[33,34] and flow cytometry (FACS),[35] atomic force microscopy (AFM),[36] and isothermal titration calorimetry (ITC).[37–39] Given the importance of the processes that they promote, the study of the complexes formed between gangliosides and AB5 toxins at different levels is very relevant. For basic research, they offer a paradigmatic model for studying the structural and thermodynamic basis of protein–carbohydrate interactions at atomic resolution. For medicinal chemistry, they may provide key insights for the structure-based design of ligands that can potentially be used to fight the above-mentioned diseases. Structure-based design of galactose-based ligands for CT and LT has been reported and recently reviewed.[27,40]

3 The Interaction with GM1

The B pentamer of CT (CTB) interacts with the soluble, monovalent oligosaccharide portion of GM1 (GM1os, **1**, Chart 1) with a strong affinity. The binding process is weakly cooperative and the dissociation constant for the monovalent interaction of one GM1os with one pentamer binding site, initially estimated as 1 μM,[39] has recently been re-evaluated by ITC and found to be 43 nM[38] at room temperature, which places it among the highest affinity protein–carbohydrate interactions described to date. In biological settings this initial binding is further amplified through positive cooperativity between the B subunits and through the multivalent interaction of the B5 pentamer with multiple copies of GM1 presented at high concentration in lipid rafts at the cell surface.[35]

A number of crystal structures of complexes between galactose-containing derivatives and CTB or LTB have been reported,[22–28] also including that for the CTB–GM1os complex.[23] In all such complexes, the orientation of galactose and the structure of the binding site are essentially identical. Comparison of the published structures of the bound and free pentamers reveals that there are only minor changes in the backbone conformations upon complexation, basically adjustments around the galactose-binding site.[4,28] The 1.25-Å resolution structure deposited for the CTB–GM1os complex[23] shows a bidentate interaction of the branched GM1os pentasaccharide (Figure 1). This type of interaction resembles a "two-fingered grip" comprising a sialic acid thumb and a Galβ(1→3)GalNAc forefinger.

Most of the contacts are given by the "finger" tips: in fact, in terms of buried surface area, the terminal Gal and Neu5Ac residues contribute with *ca.* 80% of the intermolecular contacts. There are many intermolecular hydrogen-bonding contacts, both directly between ligand and receptor and also via bridging water molecules. The terminal galactose residue in the "forefinger" is reaching a well-defined galactose-binding pocket,

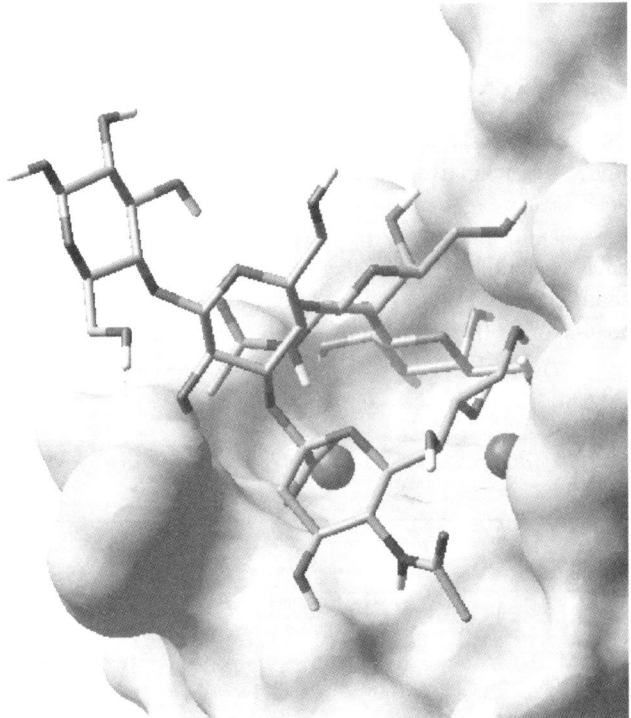

Figure 1 *CTB:GM1os complex. X-ray structure. (From Kuhn et al., J. Mol. Biol., 1998, **282**, 1043–1059)*

lined by the indole side chain of Trp-88 and shielded from the solvent. The rest of the toxin-binding site is shallow and solvent exposed. The NeuAc "thumb" interacts with a carboxylate-binding region, which includes one highly conserved crystallographic water molecule.

The comparison between the bound GM1os pentasaccharide and previously reported NMR-based solution structures of GM1os (Figure 2a) indicates[41–46] that this sugar ligand is essentially preorganized for a near lock-and-key interaction with CTB. Such preoganization appears to be the source of the unusually high binding affinity of the CTB:GM1os pair. This qualitative observation was recently reinforced by a detailed ITC study of the thermodynamics of CT binding by GM1os and fragments thereof, reported by Homans and co-workers.[38] The observed dissociation constants (Table 1) vary over 7 orders of magnitude from the nanomolar range for GM1os **1** to the high millimolar range for methyl sialoside **4**. From the results mentioned in Table 1, it is clear that the sum of $\Delta G°$ for complementary fragments of GM1os (*i.e.*, GM2os + Galβ OMe and Galβ13GalNAcβOMe + Neu5AcαOMe) falls a long way short of $\Delta G°$ for the entire ligand. Analytical treatment of these data using Jencks model[47] allows one to account for the different entropic effects at play for a bidentate ligand or for two complementary fragments binding to the receptor. Even after this treatment is applied, the data strongly suggest that presentation of the binding determinants in the appropriate orientation for toxin recognition is the key element influencing the high affinity of GM1os to CTB.

4   The Use of GM1-Glycomimetics

Oligosaccharides notoriously are conformationally mobile molecules, so why is GM1os so rigid? A comparative analysis of the conformational behavior of various ganglioside head groups suggests that branching at the Gal-II residue (see Chart 1) is responsible for much of the GM1os rigidity. Indeed, there are several studies on the conformational behavior of GM1 and other ganglioside head groups (*e.g.*, GM1, GM2, GM3, GM4, and asialo-GM1 [GA1] oligosaccharides; see Chart 1).[41–46,48–60]

The combination of NMR spectroscopy and theoretical calculations have shown that the Galβ-1,3-GalNAcβ-1,4-Gal trisaccharide (GM1, GA1) or its GalNAcβ-1,4-Gal

Table 1 *Thermodynamic parameters for GM1os and its fragments binding to CTB at 25° C[a]*

Ligand	K_d *(mM)*	$\Delta G°$ *(kJ mol^{-1})*	$\Delta H°$ *(kJ mol^{-1})*	$T \Delta S°$ *(kJ mol^{-1})*
GM1os **1**	$(4.33 \pm 0.14)\ 10^{-5}$	-41.6 ± 0.1	-72.4 ± 0.1	-30.8 ± 0.1
GM2os **2**	2.0 ± 0.2	-15.2 ± 0.2	-18.0 ± 2.0	-2.8 ± 2.0
Galβ13GalNAcβOMe **3**	7.6 ± 0.8	-12.0 ± 0.3	-42.1 ± 1.8	-30.1 ± 1.9
Neu5AcαOMe **4**	210 ± 100	-3.8 ± 1.7	-44.4 ± 35.7	-40.6 ± 34.6
GalβOMe **5**	14.8 ± 1.6	-10.4 ± 0.3	-37.4 ± 2.0	-27.0 ± 2.0
GalNAcβOMe **6**	118 ± 12	-5.3 ± 0.2	-38.4 ± 1.9	-33.1 ± 1.9

[a]From Turnbull *et al.*, *J. Am. Chem. Soc.*, 2004, **126**, 1047–1054.

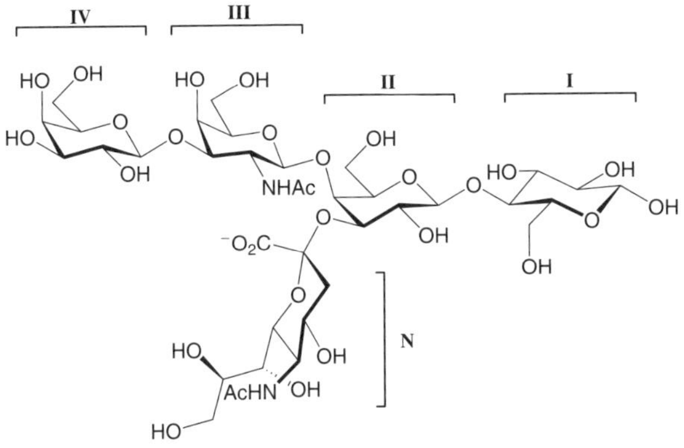

	IV	III	N	II	I
1 GM1os	Galβ1-3GalNAcβ1-4(Neu5Acα2-3)Galβ1-4Glcβ1-OH				
2 GM2os	GalNAcβ1-4(Neu5Acα2-3)Galβ1-4Glcβ1-OH				
GM3os	Neu5Acα2-3Galβ1-4Glcβ1-OH				
GM4os	Neu5Acα2-3Galβ1-OH				

	IV	III	II	I
GA1os	Galβ1-3GalNAcβ1-4Galβ1-4Glcβ1-OH			

Chart 1 *Ganglioside head groups*

fragment (GM2) always adopts one major conformation, whereas the NeuAcα-2,3-Gal fragment is more flexible and can assume the two conformations shown in Figure 2b, which differ by the orientation of the NeuAc residue, at χ –60° or χ 180°. The two conformations are almost equally represented when Gal-II O4 is not substituted, as in GM4 and GM3. In contrast, the χ 180° conformation is strongly stabilized in the Gal-II branched gangliosides, such as GM1 (Figure 2a) and GM2. This branching has also chemical and biological consequences. Several studies have shown that the Gal-branched head groups of GM1 and GM2 are less mobile and less accessible to glycosidases (both sialidases and hexosaminidases) than the unbranched structures. Moreover, distinct flexibility has been described for the internal and external sialic acid residues of disialogangliosides, such as GD1a. The low enzymatic susceptibility of GM1 and GM2 may be explained by the restricted access to the enzyme approach owing to the steric hindrance that results from the proximity of the sialic acid and GalNAc rings in the branched structure. This structural feature may also be the ultimate cause of the observed decrease of conformational mobility of the sialic acid residue.

Thus, the loss of conformational freedom in GM1 is very likely to result from the 3,4-branching at Gal-II, and this residue appears to act as the scaffold holding the two terminal Gal-IV and NeuAc moieties at the proper place and distance for optimal interaction with CT. From the above structure-based hypothesis, Bernardi *et al.*[61] have

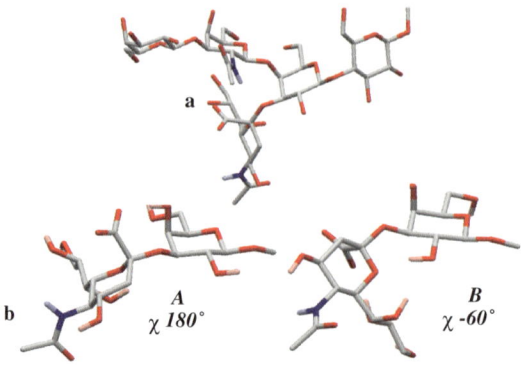

Figure 2 *(a) The experimental conformation of ganglioside GM1. The NeuAcα-2,3-Gal fragment adopts the A (χ 180°) orientation. (b) The two available conformations of the NeuAcα-2,3-Gal fragment in unbranched gangliosides (GM4 shown, A χ 180°, B χ –60°). The improper dihedral angle χ is defined as NeuAcC(=O) – NeuAcC2 – GalC3 – GalH3*

designed and prepared, using organic synthesis, a series of ganglioside mimics by replacing the scaffold element with an appropriate diol, chosen to reproduce the topological features of a 3,4-disubstituted galactose. As it will be shown below, the terminal Gal-IV moiety has always been kept in the mimics, while the scaffold and the GalNAc and NeuAc units have been replaced with simpler chemical fragments. In a long-term collaboration, the groups of Bernardi and Jiménez-Barbero have then used NMR spectroscopy assisted by molecular modeling methods to determine the structural and conformational similarity between the newly designed molecules and GM1 both in the free and CTB-bound states, and have used this information to improve the design of the mimics. In particular, NOE-based and STD NMR experiments, together with docking protocols, have allowed to propose a binding mode to CTB for the first generation of mimics and to provide simpler CTB binders. The ability of the synthetic mimics to functionally substitute ganglioside GM1 has been tested in affinity assays toward cholera toxin (CTB), giving moderate to good results.

According to the interpretation of the structural data described above, the central 3,4-disubstituted Gal-II unit of GM1 (Chart 1) acts as a scaffold element, and its topological properties are crucial in defining the proper placement of the binding determinants in GM1os. Thus, the design of the first mimic **7**[62] (Chart 2) was based on replacing the Gal-II unit with **8**[63] (DCCHD, Chart 2), a cyclohexanediol that possesses the same absolute and relative configuration of natural galactose and is locked in a single-chair conformation, as shown by molecular mechanics calculations and NMR data. According to the above rationale, the pseudo-tetrasaccharide **7** should behave as a structural mimic of GM1. Furthermore, since both CT and the related heat-labile enterotoxin of *E. coli* recognize GM1 through the terminal Gal-IV and NeuAc residues, the new mimic **7** should also be able to reproduce its binding features. A computational protocol initially allowed validation of this working hypothesis.[62] Thus, the conformation of **7** was studied both for the isolated molecule and in the binding pocket of LTB, and the predicted three-dimensional (3D) structures were

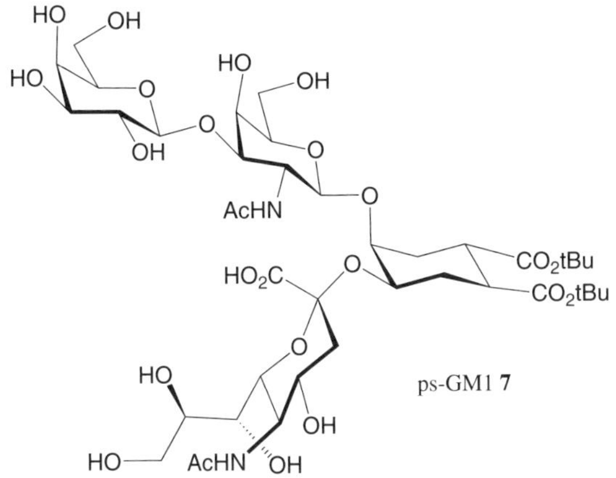

Chart 2 *The ps-GM1mimic 7 and the dicarboxy cyclohexanediol (DCCHD) scaffold 8*

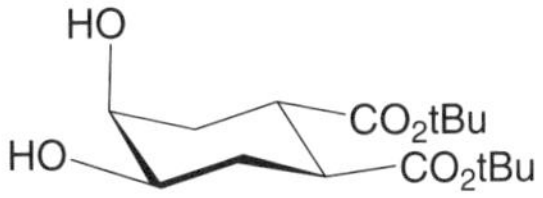

compared with those described for GM1.[46,64] The calculation predicted that the LTB:**7** complex would feature all the expected intermolecular interactions. The superimposition of the computer model of LT:**7** with the X-ray structure of CT:GM1 showed that **7** and GM1 adopt a common disposition in the toxin-binding pocket.

After synthesis of **7**, NMR studies in both D$_2$O and DMSO-d$_6$ solutions showed a characteristic set of NOE contacts, consistent with a single dominant conformation for both the Neu5Acα2-3Gal and GalNAcβ1-4Gal glycosidic linkages.[62] Indeed, the corresponding NOE distances for **7** were very similar to those observed for GM1. Binding of **7** to CT was investigated by inhibiting the GM1:CT interaction in ELISA assays and by fluorescence spectroscopy.[62] The ELISA and the fluorescence emission profiles of **7** and of the GM1 pentasaccharide **1** as reported in Figure 3 are clearly overlapping. Furthermore, both substrates appear to bind cooperatively to CTB, albeit with different cooperativity factors. Thus, this initial study validated the structural hypothesis described in the first part of this chapter, and allowed design and synthesis of a GM1 mimic that shares the exceptional CT-binding properties of the natural template, while reducing the structural complexity and the sugar-like character of GM1os. Further simplification of the pseudo-GM1 structure of **7** was pursued in the second part of the project.

Stereoselective sialylation is a major problem in synthetic carbohydrate chemistry and, indeed, sialylation of **8** represents the bottleneck of the synthesis of **7**. Hence, replacement of the Neu5Ac residue with more treatable chemical entities was envisaged as a way of considerably increasing the synthetic accessibility of the artificial ligands. In this context, simple α-hydroxyacids as mimics of the Neu5Ac moiety were examined in the series of second generation mimics **9–11** (Chart 3).[65] Initial modeling data on **9–11** showed that, although more flexible than **7**, the (*R*)-lactic acid derivative **11** and, to a minor extent, the glycolic acid derivative **9**, should be able to simultaneously fit the galactose and the carboxy-binding sites of CT using low-energy conformations. The interaction of **9–11** with CTB was then probed using fluorescence titrations. In contrast to the NeuAc-containing ligands **1** and

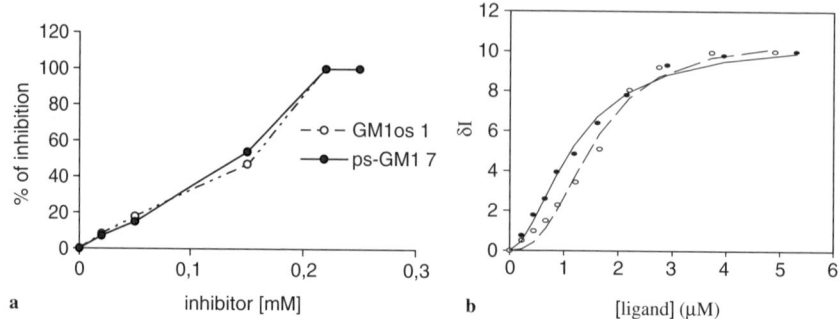

Figure 3 *CTB affinity of the pseudo-GM1 ligand 7. (**a**) Inhibition of GM1:CT binding in ELISA assays by GM1os (open circles, dashed line) and by **7** (closed circles, solid line). (**b**) Fluorescence titrations of CTB (0.5 μM) with micromolar amounts of GM1os (open circles, dashed line) or of mimic **7** (closed circles, solid line)*

7, binding of **9–11** to CTB is not cooperative, thus suggesting the involvement of additional interactions in GM1 with CTB that are not contained in the glycomimetics. Within this series **9–11**, the (*R*)-lactic acid derivative **11** displays the strongest affinity for CTB. The dissociation constants determined by nonlinear regression analysis are 667 μM for **9**, 1.1 mM for **10**, and 190 μM for **11**. For comparison, CTB has a dissociation constant of about 40 and 81 mM for galactose and lactose, respectively, while asialo-GM1 (GA1) showed no detectable binding to CT. Thus, the carboxy group of **9–11** appears to have a measurable effect on the affinity of the artificial receptors for the toxin.

Further improvement of binding in this series was sought based on the knowledge of their binding mode to CTB. Thus, NOESY- and ROESY-based NMR studies were performed on **9–11** both in the free and bound states.[66] In particular, exchange-transferred NOE and STD experiments allowed us to deduce the bioactive conformation and the binding epitope of the ligand. The NMR data were interpreted with the aid of computer models, generated and analyzed using a combination of different approaches. In the free state, the three molecules were shown to be rather flexible, especially in the hydroxyacid region. However, there is a process of conformational selection upon binding to CTB. Indeed, CTB selects a conformation similar to the global minima of the free pseudo-saccharides from the ensemble of presented conformations. No evidence of major conformational distortion was obtained, especially for the Galβ1-3GalNAc and GalNAcβ1-DCCHD linkages (see Chart 3), but just one or two out of the available staggered rotamers of the hydroxy acid side chain are selected in the bound state. The selected orientation allows fitting to the galactose-binding pocket of CT while placing the carboxy group of the hydroxy acid side chain in the carboxylate-binding region of the protein, and this determines the orientation of the methyl substituent in the complexes formed by **10** and **11**. The model of the complexes obtained as the best fit with the exchange-transferred NOE and STD experimental data are shown in Figure 4. It is clear that the methyl group in **10** sticks out in the solvent, whereas the methyl group of **11** points toward a hydrophobic area located in the vicinity of the sialic acid sidechain binding region of the CT:GM1 complex (Figure 4).

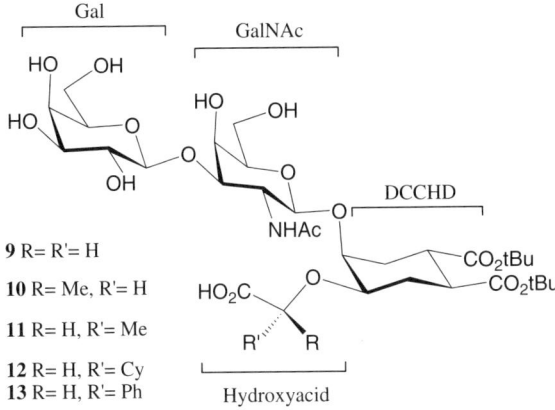

Chart 3 *Hydroxyacid mimics*

Figure 4 *Suggested binding mode of **10** and **11** to CT, based on modeling and NMR data*

The weakest STD signals were observed for the CHD moiety protons, indicating that this part of the molecule is not involved in the molecular recognition process to a significant extent. Thus, the 3D models obtained for the complexes suggested that the higher affinity for CTB of the *R*-derivative **11** relative to **9** and **10** could be because of the establishment of stabilizing van der Waals interactions between the methyl group of **11** and the toxin. From this model, it was suggested that the affinity of the pseudo-GM1-based glycomimetics might be improved by replacing the *R*-lactic moiety of **11** with appropriate hydrophobic fragments.[66]

Thus, a new generation of pseudo-GM1 ligands, **12** and **13**, which include a cyclohexyl group and a phenyl group, respectively, was prepared[67] and exhaustively studied by fluorescence titrations, NMR, and modeling.[68] In the free state, the conformational behavior of the cyclohexyl derivative **12** does not depart from the parent compound **11**: the NOE data indicated that several rotamers are contributing to the conformational equilibrium of the hydroxy acid side chain, whereas the "upper" Galβ(1→3)GalNAcβ(1→4)DCCHD fragment (see Chart 3) can be described by oscillations around one major conformation. In contrast, the phenyl derivative **13** is significantly less flexible than **12**, and a single conformation of the hydroxyacid chain can describe its conformational behavior in aqueous solution. Furthermore, the aromatic protons of **13** gave strong NOE cross peaks to the protons located at the α face of the GalNAc residue (H-1, H-3, and H-5; see Figure 5).

Thus, it appears that there is only one major solution conformation of **13**, in which the phenyl moiety stacks below the GalNAc residue, and that is depicted in Figure 5a. The origin of this conformational lock seems to arise from intramolecular van der Waals and CH-π aromatic–sugar interactions. These interactions require strict geometric and electronic features that are not achievable within the cyclohexyl derivative **12**. Although the importance of carbohydrate–aromatic interactions for the molecular recognition of oligosaccharides by the binding sites of proteins has been well documented,[69–73] this case represented a clear-cut evidence of the importance of this type of interaction for favoring a given type of 3D structure.

TR-NOESY[74–78] and STD NMR[79–81] experiments were performed to deduce the bound conformations of **12** and **13** to the CTB pentamer. For the phenyllactic acid ligand **13** the cross-peak pattern was similar to that observed for free **13** in aqueous solution. Thus, only the conformation, which displays the GalNAc/aromatic stacking, is bound by CTB. The aromatic protons, together with the Gal protons, gave rise to the most prominent STD signals (Figure 6). Thus it can be inferred that this region (aromatic ring and Gal moiety) of the glycomimetic is in more intimate contact with the CTB-binding site. Again, no STD signals were observed for the DCCHD moiety protons, indicating that this part of the molecule is not involved in the molecular recognition process to a significant extent.

In contrast, the STD spectrum of the CTB:**12** mixture did not show any peak for the pendant cyclohexane, thus indicating that this region establishes only marginal interactions with the protein, as compared to the aromatic ring of the phenyllactic derivative, **13**.

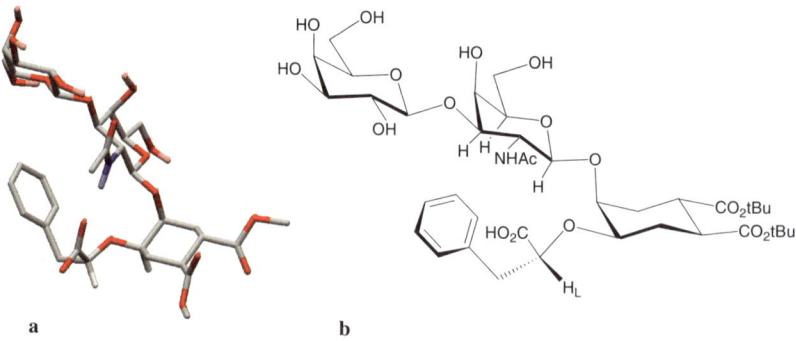

a   b

Figure 5 *Preferred solution conformation of the phenyllactic derivative **13** as deduced from ROESY spectra. (a) Computer-generated 3D-structure. (b) Structure of **13**, showing the NOE contacts observed for HL*

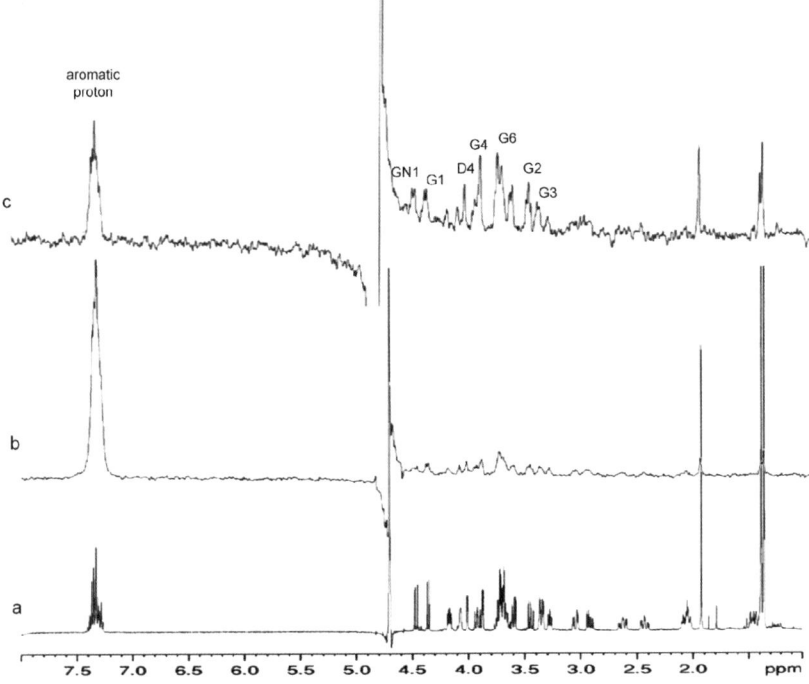

Figure 6 *(a) Reference 1D ^{1}H-NMR spectrum of **13** with CTB (ratio 25:1) in D$_2$O. (b) STD spectrum with 150 msec of presaturation of the protein envelope protons. The aromatic protons (7.35 ppm) appear to be the strongest signals. (c) Additional ligand protons from the sugar portion become evident at a longer presaturation time (1 sec)*

The 3D models of both complexes were then obtained by MC/EM and MC/SD calculations. The corresponding views are collected in Figure 7. The typical Gal/CTB interaction, which defines the structure of CTB complexes, is evident. In addition, the phenyl ring of **13** in the complex (Figure 7b) is locked between the protein and the GalNAc ring. A clear intramolecular carbohydrate–aromatic stacking interaction takes place, locating the phenyl ring in the proper position to establish additional contacts with the hydrophobic patch in the protein. In contrast, in the LT:**12** complex, the cyclohexyl ring displays more flexibility and projects away from the protein surface, with the Gal moiety establishing the major contacts with the toxin. In this case, no contacts of the GalNAc moiety with the aliphatic cyclohexyl side chain are seen. Dynamic simulations of the complexes also showed a high mobility of the cyclohexyl ring, which fluctuates in and out of the binding site. In contrast, dynamics simulations of the LT:**13** complex show that the phenyl group remains permanently in contact with the protein.[68]

Although **12** and **13** are similar in nature and activity, a detailed analysis of their behavior in solution and in the binding site of CT reveals striking differences, both in terms of conformational flexibility and of binding mode. Intra- and intermolecular interactions among the different residues strongly modulate the conformational

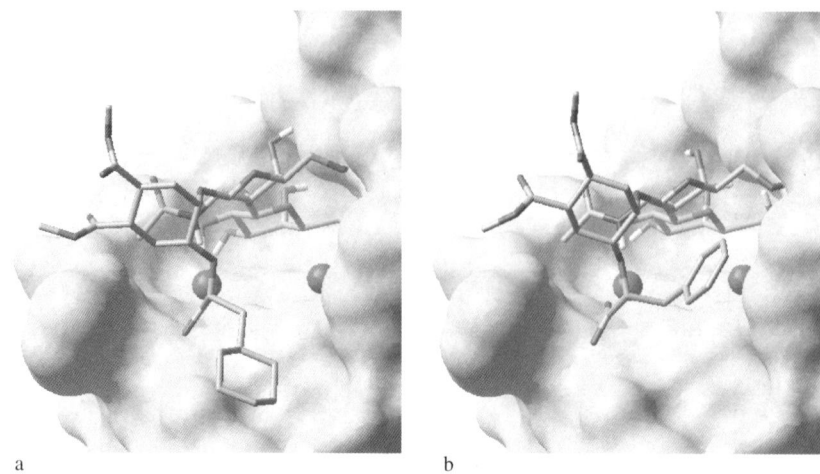

Figure 7 *(a) CT:12 complex. (b) CT:13 complex*

features of these molecules in solution and when bound to the toxin. The major differences between **12** and **13** may be ascribed to the presence of an intermolecular aromatic–carbohydrate interaction in the phenyllactic acid derivative **13**, that strongly biases its conformational behavior, severely restricting its conformational freedom. As a result of this conformational lock, the side chain of **13** is preorganized in a conformation that allows optimal interaction of the carboxy group in the carboxylate-binding region of CT. The same conformation appears to be attained by **12** in its bound state, but it has to be selected from a pool of different rotamers that are simultaneously present in solution. Thus, both the preorganization effect and a more efficient van der Waals interaction between the side chain substituent and the protein appear to concur in determining the high affinity of **13** for CT. Remarkably, the simple phenyllactic acid derivative **13** showed a 10-μM dissociation constant, which is only one order of magnitude less potent than the monovalent association of the natural ligand GM1os against the cholera toxin. The cyclohexyl derivative **12** showed a somewhat lower affinity, 45 μM, still better than the value obtained for the simple (*R*)-lactic acid compound **11** (190 μM). These experimental facts support the starting hypothesis that lipophilic substituents on the (*R*)-hydroxy acid side chain can extend toward a hydrophobic area of the toxin-binding site and improve ligand affinity, and they emphasize the importance of structure-based ligand design.

A third group of GM1 mimics was obtained to examine the role of the GalNAc residue of GM1os in the CT:GM1 complex. Modification of the hexosamine was considered because it would simplify the synthesis of the artificial receptors (GalNAc is actually made from GlcNAc), while not negatively interfering with the formation of the toxin complex, because the GalNAc of GM1 interacts with the protein only via the *N*-acetyl group. Furthermore, the 4-hydroxy group of GalNAc in the experimental CTB:GM1os complex[23] and in the computational LT:GM1[64] and LT:**7**[62] models is located outside the protein-binding pocket and fully exposed to the solvent. Assuming that a GlcNAc-containing pseudo-GM1 ligand will bind to CT with the same general mode of GM1os and **7**, it appeared likely that inversion of configuration at C4 of the hexosamine could allow new H-bond contacts to be formed between the toxin and the substrate. In principle, these may or may not compensate for the loss in complex solvation because of the burying of the hydroxy group in the binding cavity. Ligands **14** and **15** (Chart 4) were designed starting from the known GM1 mimics **7** and **11** by replacement of their GalNAc residue with the more accessible GlcNAc (C4 isomer). NMR and modeling experiments[82] indicated that the conformational properties of the equivalent pairs **7–14** and **11–15** are very similar in both the free and bound states, while emission fluorescence measurements showed that their affinity for CT is of the same order of magnitude. Competitive TRNOE experiments of GM1os with **15** also proved that the mimic **15** occupies the GM1-binding site of the toxin and allowed determination of its bound conformation.[82]

5   Multivalency

Further improvement on the design has to address cooperativity in an efficient manner. As mentioned above, the oligosaccharide part of ganglioside GM1 binds strongly to the B subunits with a K_d of 43 nM at 25 °C,[38] and the multivalent display of the GM1 molecules makes the affinity of the toxin to the cell surface even higher. Effective interference is therefore a tremendous challenge.

In recent years, the structure-based design of multivalent ligands has attracted increasing attention[5,40,83–90] with some of the most successful examples developed for targeting AB5 toxins. Indeed, impressive results have been recently reported in the design of multivalent ligands for AB_5 toxins.[40] A paradigmatic example has been presented by Bundle's group,[91] by using a designed decavalent inhibitor for Shiga-like toxins based on an asymmetric scaffold, with a pentavalent cluster of bivalent ligands that extend from a glucose core with five arms. Because each B subunit of Shiga-like toxin's B pentamer contains three receptor-binding sites of varying affinities, the inhibitor could block up to 10 of these 15 total sites upon binding, showing several million-fold gain in affinity over the monovalent trisaccharide. NMR relaxation studies on the analogous B subunit homopentamer derived from *E. coli* O157 has shown that the pentamer exists in solution in fast exchange between an axially symmetric form and a

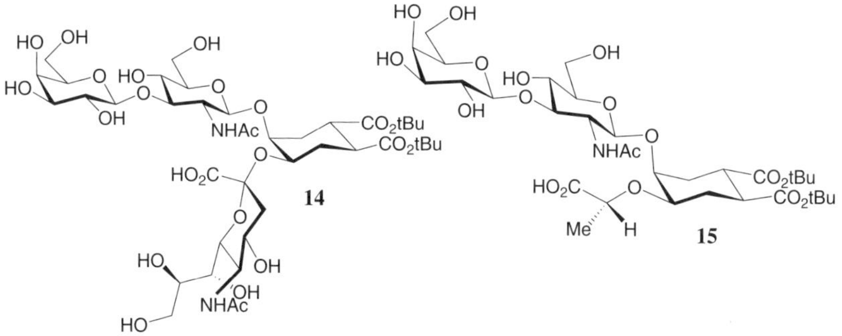

Chart 4 *The GlcNAc-containing mimics **14** and **15***

higher energy form, which is populated by *ca.* 10%. Homans and co-workers[92] showed that the binding of a bivalent ligand dimer straddled adjacent monomers and suppressed the conformational equilibrium, but with a substantial entropic cost to binding.

Polyvalent CT binders have also been developed starting from weak ligands, such as lactose or galactose.[93–97] Huge affinity enhancements (10^5 fold) compared to the monosaccharide were reported by Fan and co-workers,[94–97] who prepared and screened a series of galactose-based pentavalent ligands with linkers of various lengths. A different approach has been used by Bernardi and Pieters.[98] Dendrimers based on the 3,5-di-(2-aminoethoxy)-benzoic acid branching unit were used to attach multiple copies of the monovalent GM1 mimic **11** and the resulting polyvalent glycoconjugates (Chart 5) were used for inhibition of CT binding. Systems up to the octavalent **17** (Chart 5) were synthesized along with relevant reference compounds that contained the ligand in a monovalent format or the scaffold but not the ligand. Using a surface plasmon resonance inhibition assay, the prepared inhibitors showed good inhibition. While the monovalent GM1 mimic **11** showed the expected inhibition in the 200-μM range, the multivalent constructs led to increased binding.

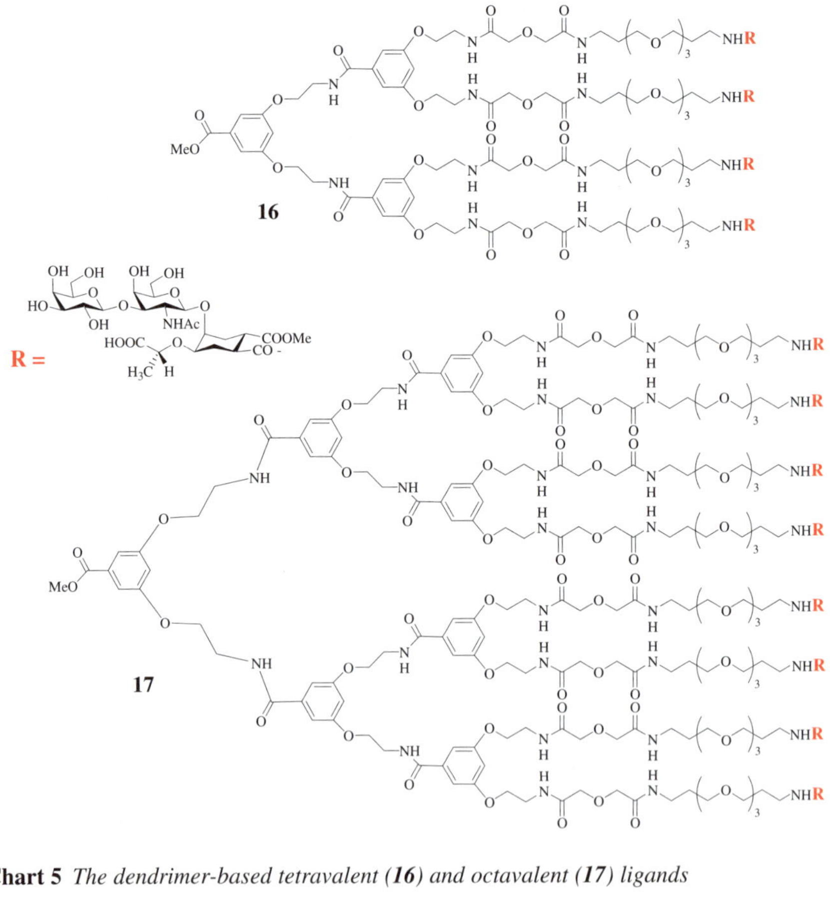

Chart 5 *The dendrimer-based tetravalent (**16**) and octavalent (**17**) ligands*

The tetravalent compound **16** was shown to be 440-fold more potent than its mono-valent counterpart. The octavalent analogue **17**, however, was the most potent com-pound as determined using an ELISA assay.[98]

This approach to high-affinity ligands for CT represents a combination of structure-based design of monovalent ligands with further enhancement by multivalent presenta-tion using dendrimers. The multivalency enhancement of 442 relative to the monovalent ligand **11** was of the same order of magnitude (263-fold) as the enhancement observed by Hol and co-workers[95] with their pentavalent version of the *m*-nitrophenyl-α-D-galac-toside ligand. The results obtained with the dendrimeric pseudo-GM1 ligands show that a pentavalent ligand design is not a crucial factor for strong multivalency binding enhancement to AB5 toxins. Precisely how the most effective compound, octavalent **17**, compares to the tetravalent **16** could not be quantified with the assays used,[98] however, its affinity was clearly higher.

Preliminary results were also obtained by Bernardi, Casnati and co-workers[99] with the divalent ligand **18** (Chart 6), prepared by hooking two units of GM1 mimic **11** onto a functionalized calix[4]arene. The size of the affinity enhancement measured by fluorescence spectroscopy was found to be 3800-fold (1900-fold per sugar mimic). The enhancement value observed is much higher than normally measured for a diva-lent ligand interacting with a polyvalent receptor,[90,100] and higher than the enhance-ment observed for the dendrimer-based tetravalent and octavalent analogues.[98] The energetics of this interesting divalent ligand are currently under investigations.

In general, the above results showed that multivalent presentation of designed lig-ands can bring their affinity closer to that required for practical application against AB5 toxins. In order to do so the compounds need to be considerably larger than monovalent ligands to be effective. Considering that the toxins reside in the intes-tinal tract, it is not a problem, in fact, it is an added bonus if the structure turns out to be too large and too polar for absorption. We are currently still in the phase where the optimal multivalent geometry for toxin neutralization, be it cyclic pentameric, dendritic, or polymeric, is not yet known, although several promising ones are now available. For the longer term concise syntheses of both simple but effective ligands

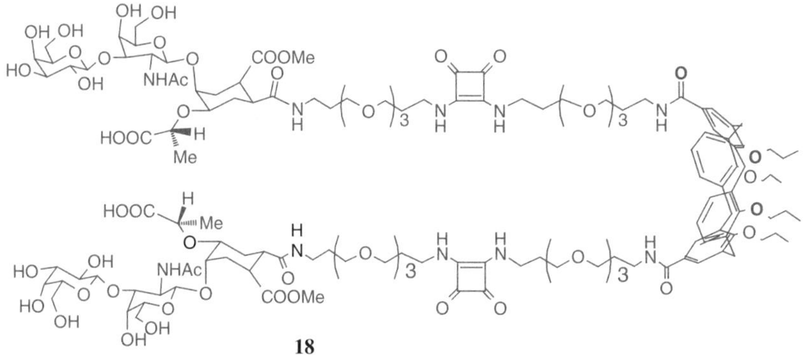

Chart 6 *The divalent calixarene-based ligand* **18**

and similarly straightforward multivalent scaffolds may bring practical intervention within reach.

References

1. Review Series. *Science*, 2001, **291**, 2337–2378.
2. C.R.Bertozzi and L.L. Kiessling, *Science*, 2001, **291**, 2357–2364.
3. H.-J. Gabius, H.-C. Siebert, S. Andre, J. Jimenez-Barbero and H. Rüdiger, *Chem. Bio. Chem.*, 2004, **5**, 740–764.
4. M. von Itzstein, W.-Y. Wu, G.B. Kok, M.S. Pegg, J.C. Dyason, B. Jin, T.V. Phan, M.L. Smythe, H.F. White, S.W. Oliver, P.M. Colman, J.N. Varghese, D.M. Ryan, J.M. Woods, R.C. Bethell, V.J. Hotham, J.M. Cameron and C.R. Penn, *Nature*, 1993, **363**, 418–423.
5. L.L. Kiessling, L.E. Strong and J.E. Gestwicki, *Annu. Rep. Med. Chem.*, 2000, **35**, 321–330.
6. P.I. Kitov and D.R. Bundle, *J. Am. Chem. Soc.*, 2003, **125**, 16271–16284.
7. N.T. Ohtomo, A. Muraoka, Y. Tashiro, Z. Zinnaka and K. Amako, *J. Infect. Dis.*, 1976, **133**(Suppl.), 31–40.
8. S. Fukuta, J.L. Magnani, E.M. Twiddy, R.K. Holmes and V. Ginsburg, *Infect. Immun.*, 1988, **56**, 1748–1753.
9. E.A. Merritt and W.G.J. Hol, *Curr. Opin. Struct. Biol.*, 1995, **5**, 165–171.
10. C.L. Gyles, *Can J. Microbiol.*, 1992, **38**, 734–746.
11. S.V. Heyningen, *Science*, 1974, **183**, 656–657.
12. A.A. Lindberg, J.E. Brown, N. Stromberg, M. Westerling-Ryd, J.E. Schultz, and K.-A. Karlsson, *J. Biol. Chem.*, 1987, **262**, 1779–1785.
13. N. Sharon and I. Ofek, *Glycoconjugate J.*, 2000, **17**, 659–664.
14. J.L. Brunton and R.J. Read, *Biochemistry*, 1998, **37**, 1777–1788.
15. D.J. Bast, L. Banerjee, C. Clark, R.J. Read and J.L. Brunton, *Mol. Microbiol.*, 1999, **32**, 953–960.
16. A.M. Soltyk, C.R. MacKenzie, V.M. Wolski, T. Hirama, P.I. Kitov, D.R. Bundle and J.L. Brunton, *J. Biol. Chem.*, 2002, **277**, 5351–5359.
17. P.G. Nyholm, G. Magnusson, Z. Zheng, R. Norel, B. Binnington-Boyd and C.A. Lingwood, *Chem. Biol.*, 1996, **3**, 263–275.
18. P.M. St. Hilaire, M.K. Boyd and E.J. Toone, *Biochemistry*, 1994, **33**, 14452–14463.
19. J.M. Richardson, P.D. Evans, S.W. Homans and A. Donohue-Rolfe, *Nat. Struct. Biol.*, 1997, **4**, 190–193.
20. P.H. Fishman, in *ADP-Ribosylating Toxins and G Proteins: Insights into Signal Transduction*, J. Moss and M. Vaughan (eds), American Society of Microbiology, Washington, DC, 1990, 127–140.
21. J. Moss and M. Vaughan, *Annu. Rev. Biochem.*, 1979, **48**, 581–600.
22. E.A. Merritt, S. Sarfaty, F. van den Akker, C.L. L'Hoir, J.A. Martial and W.G.J. Hol, *Protein Sci.*, 1994, **3**, 166–175.
23. P. Kuhn, S. Sarfaty, J.L. Erbe, R.K. Holmes and W.G.J. Hol, *J. Mol. Biol.*, 1998, **282**, 1043–1059.
24. E.A. Merritt, S. Sarfaty, I.K. Feil and W.G.J. Hol, *Structure*, 1997, **5**, 1485–1499.

25. E. Fan, E.A. Merritt, Z. Zhang, J.C. Pickens, C. Roach, M. Ahn and W.G.J. Hol, *Acta Crystallogr.*, 2001, **D57**, 201–212.

26. F. van den Akker, E. Steensma and W.G.J. Hol, *Protein Sci.*, 1996, **5**, 1184–1188.

27. J.C. Pickens, E.A. Merritt, M. Ahn, C.L.M.J. Verlinde, W.G.J. Hol and E.K. Fan, *Chem. Biol.*, 2002, **9**, 215–224.

28. E.A. Merritt, T.K. Sixma, K.H. Kalk, B.A.M. van Zanten and W.G.J. Hol, *Mol. Microbiol.*, 1994, **13**, 745–753.

29. R.G. Zhang, M.L. Westbrook, E.M. Westbrook, D.L. Scott, Z. Otwinowski, P.R. Maulik, R.A. Reed and G.G. Shipley, *J. Mol. Biol.*, 1995, **251**, 550–562.

30. S. Fukuta, J.L. Magnani, E.M. Twiddy, R.K. Holmes and V. Ginsburg, *Infect. Immun.*, 1988, **56**, 1748–1753.

31. J. Ångström, S. Teneberg and K.-A. Karlsson, *Proc. Natl. Acad. Sci. USA*, 1994, **91**, 11859–11863.

32. G.M. Kuziemko, M. Stroh and R.C. Stevens, *Biochemistry*, 1996, **35**, 6375–6384.

33. W.L. Picking, H. Moon, H. Wu and W.D. Picking, *Biochim. Biophys. Acta*, 1995, **1247**, 65–73.

34. J.A. Mertz, J.A. McCann and W.D. Picking, *Biochem. Biophys. Res. Commun.*, 1996, **226**, 140–144.

35. S. Lauer, B. Goldstein, R.L. Nolan and J.P. Nolan, *Biochemistry*, 2002, **41**, 1742–1751.

36. X.-E. Cai and J. Yang, *Biochemistry*, 2003, **42**, 4028–4034.

37. M. Masserini, E. Freire, P. Palestini, E. Calappi and G. Tettamanti, *Biochemistry*, 1992, **31**, 2422–2426.

38. W.B. Turnbull, B.L. Precious and S.W. Homans, *J. Am. Chem. Soc.*, 2004, **126**, 1047–1054.

39. A. Schoen and E. Freire, *Biochemistry*, 1989, **28**, 5019–5024.

40. E. Fan, E.A. Merritt, C.L.M.J. Verlinde and W.G.J. Hol, *Curr. Opin. Struct. Biol.*, 2000, **10**, 680–686, and references therein.

41. D. Acquotti, L. Poppe, J. Dabrowski, C.W. von der Lieth, S. Sonnino and G. Tettamanti, *J. Am. Chem. Soc.*, 1990, **112**, 7772–7778.

42. J.M. Richardson, M.J. Milton and S.W. Homans, *J. Mol. Recog.*, 1995, **8**, 358–362.

43. P. Brocca, P. Berthault and S. Sonnino, *Biophys. J.*, 1998, **74**, 309–318.

44. P. Brocca, A. Bernardi, L. Raimondi and S. Sonnino, *Glycoconjugate J.*, 2001, **17**, 283–299.

45. J.N. Scardsale, J.H. Prestegard and R.K. Yu, *Biochemistry*, 1990, **20**, 9843–9855.

46. A. Bernardi and L. Raimondi, *J. Org. Chem.*, 1995, **60**, 3370–3377.

47. W.P. Jencks, *Proc. Natl. Acad. Sci. USA*, 1981, **78**, 4046–4050.

48. L. Poppe, J. Dabrowski, C.W. von der Lieth, M. Numata and T. Ogawa, *Eur. J. Biochem.*, 1989, **180**, 337–342.

49. H.-C. Siebert, G. Reuter, R. Schauer, C.W. von der Lieth and J. Dabrowski, *Biochemistry*, 1992, **31**, 6962–6971.

50. S. Sabesan, K. Bock and R.U. Lemieux, *Can. J. Chem.*, 1984, **62**, 1034–1045.

51. Y.-T. Li, S.-C. Li, A. Hasegawa, H. Ishida, M. Kiso, A. Bernardi, P. Brocca, L. Raimondi and S. Sonnino, *J. Biol. Chem.*, 1999, **274**, 10014–10018.

52. L. Poppe, H. van Halbeek, D. Acquotti and S. Sonnino, *Biophys. J.*, 1994, **66**, 1642–1652.

53. Y. Aubin, Y. Ito, J.C. Paulson and J.H. Prestegard, *Biochemistry*, 1993, **32**, 13405–13413.

54. K.R. Barber, K.S. Hamilton, A.C. Rigby and C.W.M. Grant, *Biochem. Biophys. Acta*, 1994, **1190**, 376–384.

55. D.M. Singh, X. Shan, J.H. Davis, D.H. Jones and C.W.M. Grant, *Biochemistry*, 1995, **34**, 451–463.

56. D.H. Jones, K.R. Barber and C.W.M. Grant, *Biochemistry*, 1996, **35**, 4803–4811.

57. J. Breg, L.M.J. Kroon-Batenburg, G. Strecker, J. Montreuil and J.F.G. Vliegenthart, *Eur. J. Biochem.*, 1989, **178**, 727–739.

58. S. Sabesan, K. Bock and J.C. Paulson, *Carbohydr. Res.*, 1991, **218**, 27–54.

59. C. Mukhopadhyay and C.A. Bush, *Biopolymers*, 1994, **34**, 11–20.

60. J. Schulte, J. Lauterwein and U. Hoeweler, *Magn. Reson. Chem.*, 2000, **38**, 751–756.

61. A. Bernardi, D. Arosio and S. Sonnino, *Neurochem. Res.*, 2002, **27**, 539–545, and references therein.

62. A. Bernardi, A. Checchia, P. Brocca, S. Sonnino and F. Zuccotto, *J. Am. Chem. Soc.*, 1999, **121**, 2032–2036.

63. A. Bernardi, D. Arosio, L. Manzoni, F. Micheli, S. Pasquarello and P. Seneci, *J. Org. Chem.*, 2001, **66**, 6209–6216.

64. A. Bernardi, L. Raimondi and F. Zuccotto, *J. Med. Chem.*, 1997, **40**, 1855–1865.

65. A. Bernardi, L. Carrettoni, A. Grosso Ciponte, D. Monti and S. Sonnino, *Bioorg. Med. Chem. Lett.*, 2000, **10**, 2197–2301.

66. A. Bernardi, D. Potenza, A.M. Capelli, A. García-Herrero, F.J. Cañada and J. Jiménez-Barbero, *Chem. Eur. J.*, 2002, **8**, 4597–4612.

67. D. Arosio, S. Baretti, S. Cattaldo, S. Potenza and A. Bernardi, *Bioorg. Med. Chem. Lett.*, 2003, **13**, 3831–3834.

68. A. Bernardi, D. Arosio, D. Potenza, I. Sanchez-Medina, S. Mari, F.J. Cañada and J. Jimenez-Barbero, *Chem. Eur. J.*, 2004, **10**, 4395–4406.

69. N.K. Vyas, *Curr. Opin. Struct. Biol.*, 1991, **1**, 732–740.

70. F.A. Quiocho, *Biochem. Soc. Trans.*, 1993, **21**, 442–448.

71. W.I. Weis and K. Drickamer, *Annu. Rev. Biochem.*, 1996, **65**, 441–473.

72. S. Elgavish and B. Shaanan, *J. Mol. Biol.*, 1998, **277**, 917–932.

73. J. Jimenez-Barbero, J.L. Asensio, F.J. Cañada and A. Poveda, *Curr. Opin. Struct. Biol.*, 1999, **9**, 549–555.

74. A.A. Bothnerby and R. Gassend, *Ann. N.Y. Acad. Sci.*, 1973, **222**, 668–676.

75. P.L. Jackson, H.N. Moseley and N.R. Krishna, *J. Magn. Reson. Ser. B*, 1995, **107**, 289–292.

76. V.L. Bevilacqua, D.S. Thomson and J.H. Prestegard, *Biochemistry*, 1990, **29**, 5529–5537.

77. V.L. Bevilacqua, Y. Kim and J.H. Prestegard, *Biochemistry*, 1992, **31**, 9339–9349.

78. H. Kogelberg, D. Solis and J. Jimenez-Barbero, *Curr. Opin. Struct. Biol.*, 2003, **13**, 646–653.

79. M. Meyer and B. Meyer, *Angew. Chem. Int. Ed.*, 1999, **38**, 1784–1788.

80. J. Klein, R. Meinecke, M. Meyer and B. Meyer, *J. Am. Chem. Soc.*, 1999, **121**, 5336–5337.

81. M. Vogtherr and T. Peters, *J. Am. Chem. Soc.*, 2000, **122**, 6093–6099.

82. A. Bernardi, D. Arosio, L. Manzoni, D. Monti, H. Posteri, D. Potenza, S. Mari and J. Jimenez-Barbero, *Org. Biomol. Chem.*, 2003, **1**, 785–792.

83. J.J. Lundquist and E.J. Toone, *Chem. Rev.*, 2002, **102**, 555–578.

84. T.K. Lindhorst, *Top. Curr. Chem.*, 2002, **218**, 201–235.

85. R. Roy, *Curr. Opin. Struct. Biol.*, 1996, **6**, 692–702.

86. R.T. Lee and Y.C. Lee, *Glycoconjugate J.*, 2002, **17**, 543–551.

87. W.B. Turnbull and J.F. Stoddart, *Rev. Mol. Biotechnol.*, 2002, **90**, 231–255.

88. M.J. Cloninger, *Curr. Opin. Chem. Biol.*, 2002, **6**, 742–748.

89. R. Roy, *Trends Glycosci. Glycotechnol.*, 2003, **15**, 291–310.

90. M. Mammen, S.K. Choi and G.M. Whitesides, *Angew. Chem. Int. Ed.*, 1998, **37**, 2755–2794.

91. P.I. Kitov, J.M. Sadowska, G. Mulvey, G.D. Armstrong, H. Ling, N.S. Pannu, R.J. Read and D.R. Bundle, *Nature*, 2000, **403**, 669–672.

92. A. Yung, W.B. Turnbull, A.P. Kalverda, G.S. Thompson, S.W. Homans, P. Kitov and D.R. Bundle, *J. Am. Chem. Soc.*, 2003, **125**, 13058–13062.

93. I. Vrasidas, N.J. de Mol, R.M.J. Liskamp and R.J. Pieters, *Eur. J. Org. Chem.*, 2001, **24**, 4685–4692.

94. E.K. Fan, Z.S. Zhang, W.E. Minke, Z. Hou, C.L.M.J. Verlinde and W.G.J. Hol, *J. Am. Chem. Soc.*, 2000, **122**, 2663–2664.

95. E.A. Merritt, Z.S. Zhang, J.C. Pickens, M. Ahn, W.G.J. Hol and E.K. Fan, *J. Am. Chem. Soc.*, 2002, **124**, 8818–8824.

96. Z.S. Zhang, E.A. Merritt, M. Ahn, C. Roach, Z. Hou, C.L.M.J. Verlinde, W.G.J. Hol and E.K. Fan, *J. Am. Chem. Soc.*, 2002, **124**, 12991–12998.

97. J.C. Pickens, D.D. Mitchell, J. Liu, X. Tan, Z. Zhang, C.L.M.J. Verlinde, W.G.J. Hol and E. Fan, *Chem. Biol.*, 2004, **11**, 1205–1215.

98. D. Arosio, I. Vrasidas, P. Valentini, R.M.J. Liskamp, R.J. Pieters and A. Bernardi, *Org. Biomol. Chem.*, 2004, **2**, 2113–2124.

99. D. Arosio, M. Fontanella, L. Baldini, L. Mauri, A. Bernardi, A. Casnati, F. Sansone and R. Ungaro, *J. Am. Chem. Soc.*, 2005, **127**, 3660–61.

100. B.T. Houseman and M. Mrksich, *Top. Curr. Chem.*, 2002, **218**, 1–44.

CHAPTER 6

Retrocyclins: Miniature Lectins with Potent Antiviral Activity

ROBERT I. LEHRER

David Geffen School of Medicine at UCLA, 10833 Le Conte Avenue, Los Angeles, CA 90095, USA

1 Introduction

To date, θ-defensin peptides have been isolated only from rhesus macaques. However, intact θ-defensin (DEFT) genes exist in other Old World monkeys, lesser apes, and orangutans. Humans do not produce θ-defensin peptides despite the presence of mRNA transcripts in their bone marrow, thymus, and other tissues. Human θ-defensin peptide deficiency results from a premature stop codon in codon 17 of the signal sequence encoded in human DEFT genes. The DEFT genes of our chimpanzee and gorilla cousins harbor this premature stop codon, TAG, in exactly the same location as our genes. The mutation probably occurred 7.5 to 10 million years ago,[1] when the ancestors common to humans and orangutans set off on different paths through the African countryside.

   We think of retrocyclin-1, -2, and -3 as the θ-defensins that humans would express *if* their DEFT genes had retained their integrity. Since the retrocyclin peptides we will discuss are synthetic, our experiments are a form of molecular archeology. Originally, this work was motivated by simple curiosity. Or, to use hypothetical study section parlance (HSSP), by an inexplicable desire to fish in a lake that has been barren for millions of years using chemically synthesized virtual bait, with no assurance that anything would be caught. Fortunately, our "fishing expedition" was productive, and we will display some of its "catch" here.

2 Antimicrobial Peptides

Animals require antimicrobial peptides[2–5] because they are surrounded and outnumbered by multitudes of bacteria, fungi, and viruses. Although most of these organisms are innocuous to us, and many are helpful, others are potentially pathogenic.

Vertebrate immunity is provided by innate and adaptive mechanisms. Innate immunity is "hard-wired" and relies on gene-encoded molecules that are broadly specific, rapidly available, and ready to be deployed. Adaptive immunity is programmed by gene rearrangements. Its products are highly specific, undeniably elegant, and (there is always a catch) relatively slow to appear owing to the reliance on clonal expansion. Considering the rates at which bacteria and viruses can replicate, the importance of a speedy host response is clear. Defensin-like peptides had already existed for at least 500 million years[6] before the first glimmerings of the adaptive immune system appeared.[7] Surprisingly, the importance of innate immunity is often overlooked, even though most animal species that inhabit our planet are invertebrates and have only an innate immune system. Whether this reflects a "What have you done for us lately?" perspective or the failure of antimicrobial peptides to "bring bling" (*i.e.*, pharmaceutical profits) is unclear.

Defensins equip the innate immune system to resist microbial and viral incursions. They act directly to kill or otherwise hinder the progress of potential pathogens[5,8] and also provide signals that recruit and amplify responses provided by the adaptive immune system.[9,10] The cells of the innate immune system (*e.g.*, PMNL, macrophages, and NK cells) are produced continually, and their motility and receptors allow them to confront intruders directly. Innate immunity also depends on an array of gene-encoded effector molecules (*e.g.*, defensins, lectins, and complement). Many of these are produced continually, and others can be induced within minutes to hours. Collectively, these innate elements provide broad-ranging protection against incursions by virtually any potential pathogen. By holding the attackers at bay, they allow the adaptive immune system the days and weeks it needs to accomplish clonal expansion – a crucial but slow step necessary to provide effective concentrations of its pathogen-specific antibodies or adequate numbers of target-specific T cells. Although it is convenient to draw distinctions between these systems, they in fact complement each other and their borders can be indistinct.

3 Defensins

Defensins are cysteine rich, cationic, and amphipathic peptides that play multiple roles *in vivo*, including many that convey resistance to infection. Humans may express as many as 40 different defensins that belong to two subfamilies: α-defensins and β-defensins.[5] All α- and β-defensin molecules have six conserved cysteines that form three intramolecular disulfide bonds. Four of the six human α-defensins were initially discovered in neutrophils (a synonym for PMNL) and were named human neutrophil peptides (HNPs).[11,12] The sequences of HNP-1 to HNP-3 differ only by a single amino acid (Figure 1). HNP-1 to HNP-3 account for 5–7% of the *total* protein of human PMNL and more than 30% of the total protein stored in their primary (azurophil) cytoplasmic granules.[13] Other human white blood cells that express HNP-1 to HNP-3 include NK ("natural killer") cells and certain T lymphocytes.[14,15] Cysteines-1 and -6 (the amino-terminal and carboxy-terminal cysteines) of α-defensins are disulfide bonded to each other (Figure 1), effectively rendering these molecules macrocyclic. All human α-defensin genes reside on the short arm of chromosome 8, close to the telomere at 8p22-23. Their closest neighbors in this locus

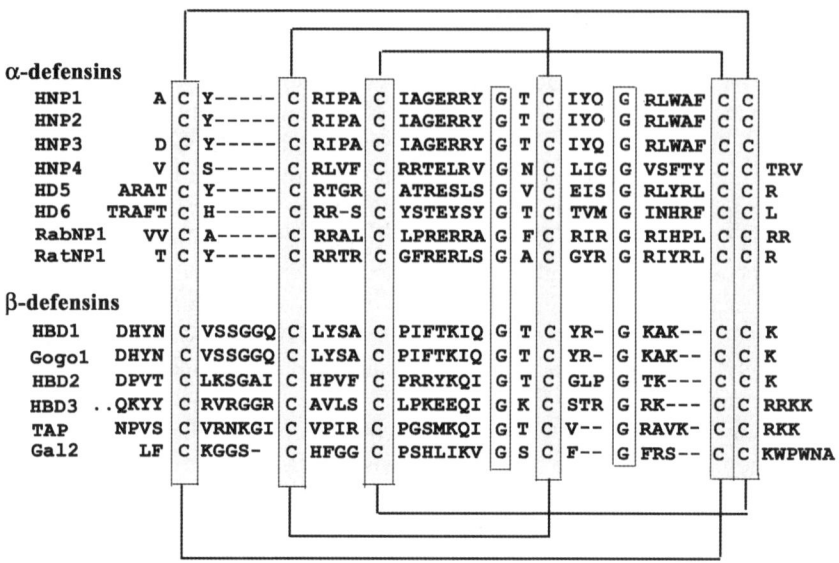

Figure 1 *Sequences of selected α- and β-defensins. Sequences of six human α-defensins (HNP-1 to -4 and HD 5 and 6) and three human β-defensins (HBD-1 to HBD-3) are shown. Cysteine [C] residues are boxed, and their disulfide connectivity is shown. Abbreviations: RabNP1, rabbit neutrophil peptide-1; RatNP-1, rat neutrophil peptide-1; Gogo1, Gorilla gorilla β-defensin-1; TAP, tracheal antimicrobial peptide, a bovine β-defensin; Gal2, Gallinacin-2, a β-defensin of chicken PMNL. Gaps (-) were introduced to maximize alignments*

include several β-defensin genes and several θ-defensin pseudogenes.[16–18] α-Defensins have been identified only in certain mammals, including rabbits, rodents, and primates, and β-defensins in these, and also in many ruminants, dogs, birds, and snakes.[16,17,19]

The human genome contains at least 30 different β-defensin genes, some of which exist in as many as 12 copies owing to gene reduplication.[20,21] β-Defensin peptides are usually larger than α-defensins, and their cysteines show different spacing and pairing (Figure 1). Many β-defensin peptides contain a short helical domain near their amino terminus.[22] A similar helical domain is present in the defensin peptides of plants,[23,24] invertebrates,[25] and fungi.[6] The β-defensin genes of vertebrates existed before α- and θ-defensin genes appeared; however, their relationship to the older β-defensin-like genes of invertebrates and plants and fungi is unknown.[6] Recent reviews on the various defensins mentioned above provide more information about them.[5,25–30]

4   Theta (θ)-Defensins

4.1   Presence in Nonhuman Primates

Figure 2 shows the structures of RTD-1 and two retrocyclins. RTD-1 was isolated from rhesus macaque PMNLs, and later from rhesus bone marrow. Although alternative splicing relegated "One gene, one protein" to the history division of molecular biology, θ-defensins also modify the dictum, but in a different way, to

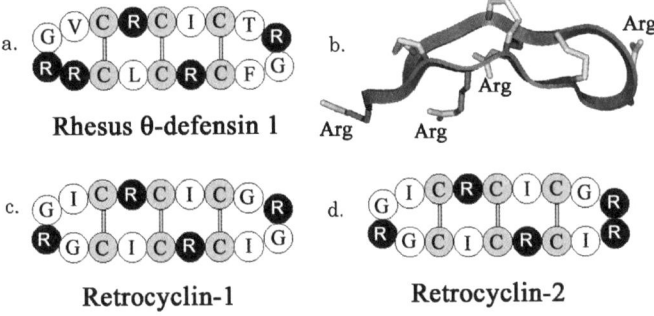

Figure 2 *Theta (θ) defensins. Panels a, c, and d illustrate the salient features of three
θ-defensins. These include a cyclic peptide backbone, the presence of six cysteines
(C) and three evenly spaced intramolecular disulfide bonds. The multiple arginine
(R) residues are distributed, so that, unlike protegrins, θ-defensins are not amphipathic. Panel b is a somewhat more realistic model of retrocyclin-1, based on NMR
studies of RTD-1 and molecular modeling*

"Two genes, a little magic, one peptide." Each of the two genes that collaborate to
produce a single θ-defensin peptide molecule is a mutated α-defensin gene. A
normal α-defensin gene encodes a prepropeptide of approximately 100 amino acid
residues, with a signal sequence, an anionic propiece, and a C-terminal defensin
domain. DEFT genes are mutated α-defensin (DEFA) genes. The mutation is a premature stop codon that terminates translation after residue 12 of the α-defensin,
which would otherwise contain 29–35 residues (Figure 1). How the subsequent
magic is performed has not been publicly disclosed, but it is clearly not an illusion.
Rhesus cells process the prepropeptide, discarding three residues and salvaging nine
– including three cysteines. Then (watch closely, because this is the magical part),
they splice two nonapeptides and form the tridisulfide ladder shown in Figure 2 and
a seamless cyclic backbone. The rhesus macaque has DEFT genes that encode (at
least) two slightly different nonapeptides. Since it can incorporate identical or different nonapeptides into a θ-defensin molecule, two different DEFT genes can (and
do) give rise to three different θ-defensin peptides: RTD-1, RTD-2, and RTD-3.[31,32]

4.2 Why Humans Lack Them

We adopted a twofold approach to address this issue. One approach involved searching the human genome database for clues, and the other approach was to question
close relatives or – to be more precise – their DNA. Searching the human genome
database revealed six human DEFT pseudogenes, of which five resided on chromosome 8p23. One gene had been translocated to chromosome 1. All of the human
DEFT genes were defective. Those on 8p23 had been converted into pseudogenes by
a mutation that converted codon 17 of the signal sequence into a "tag" stop codon.
Human DEFT pseudogenes and human α-defensin genes had identical 3-exon
layouts. Exon-1 (63-bp) of the DEFT pseudogenes encoded a 5' untranslated region,
exon-2 (187-bp) encoded the mutated signal sequence, and end exon-3 (246-bp)
encoded the propiece and 12 residues from the C-terminal defensin domain. If the
human DEFT pseudogenes had remained functional, they would have contributed

two nearly identical nonapeptides (RCICGRGIC and RCICGRRIC) for inclusion into θ-defensins. Thus, the innate immune system of humans would have included at least three θ-defensins in addition to its six α-defensins and more numerous β-defensins. These peptides would correspond to retrocyclin-1, net charge +4; retrocyclin-2, net charge +5; and retrocyclin-3, net charge +6. Both retrocyclin-1 and retrocyclin-2 are illustrated in Figure 2. Retrocyclin-3, which is not shown, is a homodimer composed of tandem RCICGRRIC nonapeptides.

Having obtained a description of the missing peptides, and suspecting that their disappearance might have been an "inside job," we rounded up the relatives, who included five prosimians, six New World monkeys, five Old World monkeys, 1 lesser ape (the siamang), and four great apes (orangutan, gorilla, chimpanzee, and bonobo). Their DNA was probed with nondegenerate DNA primers based on the human θ-defensin cDNA sequence obtained from the human bone marrow. We found seven homologous defensin genes in the Sumatran orangutan, *Pongo pygmaeus abeli*. Two were intact α-defensin genes, four were intact θ-defensin genes, and one was a retrocyclin-like DEFT pseudogene. The orangutan's DEFT genes showed about 90% sequence identity to an orangutan α-defensin gene. Gorillas, chimpanzees, and bonobos had defective DEFT genes (Figure 3). In all of these defective DEFT genes, codon 17 of the signal sequence was a stop codon ("tag"), whose presence would abort translation of the θ-defensin propeptide. Codon 17 of the orangutan's single DEFT pseudogene also harbored the "tag" mutation.

In contrast, in the intact DEFT genes of Old World monkeys, codon 17 was "cac" or "cat," both signifying histidine. In four intact orangutan DEFT genes, codon 17 had mutated to "cag" (glutamine). Thus, the lack of θ-defensin peptides in humans may have transpired in two ways. If the codon 17 "tag" mutation occurred before the last common ancestor of the orangutan and humans diverged, our lineage may have been left with the mutated and defective allele. Alternatively, the mutation could have occurred independently in orangutans and in the last common ancestor of chimpanzees, gorillas, and hominids. By either scenario, our ability to make θ-defensin peptides was lost at least 7.5 million years ago.

We used the findings from our phylogenetic survey to construct the consensus θ-defensin sequence shown in Figure 4. Eleven of the 18 residues were invariant, including all 6 cysteines, 4 arginines, and 1 glycine. Although each β-turn contained a conserved arginine, most of the other residues were variable. This variability contrasted with the β-sheet domains, which were either invariant or contained conservative substitutions.

4.3   Resemblance to Protegrins

When we first became aware of θ-defensins, we were impressed by their considerable resemblance to protegrins, a family of unusually potent antimicrobial peptides found in porcine (but not human) PMNL. The structures of protegrin (PG)-1 and RTD-1 are compared in Figure 5. Pigs express at least five different protegrin peptides (PG-1 to PG-5). Four of these peptides contain 18 amino acid residues, and one (PG-2) has 16 residues. Protegrins have exceptionally potent antibacterial and antifungal properties, but they are also fairly cytotoxic and cause the red blood cells of humans, rabbits, and mice (but not those of ruminants or pigs) to undergo hemolysis. In contrast, retrocyclins-1 and -2 are

SIGNAL SEQUENCE

CODON	12	13	14	15	16	17	18	19	20	21	22

PRIMATES THETA DEFENSIN PSEUDOGENES

	12	13	14	15	16	17	18	19	20	21	22
Human	ctc	ctg	gtg	gcc	ctg	tag	gct	cag	gcg	gag	cca
Gorilla	ctc	ctg	gtg	gac	ctg	tag	gct	cag	gcg	gag	cca
Bonobo	ctc	ctg	gtg	gcc	ctg	tag	gct	cag	gca	gag	cca
Chimpanzee	ctc	ctg	gtg	gcc	ctg	tag	gct	cag	gca	gag	cca
Orangutan	ctc	ctg	gtg	gcc	ctg	tag	gct	gag	gca	gag	cca

ORANGUTAN THETA AND ALPHA-DEFENSIN GENES

DEFT-1	ctc	gtg	gtg	gcc	ctg	cag	gct	cag	gcg	gag	cca
DEFT-2	ctc	gtg	gtg	gcc	ctg	cag	gct	cag	gcg	gag	cca
DEFT-3	ctc	gtg	gtg	gcc	ctg	cag	gct	cag	gcg	gag	cca
DEFT-4	ctc	ctg	gtg	gcc	ctg	cag	gct	cag	gca	gag	cca
DEFA-1	ctc	ctg	gtg	gcc	ctg	cag	gct	cag	gca	gag	cca

OLD WORLD MONKEYS THETA DEFENSIN GENES

Rhesus macaque	ctc	ctg	gtg	gcc	ctg	cac	gct	cag	gca	gag	gca
Pigtail macaque	ctc	ctg	gtg	gcc	ctg	cat	gct	cag	gca	gag	gca
Colobus monkey	ttc	ctg	gtg	gcc	ctg	cac	act	cag	gca	gag	gca

Figure 3 *The slippery slope to a silenced signal sequence. Partial signal sequences of 13 primate defensin genes are shown and aligned. The first five sequences are from θ-defensin genes that have been silenced by the premature "tag" stop codon. Orangutan DEFA-1 is an intact α-defensin gene. The remaining sequences are from DEFT genes. In the intact DEFT genes, codon 17 signifies histidine (cac, cat) or glutamine (cag). Bonobos are also called pygmy chimpanzees*

Consensus θ-Defensin

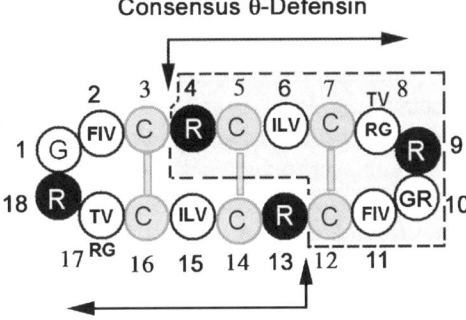

Figure 4 *Consensus θ-defensin sequence. The figure represents 17 sequences from 9 different primate species. All six cysteines, four arginines, and one glycine were invariant. Substantial variation existed in the β-turn, but β-sheet regions were highly conserved. Since θ-defensins are circular, residue numbering is somewhat arbitrary. Here, it represents the usual layout of the linear precursor that we synthesize. The short arrows identify where the tandem nonapeptides would be spliced in vivo, and the long arrows show the direction of the peptide backbone, pointing from N→C termini. One nonapeptide is shaded*

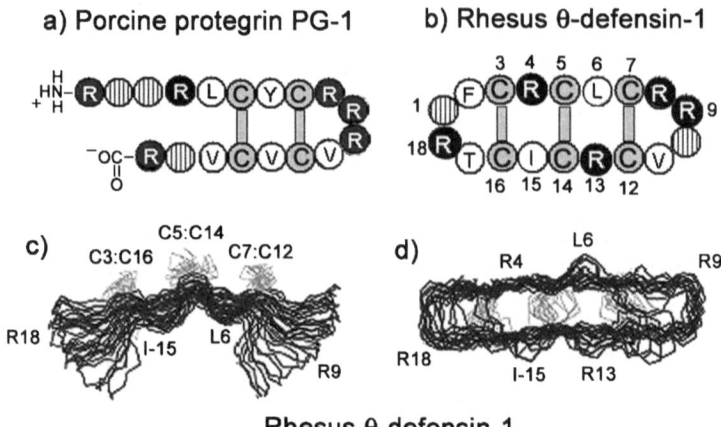

a) Porcine protegrin PG-1   b) Rhesus θ-defensin-1

c)   d)

Rhesus θ-defensin-1

Figure 5 *θ-defensins and protegrins: similarities and differences. Panel a shows PG-1, an unusually potent antimicrobial peptide found in the PMNL of pigs. Like RTD-1 (panel b) PG-1 is an arginine-rich octadecapeptide whose two antiparallel β-sheets are connected by a β-turn and crossconnected by cysteine disulfide bonds. Unlike RTD-1, the peptide backbone of PG-1 is not cyclic. The arginine residues of PG-1 are clustered away from the apolar β-sheet residues (C3 to C7 and C12 to C16) imparting a highly amphiphilic character to PG-1 that is not found in θ-defensins. Panels c and d depict 20 of the lowest energy NMR structures of RTD-1 superimposed over the backbone atoms of its β-sheets, and shows that both β-turns manifest great flexibility, relative to the β-sheet region.[48] The interaction of RTD-1 with lipid bilayers were examined, and revealed several unusual features that could explain why θ-defensins lack the cytotoxicity and hemolytic activity seen with PG-1 and many other antimicrobial peptides[49]*

nonhemolytic and only minimally cytotoxic to human red cells. Retrocyclins and RTDs have moderate antibacterial properties, generally equivalent to those of full-length human α-defensins and substantially less than those exerted by porcine protegrins.

In vitro studies of PG-1 and RTD-1 in model systems have suggested a basis for their many functional differences. Huang and co-workers used oriented circular dichroism and X-ray lamellar diffraction to examine the actions of these peptides on model phospholipid membranes. PG-1 bound and embedded itself into the headgroup region of the lipid bilayer (the S state) and, accepting the protegrin molecules in this location, caused the bilayer to thin considerably. On attaining a critical protegrin-to-phospholipid ratio, PG-1 molecules changed their orientations, inserted into the membrane, and formed transmembrane toroidal pores. Membrane-bound RTD-1 also initially oriented with the plane of its β-sheet backbone parallel to the plane of the bilayer. However, it induced only mild membrane thinning, and – although its orientation also changed as the peptide-to-phospholipid concentration increased – it did not lead to membrane insertion.

4.4 *In Vivo* Expression and Chemical Synthesis

We examined the expression of θ-defensin mRNA in 14 different human tissues. It was most prominent in bone marrow and also occurred in skeletal muscle, spleen, thymus, and testis. Which nonmyeloid tissues and cells of nonhuman primates express θ-defensins has not been reported. Whether θ-defensin expression is constitutive or inducible remains to be answered.

Because our magical powers are limited, we did not attempt to imitate the natural processes that rhesus cells use to manufacture θ-defensins *in vivo*. Instead, we would synthesize a linear 18-residue peptide, release it from the resin, reduce it with dithiothreitol, and purify it by HPLC. Then, we dissolved the peptide at a low concentration (to minimize oligomerization), allowed it to oxidize in room air, and repurified it by further HPLC. Then, we dissolved the now-oxidized peptide in dimethyl sulfoxide and cyclized it with 1-ethyl-3-(3-dimethylaminopropyl)carbodiimide·HCl and *N*-hydroxybenzotriazole. Finally, we purified the cyclic product by HPLC and determined peptide yields and concentrations by amino acid analysis. We monitored each step along the way by MALDI-MS.

We analyzed the DEFT gene and pseudogene sequences of nine primates, concentrating on the nonapeptides that would be incorporated into a mature θ-defensin molecule.[12,29] Seven of these nine residues were highly conserved, and only two showed considerable inter-species variability. The effects of this are evident in the consensus theta-defensin peptide illustrated in Figure 4. From the solution structure of RTD-1,[28] these variable residues are in the turns and the more conserved or invariant residues constitute the antiparallel β-strands. The other illustrations show retrocyclin-1, -2, and -3 – the hypothetical ancestral hominid θ-defensins.

4.5 Antimicrobial Properties

Human α-defensins are pore-forming peptides that kill bacteria and fungi by targeting their membranes and causing them to lose structural and functional integrity. These defensins work best in close-combat situations, where very high defensin concentrations can attack a small number of bacteria. To establish these odds, PMNL store α-defensins such as HNP-1 to HNP-3 in tiny intracellular depots called azurophil granules. Many hundreds of these granules exist in the cytoplasm of each PMNL, and their defensin concentrations are extremely high, about 50 mg ml^{-1}. PMNL are motile cells with receptors that can guide them to sites of infection. PMNL are phagocytes that sequester the microbes they ingest within membrane-bounded cytoplasmic compartments called phagosomes. Most azurophil granules deliver their contents directly to these phagosomes, establishing exactly the conditions that allow defensins to work most effectively. HNP-1 to HNP-3 are ineffective in the extracellular milieu for two main reasons. First, serum generally contains very low concentrations of α-defensins (<0.1 μg ml^{-1}). Second, the concentrations of NaCl and divalent cations in serum prevent defensin-mediated antimicrobial activity. θ-defensins generally resemble α-defensins in their ability to kill bacteria; however, they may be able to tolerate higher NaCl concentrations before losing their antimicrobial activity.

4.6 Activity Against HIV-1 and Herpes Simplex

Nearly 40 million people worldwide are living with HIV-1 infection. If left untreated, HIV progressively impairs the immune system and allows an array of opportunistic pathogens to emerge and wreak havoc. To initiate infection, every virus must enter a host cell and co-opt its biosynthetic machinery to manufacture copies of the virus. Usually, HIV-1 only infects cells with a surface receptor called CD4. This receptor normally promotes interactions between T cells and cells that present antigens to it, and it also has signaling functions. Target cells that display CD4 on their surface must also display a second receptor molecule on their surface to be subject to infection by HIV-1. Usually, this "coreceptor" is either CXCR4 or CCR5. HIV-"M-tropic" strains of HIV-1 use CCR5 as the coreceptor, "T-tropic" strains use CXCR4, and dual-tropic strains can use either CXCR4 or CCR5. M and T signify macrophage and T cell, respectively.

Our initial studies revealed that retrocyclin-1 protected CD4+ human lymphocytes from infection by *both* the T-tropic IIIB strain of HIV-1 and the M-tropic JR-CSF strain.[33] Unlike α-defensins,[34] retrocyclin did not directly inactivate HIV-1;[33] yet it could inhibit proviral DNA formation, indicating that its protective effect occurred early in the infection process.[33,35] We prepared a fluorescent analogue of retrocyclin-1 and observed patch-like fluorescent aggregates on the surface of CD4+ cells by confocal microscopy, but no evidence of internalized peptide. This was consistent with the behavior of RTD-1 in model membrane bilayers, as studied by circular dichroism and X-ray diffraction.[36] Here, the peptide oriented its backbone ring parallel to the plane of the bilayer, causing some membrane thinning without inserting into the bilayer.

The ability of retrocyclin-1 to protect against M- and T-tropic strains of HIV-1 suggested that it might bind either gp120 on the surface of HIV-1, or CD4, its primary receptor on the target cell surface. When we studied this by surface plasmon resonance (SPR), we found to our surprise that it bound both CD4 and gp120 and bound them with similar affinity: a K_d of 35.4 nM for gp120 and 31 nM for CD4.[37] To make matters even more complex, retrocyclin-1 bound galactosyl ceramide, a cell-surface glycolipid and suspected HIV-1 receptor,[38,39] with a K_d of 24.1 nM. How would we resolve this paradox of binding that, on the surface, appeared nonspecific yet was also demonstrably of high affinity. Ironically, enlightenment came from a phrase that was formulated over a century ago by Paul Ehrlich, who used it to explain immunological specificity: "side chains!" Perhaps the peptides were recognizing carbohydrate molecules. Both gp120 and CD4 are glycoproteins and galactosyl-ceramide is a glycolipid. Several lines of evidence that strongly supported this hypothesis were provided by SPR. For example, the binding of retrocyclin-1 to gp120 and CD4 was greatly reduced when the glycoproteins were enzymatically deglycosylated, and their binding to bovine serum albumin was enhanced when this nonglycosylated protein was covalently coupled to various carbohydrates.

Cyclic retrocyclin-1 did not bind gp120 or CD4 effectively if its tridisulfide ladder was disrupted by reduction and alkylation. The noncyclic, octadecapeptide precursor of retrocyclin-1 was also ineffective despite containing the three disulfide bonds. Taken together, these findings indicate that retrocyclins require both a cyclic backbone and a tridisulfide ladder to manifest lectin-like behavior.

Norelle Daly and David Craik (University of Queensland, Brisbane, Australia) have defined the solution structure of retrocyclin-2 by NMR (Protein Data Bank, Accession number 2ATG) and have provided insights into the ability of these peptides to self-assemble into structured oligomers. We have made and tested many analogues of retrocyclin-1 with one or more amino acid substitutions, and their use allowed us to identify the residues that participate in carbohydrate binding. These studies, yet unpublished, will be described elsewhere.

For any readers especially interested in our work on HIV-1 or phylogenetic aspects of θ-defensins, we will summarize a few additional publications. Our phylogenetic studies[1,40] found six human DEFT genes, five on chromosome 8p23 and one on chromosome 1. All six pseudogenes, as well as their homologues in chimpanzees and gorillas, contained the same premature stop codon. Intact DEFT genes were identified in several Old World monkeys, a gibbon, and orangutans. No DEFT genes were found in six New World monkeys and five prosimians. These and other findings suggest that DEFT genes and θ-defensins arose in Old World monkeys by mutation of a preexisting α-defensin gene. Although intact DEFT genes survive in some nonhuman primates, our hominid ancestors lost their ability to produce θ-defensins after the orangutan and hominid lineages diverged. It is possible that this mutation made our species more susceptible to infection by HIV-1, but this hypothesis would be difficult or impossible to test.

The activity of retrocyclin-1 and several analogues was tested against primary HIV-1 isolates by investigators at the Communicable Disease Center in Atlanta.[41] The viruses included R5 and R5X4 strains of subtypes (clades) A–D, CRF-01_AE, and recombinant strains. The analogues differed from retrocyclin-1 by a single amino acid substitution, for example, gly4→tyr in RC-106 and Arg9→Lys9 in RC-101. RC-106 showed much reduced activity against HIV-1. Although its modification was chemically conservative, RC-101 was significantly more potent than retrocyclin-1 across the panel of isolates. In SPR binding studies, RC-101 bound recombinant gp120 from HIV-1 LAV/IIIB (clade B) with a K_d of 30–35 nM. It bound gp120 from two circulating recombinant (CRF-01_AE) strains with K_d values of 200–750 nM. We do not know if these different binding affinities for gp120 arise from qualitative or quantitative differences in gp120 glycosylation; from differences in the number, density, or distribution of their charged amino acids (R, K, E, and D); or from a combination of these factors. Nor do we know if such differences affect the relative susceptibility of different HIV-1 isolates to defensins. Given the extensive glycosylation of gp120 (over half of its mass consists of carbohydrate) and the substantial sequence heterogeneity between (and within) isolates, this will not be so simple to determine. The factor(s) responsible for improving the performance of RC-101 over that of RC-100 are under investigation. Whatever they may be, the finding that a relatively minor amino acid change enhanced the anti-HIV properties of retrocyclin-1 is testimony that it can be used as a design platform to create novel therapeutics. Our laboratory notebooks hold additional testimonials of this type. It is also noteworthy that RC-112, an *enantio*-analogue of retrocyclin-1 composed exclusively of D-amino acids, was more effective than retrocyclin-1 against a panel of 30 different primary isolates of HIV-1,[42] even though it bound gp120 and CD4 with lower affinity.

It remains to be determined if the improved activity of RC-112 results from resistance to degradation by surface-associated or secreted proteases of the target cells, or from other factors. θ-defensins composed exclusively of D-amino acids merit consideration as potential topical microbicides for application to sites, for example, the vagina or rectum, that contain proteases produced by the host or the resident microbial flora.

Although humans lack θ-defensin peptides, six human α-defensins show activity against HIV-1.[43–45] We compared the ability of six θ-defensins (hominid retrocyclin-1 to -3 and RTDs 1 to 3) and four human α-defensins (HNP-1 to -4) to bind gp120 and CD4 and to protect J53-BL indicator cells from primary HIV-1 isolates that varied in subtype and coreceptor usage. The most potent θ-defensin was retrocyclin-2. It bound with exceptionally high affinity to gp120 (K_d, 9.4 nM) and CD4 (K_d, 6.87 nM), and its effectiveness against subtype B isolates (IC_{50}, 1.05 ± 0.28 μg ml^{-1}; 520 ± 139 nM) was approximately twice as great as that of HNP-1 on a molar basis. We also showed, for the first time, that human α-defensins, HNP-1 to HNP-3, were also lectins that bound with high affinity to gp120 (K_d range, 15.8–52.8 nM) and CD4 (K_d range, 8.0–34.9 nM). Several serum proteins also bound these α- and θ-defensins and competed with their ability to bind gp120. However, even the low concentrations of α-defensins in normal human serum should suffice to bind over half of the gp120 spikes on HIV-1 and an even higher percentage of cell surface CD4 molecules.

We tested the ability of 20 synthetic θ-defensins to protect cells from infection by type 1 and type 2 herpes simplex viruses (HSV-1 and -2, respectively). The peptides included RTD-1 to RTD-3, retrocyclin-1 to -3, and 14 retrocyclin analogues, including the retro, enantio, and retroenantio forms of retrocyclin-1. Retrocyclin-1 and -2 and RTD-3 protected cervical epithelial cells from infection by both HSV serotypes, but only retrocyclin-2 did so without causing cytotoxicity or requiring preincubation with the virus. SPR studies revealed that retrocyclin-2 bound to immobilized HSV-2 glycoprotein B (gB2) with high affinity (K_d, 13.3 nM) and that it did not bind to enzymatically deglycosylated gB2. Temperature shift experiments indicated that retrocyclin-2 and human α-defensins HNP-1 to HNP-3 protected human cells from HSV-2 by different mechanisms. Retrocyclin-2 blocked viral attachment, and its addition during the binding or penetration phases of HSV-2 infection markedly diminished nuclear translocation of viral VP16 and expression of viral ICP4. In contrast, HNP-1 to -3 had little effect on binding or entry[46] but could both directly inactivate cell-free virus and act at a postentry step.[48,49] α- and θ-defensins behave in much the same way on HIV-1 infection. Whereas retrocyclins block the entry of HSV, α-defensins can directly damage them and also block a postentry step.[34,43] From these considerations, the possibility that α- and θ-defensins would exert additive or synergistic effects deserves exploration.

4.7   Mechanism of Activity Against Influenza A

The effects of retrocyclin-2 on influenza virus infection mediated by viral hemagglutinin (HA) have been examined in considerable detail.[50] Entry of influenza virus occurs in three stages. First, interactions between HA and sialic acid receptors at the cell surface mediate binding. Following this, the virus is internalized by endocytosis, after which the viral envelope fuses with the endosomal membrane, allowing virions to enter the cell. Bound HIV and herpes virus enter cells in a different way, first binding to cell surface and then directly fusing with the plasma membrane.

RC2-mediated inhibition of influenza virus did not result from changes in binding or from inhibition of clathrin-mediated endocytosis. Intact disulphide bonds were as necessary for activity against influenza as they were for activity against HIV-1. Retrocyclins did not directly inactivate influenza virions and they showed minimal, if any, cytotoxicity for the various target cells used in the study. Similar observations were also made in our studies with other viruses.

Fusion mediated by viral glycoproteins is a multistep process that proceeds through a transient stage of membrane hemifusion to complete fusion detectable as content mixing between aqueous volumes initially separated by the membranes. Because binding of influenza virus to cell-surface receptors and its subsequent membrane fusion in acidified endosomes are mechanistically uncoupled, the effects of retrocyclins on viral binding and fusion could be assessed separately.

The earliest intermediate in HA-mediated fusion is restricted hemifusion (RH) – a situation in which further lipid mixing through a hemifusion site is restricted by HA or other cell-surface molecules. Retrocyclins inhibited RH and, if added after RH had already occurred, they also prevented its progression to complete fusion. When viral fusion proteins are conformationally activated, the lipid bilayers of viral and target cell membrane are separated by at least 10 nm, and the contact zone is crowded with membrane proteins (including fusion proteins and receptors) that cover both of the membranes. To allow tight bilayer contact, hemifusion, and fusion, pore-opening membrane proteins must be displaced from the future fusion site. RC2-mediated cross-linking of surface glycoproteins evidently frustrates this process by blocking the displacement of proteins from the fusion site. RC2 inhibits not only hemifusion but also subsequent fusion stages suggesting that fusion completion downstream of RH intermediates requires further expansion of hemifusion structures and, thus, protein-depleted membrane patches.

During these experiments, we found that RC2 also inhibited fusion mediated by the dissimilar glycoproteins of Sindbis virus and baculovirus. This provides more evidence that the antiviral effects of retrocyclins do not arise from interactions specific to any one class of viral or cellular (co)receptors or viral fusion proteins.

The mechanism elucidated for retrocyclin-2 and influenza A virus exemplifies a new antiviral mechanism – one that depends on the ability to form a barricade of crosslinked and immobilized cell-surface glycoproteins that can block viral fusion and entry. This mechanism is not unique to RC2, since we found that β-defensin-3 (HBD-3) and mannan-binding lectin (MBL) also used this mechanism to block viral fusion. As lectins are increasingly being thought of as potential antiviral therapeutics, it would not be surprising to learn that other lectins, in addition to defensins and MBL, can erect cell-surface barricades to prevent viral entry.

5   Concluding Remarks

There are many ways to look at θ-defensins. While a rhesus macaque might consider them to be evidence of "intelligent design," we see them as evidence of an interesting and potentially useful molecular design. Human α-defensins, about which much is known, are useful *in vivo* because they are delivered by motile leukocytes to sites challenged by pathogens. Human cells neither produce nor deliver θ-defensins. The peptides can be produced by chemical synthesis, but only at a cost that would make their systemic use prohibitively expensive. What is most needed to develop defensin therapeutics is the establishment of a large-scale manufacturing procedure to produce GMP-quality defensins in a cost-efficient manner. With adequate resources, and 21st-century biotechnology, this should be possible. Only then can the true potential of defensin therapeutics be explored and perhaps realized.

References

1. T.X. Nguyen, A.M. Cole and R.I. Lehrer, *Peptides*, 2003, **24**, 1647.
2. R.I. Lehrer and T. Ganz, *Curr. Opin. Immunol.*, 1999, **11**, 23.
3. R.I. Lehrer and T. Ganz, *Curr. Opin. Hematol.*, 2002, **9**, 18.
4. R.I. Lehrer and T. Ganz, *Curr. Opin. Immunol.*, 2002, **14**, 96.
5. R.I. Lehrer, *Nat. Rev. Microbiol.*, 2004, **2**, 727.
6. P.H. Mygind, R.L.S.K. Fischer, M.T. Hansen, C.P. Sonksen, S. Ludvigsen, D. Raventos, S. Buskov, B. Christensen, L. De Maria, O. Taboureau, D. Yaver, S.G. Elvig-Jorgensen, M.V. Sorensen, B.E. Christensen, S. Kjaerulff, N. Frimodt-Moller, R.I. Lehrer, M. Zasloff and H.H. Kristensen. *Nature*, 2005, **437**, 975.
7. J.J. Marchalonis, M.K. Adelman, S.F. Schluter and P.A. Ramsland, *Dev. Comp. Immunol.*, *in press*, 2005.
8. R.I. Lehrer, A.K. Lichtenstein and T. Ganz, *Annu. Rev. Immunol.*, 1993, **11**, 105.
9. D. Yang, A. Biragyn, L.W. Kwak and J.J. Oppenheim, *Trends Immunol.*, 2002, **23**, 296.
10. J.J. Oppenheim, A. Biragyn, L.W. Kwak and D. Yang, *Ann. Rheum. Dis.*, 2003, **62** (Suppl 2), ii17.
11. T. Ganz, M.E. Selsted, D. Szklarek, S.S. Harwig, K. Daher, D.F. Bainton and R.I. Lehrer, *J. Clin. Invest*, 1985, **76**, 1427.
12. M.E. Selsted, S.S. Harwig, T. Ganz, J.W. Schilling and R.I. Lehrer, *J. Clin. Invest*, 1985, **76**, 1436.
13. W.G. Rice, T. Ganz, J.M. Kinkade Jr., M.E. Selsted, R.I. Lehrer and R.T. Parmley, *Blood*, 1987, **70**, 757.
14. D. Trabattoni, S.L. Caputo, G. Maffeis, F. Vichi, M. Biasin, P. Pierotti, F. Fasano, M. Saresella, M. Franchini, P. Ferrante, F. Mazzotta and M. Clerici, *J. Acquir. Immune. Defic. Syndr.*, 2004, **35**, 455.
15. B. Agerberth, J. Charo, J. Werr, B, Olsson, F. Idali, L. Lindbom, R. Kiessling, H. Jornvall, H. Wigzell and G.H. Gudmundsson, *Blood*, 2000, **96**, 3086.
16. P.M. Aldred, E.J. Hollox and J.A. Armour, *Hum. Mol. Genet.*, 2005, **14**, 2045.
17. R.M. Linzmeier and T. Ganz, *Genomics*, 2005, **86**, 423.
18. M. Boniotto, M. Ventura, J. Eskdale, S. Crovella and G. Gallagher, *Genet. Test.*, 2004, **8**, 325.
19. C. Zhao, T. Nguyen, L. Liu, R.E. Sacco, K.A. Brogden and R.I. Lehrer, *Infect. Immun.*, 2001, **69**, 2684.
20. T. Scheetz, J.A. Bartlett, J.D. Walters, B.C. Schutte, T.L. Casavant and P.B. McCray, Jr., *Immunol. Rev.*, 2002, **190,** 137.
21. B.C. Schutte, J.P. Mitros, J.A. Bartlett, J.D. Walters, H.P. Jia, M.J. Welsh, T. L. Casavant and P.B. McCray, Jr., *Proc. Natl. Acad. Sci. USA*, 2002, **99**, 2129.
22. D.M. Hoover, O. Chertov and J. Lubkowski, *J. Biol. Chem.*, 2001, **276**, 39021.
23. F. Fant, W. Vranken, W. Broekaert and F. Borremans, *J. Mol. Biol.*, 1998, **279**, 257.
24. F.T. Lay, H.J. Schirra, M.J. Scanlon, M.A. Anderson and D.J. Craik, *J. Mol. Biol.*, 2003, **325**, 175.

25. P. Bulet and R. Stocklin, *Protein Pept. Lett.*, 2005, **12**, 3.
26. M.S. Castro and W. Fontes, *Protein Pept. Lett.*, **12**, 2005, 13.
27. S. Crovella, N. Antcheva, I. Zelezetsky, M. Boniotto, S. Pacor, M.V. Verga Falzacappa and A. Tossi, *Curr. Protein Pept. Sci.*, 2005, **6**, 7.
28. F.T. Lay and M.A. Anderson, *Curr. Protein Pept. Sci.*, 2005, **6**, 85.
29. H.G. Sahl, U. Pag, S. Bonness, S. Wagner, N. Antcheva and A. Tossi, *J. Leukoc. Biol.*, 2005, **77**, 466.
30. M.E. Selsted and A.J. Ouellette, *Nat. Immunol.*, 2005, **6**, 551.
31. L. Leonova, V.N. Kokryakov, G. Aleshina, T. Hong, T. Nguyen, C. Zhao, A.J. Waring and R.I. Lehrer, *J. Leukoc. Biol.* 2001, **70**, 461.
32. D. Tran, P.A. Tran, Y.Q. Tang, J. Yuan, T. Cole and M.E. Selsted, *J. Biol. Chem.*, 2002, **277**, 3079.
33. A.M. Cole, T. Hong, L.M. Boo, T. Nguyen, C. Zhao, G. Bristol, J.A. Zack, A.J. Waring, O.O. Yang and R.I. Lehrer, *Proc. Natl. Acad. Sci. USA*, 2002, **99**, 1813.
34. T.L. Chang, J. Vargas, Jr., A. DelPortillo and M.E. Klotman, *J. Clin. Invest.*, 2005, **115**, 765.
35. C. Munk, G. Wei, O.O. Yang, A.J. Waring, W. Wang, T. Hong, R.I. Lehrer, N. R. Landau and A.M. Cole, *AIDS Res. Hum. Retroviruses*, 2003, **19**, 875.
36. T.M. Weiss, L. Yang, L. Ding, A.J. Waring, R.I. Lehrer and H.W. Huang, *Biochemistry*, 2002, **41**, 10070.
37. W. Wang, A.M. Cole, T. Hong, A.J. Waring and R.I. Lehrer, *J. Immunol.*, 2003, **170**, 4708.
38. N. Yahi, S. Baghdiguian, H. Moreau and J. Fantini, *J. Virol.*, 1992, **66**, 4848.
39. S. Bhat, R.V. Mettus, E.P. Reddy, K.E. Ugen, V. Srikanthan, W.V. Williams and D.B. Weiner, *AIDS Res. Hum. Retroviruses*, 1993, **9**, 175.
40. A.M. Cole, W. Wang, A.J. Waring and R.I. Lehrer, *Curr. Protein Pept. Sci.*, 2004, **5**, 373.
41. S.M. Owen, D.L. Rudolph, W. Wang, A.M. Cole, A.J. Waring, R.B. Lal and R.I. Lehrer, *AIDS Res. Hum. Retroviruses*, 2004, **20**, 1157.
42. S.M. Owen, D. Rudolph, W. Wang, A.M. Cole, M.A. Sherman, A.J. Waring, R.I. Lehrer and R.B. Lal, *J. Pept. Res.*, 2004, **63**, 469.
43. T.L. Chang and M.E. Klotman, *AIDS Rev.*, 2004, **6**, 161.
44. W. Wang, S.M. Owen, D.L. Rudolph, A.M. Cole, T. Hong, A.J. Waring, R.B. Lal and R.I. Lehrer, *J. Immunol.*, 2004, **173**, 515.
45. Z. Wu, F. Cocchi, D. Gentles, B. Ericksen, J. Lubkowski, A. Devico, R.I. Lehrer and W. Lu, *FEBS Lett.*, 2005, **579**, 162.
46. B. Yasin, W. Wang, M. Pang, N. Cheshenko, T. Hong, A.J. Waring, B.C. Herold, E.A. Wagar and R.I. Lehrer, *J. Virol.*, 2004, **78**, 5147.
47. K.A. Daher, M.E. Selsted and R.I. Lehrer, *J. Virol.*, 1986, **60**, 1068.
48. M. Trabi, H.J. Schirra and D.J. Craik, *Biochemistry*, 2001, **40**, 4211.
49. P.M. Abuja, A. Zenz, M. Trabi, D.J. Craik and K. Lohner, *FEBS Lett.*, 2004, **566**, 301.
50. E. Leikina, H. Delanoe-Ayari, K. Melikov, M.S. Cho, A. Chen, A.J. Waring, W. Wang, Y. Xie, J.A. Loo, R.I. Lehrer and L.V. Chernomordik, *Nat. Immunol.*, 2005, **6**, 995.

CHAPTER 7

C-type Lectin Receptors that Regulate Pathogen Recognition Through the Recognition of Carbohydrates

SANDRA J. VAN VLIET, CHRISTIAN H. GRÜN AND YVETTE VAN KOOYK

Department of Molecular Cell Biology & Immunology, VU Medical Center, Amsterdam, The Netherlands

1 Introduction

C-type lectin receptors are expressed on a variety of immune cells, such as antigen presenting cells (APC) and endothelial cells, which are involved in pathogen recognition. Both APC and endothelial cells play a central role in antigen presentation and induction of immunity, but on the other side are key players in the maintenance of peripheral tolerance to self-antigens. A well-known function of C-type lectins is to recognize carbohydrate moieties on antigens, which subsequently are internalized into the cells by receptor-mediated uptake, allowing enhanced processing on MHC molecules and presentation of antigen to T cells.

In this review we will discuss the functional similarities and differences between the following C-type lectin receptors: dendritic cell-specific intercellular adhesion molecule 3-grabbing nonintegrin (DC-SIGN), liver/lymph node-specific ICAM-3-grabbing nonintegrin (L-SIGN), macrophage galactose-type lectin (MGL), and asialoglycoprotein receptor (ASGP-R), in which we focus on their expression profile and carbohydrate and pathogen specificity.

2 Expression Patterns of C-type Lectins in the Human Body

A well-balanced immune system is crucial to the survival of the host. The human body is continuously attacked by pathogens from the external environment. These pathogens need to be eliminated before they cause major infections and potentially kill their host. For this reason the skin and all peripheral mucosal sites, which are in close contact with the outside world, are seeded with specialized APC, such as dendritic cells (DC) and macrophages (MØ). These APC express several pattern recognition receptors (PPRs), specific for integral structural components of the invading pathogen, such as viral or bacterial DNA, lipopolysaccharide (LPS) or certain carbohydrate structures. These components are commonly referred to as pathogen-associated molecular patterns (PAMPs). Upon recognition of a pathogen via its PAMPs, the APC will become activated – a process called maturation – and migrate towards the draining lymph node where the pathogenic antigens are presented to T cells, thereby initiating an adaptive immune response.[1]

Meanwhile, the immune system has to protect the host from potentially damaging immune reactions and remain unresponsive towards innocuous antigens and self-glycoproteins or tissues. The liver is one of the organs involved in maintaining this peripheral tolerogenic status. Many foreign antigens derived from food components and commensal bacteria pass the liver via the portal vein. These antigens are not harmful and therefore need to be cleared and thereby ignored by the immune system. The blood vessels of the liver are covered with specialized liver sinusoidal endothelial cells (LSEC), which display several features of APC and are capable of presenting antigens to T cells. In contrast to professional APC, LSEC induce tolerance instead of immunity.[2] The liver hepatocytes are, furthermore, involved in the clearance and uptake of glycoproteins and hormones from circulation.

As mentioned earlier, C-type lectins recognize carbohydrates on glycoproteins or glycolipids via their carbohydrate recognition domain (CRD). C-type lectins can function as either clearance receptors or genuine PPRs (Table 1). After recognition, carbohydrate antigens are internalized, and for some C-type lectins internalization results in antigen presentation on the cell surface in MHC class I or II, or in CD1 – the molecule for presentation of lipid antigens.[3–5]

2.1 DC-SIGN (CD209)

The C-type lectin DC-SIGN was originally described as a marker for DC in skin and mucosal tissues in the gastrointestinal and genital tract.[6] In lymph node, DC-SIGN positive immature DC can be found in the T-cell areas and along the outer zones of the paracortex.[7] DC-SIGN positive precursor DC constitute a small percentage of circulating DC in blood.[8] DC-SIGN is also highly expressed in the human placenta on Hofbauer cells, a cell type that was originally defined as a specialized type of MØ, and in the spleen on ellipsoids, the only site in the human body where the

Table 1 *Expression pattern of several human C-type lectins*

C-type lectin	Expression pattern
DC-SIGN	DC, MØ
L-SIGN	LSEC, LN endothelium
ASGP-R	Hepatocytes, epithelial cells (thyroid, kidney)
MGL	Immature DC, MØ

bloodstream is in direct contact with the underlying tissue without a separate endothelial layer. Further, immuno-histochemical analysis identified DC-SIGN expressing tissue resident decidual and alveolar MØ populations.[9,10]

2.2   L-SIGN (DC-SIGNR, CD299)

The closely related homologue of DC-SIGN, L-SIGN or DC-SIGNR (*i.e.*, DC-SIGN *related*) is not expressed by DC or other myeloid cells. Instead, L-SIGN is exclusively found on specialized sinusoidal endothelial cells in the outer zone of the paracortex and in the liver on LSEC.[7,11]

2.3   ASGP-R

The asialoglycoprotein receptor, (ASGP-R) was the first mammalian member of the C-type lectin family to be discovered in the mid-1960s.[12] Similar to L-SIGN, ASGP-R is highly expressed in the human liver, though not on LSEC but on liver parenchymal cells. ASGP-R is distributed in a polarized fashion on hepatocytes, where it is found on the basolateral membrane facing the capillaries but is virtually absent from the apical membrane facing the bile canaliculi.[13] ASGP-R is expressed at low levels in the thyroid and on proximal tubular epithelial cells in the kidney.[14,15]

2.4   MGL (CD301)

The galactose-type C-type lectin MGL (also called DC-ASGPR or HML[16,17]) is expressed by human DC and MØ at an intermediate stage of differentiation from monocytes.[18] MGL positive immature DC and MØ are most abundantly found in the dermis of the skin, but MGL expression is present on DC-like cells in the T-cell areas of human lymph node and tonsil as well.

3   Binding of C-type Lectins to Carbohydrates

Animal lectins are carbohydrate-binding proteins with a principal function in the recognition of carbohydrates present at cell surfaces, attached to circulating proteins, and present in extracellular matrices. By binding specifically to the carbohydrates, lectins mediate biological events, such as cell–cell adhesion, host–pathogen recognition, serum glycoprotein turnover and innate immune responses. C-type lectins form a superfamily of lectins, whose binding to carbohydrate ligands depends on the presence of Ca^{2+} ions. Loss of the ion results in conformational changes within the binding site, resulting in the loss of binding function. C-type lectins have a primary binding site that is selective for a single monosaccharide residue. The affinity for monosaccharides, however, is usually weak, with association constants in the millimolar range.[19] Secondary binding sites exist adjacent to the primary sites that interact with neighbouring monosaccharide residues present in the carbohydrate. These secondary binding sites, on the one hand, strongly enhance the binding of carbohydrates and, on the other, fine-tune lectin specificity for certain types of linkages and substitutions. Interactions with the secondary binding sites involve amino acids rather than coordination by Ca^{2+} ions, meaning that the interactions are rather weak compared to those of the primary binding site.

The selectivity for *mono*saccharides is not always high, and especially variations at C-2 are accepted by many lectins. Hence, mannose-binding lectins may bind mannose (which bears an axial OH-2) as well as glucose (with equatorial OH-2). Owing to the structural similarity of mannose and fucose, both monosaccharides serve as acceptors for many mannose-binding lectins. Furthermore, lectins that recognize galactose may also bind to *N*-acetyl-galactosamine with high affinity.

3.1 Interactions

Binding of carbohydrates occurs on shallow depressions on the surface of lectins rather than in the interior of the protein. Four different types of interactions between the lectin and the carbohydrate ligand can be distinguished that together define the selectivity and binding affinity for certain ligands: coordination by Ca^{2+} ions, hydrogen bond formation, hydrophobic interactions and van der Waals forces. Coordination by Ca^{2+} ions forms the key interaction for C-type lectins. Binding is strong and involves two vicinal hydroxyl groups of a monosaccharide residue. The Ca^{2+} ion also forms a number of coordination bonds with the protein via the side chains of certain amino acid residues. The amino acids that are involved in binding to Ca^{2+} are highly conserved in the C-type lectin superfamily: in mannose-binding lectins, these are two glutamic acid and two asparagine residues, of which two are combined in the sequence Glu-Pro-Asn. In galactose-binding C-type lectins, the glutamic acid is replaced by a glutamine in this motif and the asparagine by an aspartic acid, forming the sequence Gln-Pro-Asp. Replacing these amino acids by site-specific mutagenesis can drastically change the binding specificity of the lectin. When the glutamic acid and the asparagine residues in mannose-binding protein MBP-A were replaced by glutamine and aspartic acid, respectively, galactose became the favoured ligand instead of mannose.[20] The amino acids that coordinate with Ca^{2+} also form hydrogen bonds with the carbohydrate. These hydrogen bonds are formed between the hydroxyl groups of the carbohydrate and amines, hydroxyls, and oxygen of the protein.[19] A number of surrounding amino acids are also involved in hydrogen-bond formation, thereby stabilizing the binding. In addition to hydrogen bonds that are formed by direct interaction between the carbohydrate and the protein, water-mediated bridges can be formed. Hydrophobic interactions occur between the cyclic surface of a monosaccharide residue and hydrophobic parts of the protein. This interaction involves stacking of a monosaccharide plane parallel to the hydrophobic side chains of the aromatic amino acids phenylalanine, tyrosine and tryptophan. In addition, substitutions on monosaccharides such as the methyl group in *N*-acetyl hexosamines can interact with aromatic residues of the protein. Van der Waals forces, finally, are usually very weak. However, owing to their frequent occurrence, their contribution in binding is significant.

3.2 DC-SIGN (CD209)

Dendritic cell-specific intercellular adhesion molecule 3-grabbing nonintegrin (DC-SIGN) is a type II transmembrane protein that, on the basis of its structure, belongs to the C-type lectin family. It contains a short cytoplasmic N-terminal domain with several intracellular sorting motifs, an extracellular stalk of seven complete and one partial tandem repeat, and a C-terminal lectin or CRD.[6,21] The DC-SIGN gene is located on the human chromosome 19p13, in the proximity of a gene encoding a closely related protein, L-SIGN (discussed later), which shares 77% sequence identity with DC-SIGN.[22] DC-SIGN is exclusively expressed on DC and was identified by its interaction with T cells via the intracellular adhesion molecule ICAM-3.[6] The inhibition of DC-SIGN function by the yeast carbohydrate mannan suggests that the interaction with its ligands is mediated by mannose-like carbohydrates.[6] Indeed, DC-SIGN functions as a mannose receptor recognizing mannose monosaccharides as well as mannose present in high-mannose-containing glycoconjugates. It was shown that each individual CRD possesses high affinity for mannose-containing oligosaccharides, and the CRD-containing extracellular domains of four DC-SIGN molecules form a tetramer.[23] Inhibitory studies using methyl glycosides showed that inhibition by methyl α-galactoside is approximately 30-fold less effective than inhibition by methyl α-mannoside, indicating that DC-SIGN distinguishes between monosaccharides through the relative positions of the hydroxyl groups on C-3 and C-4.[23] Furthermore, a hydroxyl group at C-2 in axial position is preferred to an equatorial hydroxyl, which explains why methyl α-glucoside is a relatively weak inhibitor. The authors further showed that L-fucose is a good inhibitor that binds better than mannose. Even better binding is obtained by using multivalent carbohydrate ligands, such as the naturally occurring high-mannose N-glycans, which may be explained by the tetramerization of the CRDs. Crystallography studies on the CRD of DC-SIGN in complex with mannose-containing oligosaccharides showed that the oligosaccharide ligand forms a bridge between two CRDs.[24] The CRD shows a folding typical to C-type lectins with two α-helices and five β-strands. The data show that the equatorial hydroxyl groups at C-3 and C-4 of (1,3)-linked α-mannose interact with Ca^{2+}. The interaction is further stabilized by hydrogen bonds with asparagine (to OH-3), glutamic acid (to OH-4) and lysine (to OH-6) residues. Remarkably, DC-SIGN recognizes internal mannose residues, whereas common mannose-binding lectins have a preference for terminal mannose residues. This has consequences for the binding to high-mannose oligosaccharides, as will be discussed later. Interaction of the carbohydrate with the protein occurs via the formation of hydrogen bonds with a serine residue. Another serine interacts with hydroxyl groups of mannose via a water-mediated hydrogen bond. Other amino acids that are involved in binding to the carbohydrate include glutamic acid, aspartic acid, and phenylalanine residues. The latter appears to play a key role in binding, because it discriminates between internal α-Man and β-Man during binding of N-glycans: The crystallographic data combined with molecular modelling of high-mannose glycans showed that the internal mannose residue of the branched trisaccharide Manα1-3[Manα1-6]-Man can only bind if all mannose residues are in the α-anomeric configuration. This structure is only found in high-mannose oligosaccharides (Table 2).

As mentioned earlier, inhibition studies that were performed with several monosaccharides and oligosaccharides showed strong inhibitory activity for fucose, which is not uncommon for mannose-binding lectins because of the structural similarities between mannose and fucose. In fact, L-fucose, if rotated by 180°, displays a conformation similar to that of D-mannose with the hydroxyls of fucose at positions C-2, C-3, and C-4 overlapping with those of mannose at C-4, C-3, and C-2, respectively. Therefore, many mannose-binding lectins bind free fucose at the equatorial OH-2 and OH-3. In several

Table 2 *Lectin binding specificity*

Lectin	Binding specificity
DC-SIGN	Internal mannose in high-mannose N-glycans
	Fucose in Lewis-type structures
L-SIGN	Internal mannose in high-mannose N-glycans
	Lewis^a, Lewis^b, Lewis^y
ASGP-R	Terminal Gal, GalNAc
MGL	Terminal GalNAc (cellular, human)
	Terminal Gal, Lewis^x (mouse)
	Terminal GalNAc, GalNAc substituted at C-6 (recombinant)

studies it was shown that DC-SIGN can indeed bind to fucose-containing glycans, especially to Lewis-type blood group epitopes found on a number of pathogens (Table 2).[25,26] Binding to DC-SIGN was shown for all Lewis-type structures (Le^x, Le^a, Le^y, and Le^b present as neoglycoconjugates on PAA beads) without noticeable differences in binding specificity, whereas sialylation of Le^x and Le^a completely revokes binding.[25] Single α-L-fucose also binds, albeit to a lesser extend, indicating that the backbone to which fucose is linked may also play a role in binding. By screening a glycan array with over 130 different glycan structures, the group of Drickamer in collaboration with the Consortium for Functional Glycomics further refined the binding specificity of DC-SIGN. Glycans were covalently linked to biotin and were then immobilized on a streptavidin-coated plate. By incubating the carbohydrates with fluorescently labelled extracellular domains of DC-SIGN, relative quantification of the binding to various glycans was obtained.[27] The screening revealed the earlier-observed binding specificity for N-linked high-mannose oligosaccharides, with binding specificities increasing with increasing numbers of mannose residues present in the structure. Besides binding to these mannose-containing carbohydrates, DC-SIGN also showed high affinity for a number of different, terminal fucose-containing carbohydrates, most of them containing a Lewis-type structure.

This dual-binding specificity for DC-SIGN was investigated by two independent studies based on molecular modelling and on crystallography.[27,28] Both studies show that during binding of Lewis^x to the CRD of DC-SIGN, the Ca^{2+} in the primary carbohydrate-binding site interacts with the hydroxyl groups at C-3 and C-4 of fucose. As discussed earlier, binding of mannose to Ca^{2+} also coordinates with the hydroxyl groups at C-3 and C-4. However, in mannose, both hydroxyl groups are equatorially linked, whereas in fucose, the hydroxyl at C-3 is axial and that at C-4 is equatorial. The crystal structure of DC-SIGN in complex with a Lewis^x-containing oligosaccharide showed that in order to preserve full coordination by Ca^{2+}, the fucose residue is tilted, resulting in a loss of contact with Lys^{368}, which plays an important role in mannose binding.[27] The complex is stabilized by interaction with another amino acid residue, Val^{351}, which forms van der Waals contacts with the hydroxyl at C-2. The importance of Val^{351} for binding was shown by replacing the amino acid by Gly, which abrogates binding of fucose.[28,29] Importantly, mutation of Val^{351} into a different amino acid does not affect binding to mannose-containing oligosaccharides,[27] and it also does not affect binding to HIV gp120 glycoprotein.[11,29]

The molecular modelling studies gave similar results, though the hydrogen bond with OH-4 may be formed with Glu[347] instead of Glu[354].[28] Molecular modelling also makes clear that the terminal galactose residue in Lewis-type structures interacts with Glu[358] by forming a hydrogen bond with OH-6. This interaction enhances binding to Lewis-type structures compared to free fucose, which is in agreement with earlier studies on fucose binding.[25]

3.3   L-SIGN (DC-SIGNR, CD299)

Liver/lymph node-specific ICAM-3-grabbing nonintegrin (L-SIGN) is a close homologue of DC-SIGN, which shares 77% amino acid sequence identity[11,22] and was therefore originally designated DC-SIGN-related protein, or DC-SIGNR. However, the expression pattern of L-SIGN is strikingly different from that of DC-SIGN: in contrast to DC-SIGN, L-SIGN is not expressed by DC *in vivo* or by monocyte-derived DC cultured *in vitro*;[11,30] rather, L-SIGN expression is found in liver on LSEC and in lymph node.[11,30] The homology in amino acid sequence with DC-SIGN is reflected on the binding characteristics of L-SIGN. Both proteins are constituted as a tetramer and the individual CRDs bind with high affinity to mannose and mannose-containing carbohydrates, whereas their interaction with glucose is weak.[23] Despite the strong homology between DC-SIGN and L-SIGN and the similarities in affinity for mannose-containing ligands, striking differences were observed in the binding to fucose-containing carbohydrates. Whereas DC-SIGN shows high binding specificity to Lewis[x], L-SIGN does not recognize this ligand.[28] How can these differences in binding specificity be explained? Binding specificities of lectins are defined by certain conserved amino acid motifs in the CRD. For mannose-binding C-type lectins the Glu-Pro-Asn motif is highly conserved, whereas for galactose-binding the Gln-Pro-Asp sequence is essential. Both DC-SIGN and L-SIGN contain the Glu-Pro-Asn motif, which determines their affinity for mannose-containing carbohydrates.[31] However, several other amino acid residues in the CRD determine the binding specificity of the lectin; each lectin may recognize a unique branching and positioning of mannose or fucose residues on a given pathogen or cell surface structure. As discussed earlier, the binding of DC-SIGN to fucose is realized via interaction with a valine residue. This amino acid plays an essential role in binding to fucose, because replacing the valine for another amino acid completely abrogates binding.[29] Molecular modelling and crystallography studies showed that the valine forms a hydrophobic pocket that is required to fit Lewis[x] into DC-SIGN and to stabilize binding. However, in L-SIGN the valine is replaced by a hydrophilic serine, resulting in complete loss of this specific interaction, and therefore L-SIGN is unable to bind Lewis[x]. To investigate whether a hydrophobic region in the CRD of L-SIGN can be introduced that enhances binding to Lewis[x], the serine residue was replaced by a valine by means of site-specific mutagenesis, equivalent to the valine in the CRD of DC-SIGN.[28] Binding studies showed that this mutation dramatically increases recognition of fucose-containing carbohydrates by L-SIGN without compromising its binding to mannose-containing carbohydrates.[27,28] Remarkably, although Lewis[x] does not bind to wild-type L-SIGN, binding studies showed high affinity of L-SIGN for the fucose-containing Lewis[a], Lewis[y], and Lewis[b] oligosaccharides, with affinities similar to DC-SIGN (Table 2).[28] Again, molecular modelling was used to gain insight into the different aspects of ligand binding by L-SIGN. When Lewis[a] is docked in L-SIGN, the same serine residue that abrogates binding of Lewis[x] can form a hydrogen bond with OH-6 of the GlcNAc residue in Lewis[a], thereby restoring parts of the contacts that were initially lost (van Liempt, unpublished data). Similarly, Lewis[b] and Lewis[y] were docked in L-SIGN, but in contrast to Lewis[a], no additional contacts were observed. Why they bind to L-SIGN remains elusive.

3.4 ASGP-R

The asialoglycoprotein receptor (ASGP-R) is a C-type lectin that mediates the endocytosis and degradation of desialylated glycoproteins. The lectin is expressed by liver parenchymal cells in mammals and is composed of major and minor subunits. In humans, two subunits are present: the 40 kDa H1 (major) and the 48 kDa H2 (minor) that share 55% amino acid sequence identity. Each subunit is a type II transmembrane protein, which contains a short N-terminal cytoplasmic domain of 40 amino acids, a hydrophobic single-pass transmembrane domain and an extracellular stalk region to which a carboxy-terminal CRD is linked.[32,33] The CRD depends on Ca^{2+} for binding of carbohydrate ligands, and therefore the lectin belongs to the C-type lectin superfamily. The lectin has binding affinity for terminal, non-reducing galactose and *N*-acetyl-galactosamine residues (Table 2). Relative binding affinity to several ligands was determined by a competitive binding assay, which showed that Gal and GalNAc have equally binding activity. When using the methyl glycoside derivatives of these monosaccharides, however, binding affinity of methyl-α-GalNAc increases over ten times, whereas the binding affinity for methyl-Gal hardly changes.[34] As is common to C-type lectins, high affinity and specificity of ligand binding are achieved by multimerization of the CRDs. In ASGP-R, oligomerization occurs via the stalk segments of the two subunits, in such a way that a tetramer composed of two CRDs from subunit H1 and two from H2 is formed. Owing to the multimerization, multivalent ligands are bound with much higher affinity than monovalent carbohydrate ligands. The preferred ligand is a tri-antennary glycan with terminal galactose, which binds with an affinity approximately six orders of magnitude higher than monovalent galactose.[19] The crystal structure of the CRD from subunit H1 has revealed that the structure contains six long and two short β-strands and two α-helices that define a protein fold similar to that of CRDs of other mammalian lectins.[33] The CRD contains three Ca^{2+} binding sites, of which one is located in the glycine-rich loop that forms the carbohydrate-binding site. Attempts to crystallize the CRD of ASGP-R in complex with a carbohydrate ligand failed so far, probably owing to the low binding constant of the carbohydrate ligand.[33] Therefore, the structural basis for galactose-binding by ASGP-R is missing. Nevertheless, by combining the crystal structure of H1 with data obtained from closely related proteins, Meier and co-workers provided a model for lectin–carbohydrate interaction of ASGP-R. They made use of previously obtained results from site-specific mutagenisis of mannose-binding protein MBP-A, which resulted in binding properties remarkably similar to those of ASGP-R. The introduction of a motif denoted QPDWG in the CRD of MBP-A, replacing EPNH by QPDW, and the insertion of a glycine-rich segment HGLGG resulted in binding specificity and affinity similar to that of rat ASGP-R.[35] The crystal structure of this mutated CRD of MBP-A in complex with β-methyl galactoside and *N*-acetyl-galactosamine was resolved, giving a general structural basis for galactose-binding of lectins.[36] By superimposing the structure of the CRD binding site of H1 with that of the QPDWG mutant of MBP-A, Meier and co-workers were able to generate a model showing various interactions between the CRD of H1 and galactose.[33] The model indicates that binding of Ca^{2+} to Gal or GalNAc is coordinated via the hydroxyl groups at C-3 and C-4. Furthermore, hydrogen bonds are formed with glutamine, aspartic acid, glutamic acid, and asparagine residues. Binding of the carbohydrate ligand to the CRD is further enhanced by hydrophobic interactions of the apolar pyranose ring with the indole side chain of a tryptophan residue.

3.5 MGL (CD301)

Macrophage galactose-type lectin (MGL, also called DC-ASGPR or HML)[16,17,37] is most closely related to the ASGP-R lectin. In contrast to ASGP-R, MGL is expressed on human and mouse immature DC and MØ in skin and lymph node, rather than by liver parenchymal cells.[18] Mice contain two functional copies of the MGL gene, mMGL1 and mMGL2, whereas in humans only one MGL gene is found. mMGL1 and mMGL2 have different carbohydrate specificities for Lewis[x] and α/β-GalNAc structures, respectively.[38] Basic screening identified the monosaccharides Gal and GalNAc to be carbohydrate ligands for human MGL (Table 2).[17] By screening a glycan array using fluorescently labelled, recombinant human MGL, the binding specificity of MGL was further refined.[39] The glycan array showed that recombinant MGL binds with high affinity to monovalent or multivalent α- and β-GalNAc. Binding to galactose was not observed, which indicated that recombinant MGL prefers GalNAc for binding (Table 2). Screening of a large number of oligosaccharide structures confirmed the strong preference for GalNAc. Most notably, all glycans containing terminal GalNAc bind to MGL, regardless to which structure the GalNAc is substituted. Substitution on position 3 of the α-GalNAc, as found in the core 1–4 structures of O-glycans, abrogated MGL binding. Remarkably, when an α-GlcNAc residue is attached to the 6-position of GalNAc, binding is not affected, indicating that the primary interaction of MGL, similar to that of ASGP-R, may occur with OH-3 and OH-4 of GalNAc.

4 C-type Lectin–Pathogen Interactions

4.1 C-type Lectins in Viral Infections

Many enveloped viruses actively target C-type lectins during their lifecycle. C-type lectins can function as direct entry receptors, where expression of the C-type lectin is required for the virus to enter the cells of the host.[40] Other viruses use C-type lectins as transmission receptors. Therefore, expression of the C-type lectins is not sufficient for entry; instead, it concentrates virus particles at the cell surface and transmits these particles extremely efficiently to host cells carrying the necessary entry receptor. Often the actual infection is greatly enhanced by the C-type lectin, especially at low virus titers, which is often the case at the moment of infection.[41]

C-type lectins on APC can function as PPR. Recognition of a virus by the C-type lectin results in internalization of the virus. Many C-type lectins contain special internalization motifs within their cytoplasmic tails, thereby targeting internalized cargo to late endosomes or lysosomes. The efficient uptake and targeting of glycosylated viral antigens can potentially enhance presentation of virus-derived peptides to naïve T cells.[42]

The human immunodeficiency virus (HIV) is the causative agent of acquired immunodeficiency syndrome (AIDS). HIV is transmitted via sexual intercourse or via blood–blood contact. HIV affects about 40 million people worldwide, and it is estimated that 8000 people die of AIDS each day (World Health Organization, 2004, www.who.int). HIV belongs to the family of retroviruses, which are characterized by containing an envelope and positive-stranded RNA. These viruses are called retroviruses, because they encode an RNA-dependent DNA polymerase (reverse transcriptase) and replicate via a DNA intermediate, which is subsequently integrated in the host genome.[43]

The primary entry and fusion receptor for HIV is the CD4 molecule, present on T cells. Binding of the envelope glycoprotein gp120 to CD4 induces a conformational change within the gp120, necessary for the recognition of the co-receptor CCR5 or CXCR4 (depending on the HIV strain), thereby allowing the HIV envelope to fuse with the T cell membrane.[44,45]

The gp120 molecule is heavily glycosylated with more than 20 glycosylation sites and contains predominantly complex-type bi-, tri- and tetra-antennary sialo-oligosaccharides, occasionally with bisecting *N*-acetylglucosamine residues, and high-mannose-type glycans containing 5–9 mannoses. The high-mannose glycans form a cluster on the gp120 surface, while the complex glycans form another cluster with little structural overlap.[46,47]

The C-type lectins DC-SIGN and L-SIGN both recognize HIV gp120 through the high-mannose glycans, although the gp120 protein backbone may also contribute to binding.[11,29,41] Neither DC-SIGN nor L-SIGN serve as classical entry receptors for HIV. DC-SIGN, however, is expressed by DC in the mucosal layers of the genital tract. During the initial HIV infection, the mucosal DC pick up the viral particles via DC-SIGN and transport them to the draining lymph node where the virus is efficiently transmitted to CD4 positive T cells.[48] This process of *trans*-infection allows for minute amounts of virus to enter T cells that normally would not be infected without the assistance of DC-SIGN. DC-SIGN positive blood DC may be involved in the transmission of blood-borne HIV to T cells.[8]

Functional studies have shown that L-SIGN also can function as a *trans*-receptor for HIV, hinting at a possible role for LSEC in the transmission of blood-borne HIV to passing CD4 T cells.[11] HIV targets the C-type lectins DC-SIGN and L-SIGN; still other viruses, such as filoviruses or hepatitis C virus (HCV), target even more C-type lectins at multiple sites in the body.

The Marburg and Ebola viruses belong to the family of filoviruses. Filoviruses are filamentous, enveloped, negative-stranded RNA viruses, which are transmitted by mosquitoes. These agents cause haemorrhagic fevers, which are fatal in up to 50–90% of the cases.[49] Target organs in filovirus infection are the liver, spleen, lymph nodes and lungs. Marburg and Ebola viruses are endemic in Africa.

Both Ebola and Marburg viruses contain one glycoprotein, termed GP, in their envelope. The GP of 681 amino acids contains 19 N-linked glycosylation sites and several clusters of hydroxylamine acids that serve as attachment sites for O-glycosylation.[50] N-linked oligosaccharides include both high-mannose and hybrid-type glycans, as well as bi-, tri- and tetra-antennary complex-type glycans. The O-glycan clusters contain neutral mucin-type glycans, although this region has a high variability and is specific for a particular isolate.[51] GP is characterized by a complete lack of terminal sialic acids.[52,53]

Recently, several C-type lectins on different cell types were reported as attachment receptors for filoviruses. Phagocytic mononuclear cells, such as monocytes, DC and MØ, are the initial targets in filovirus infection. Indeed DC-SIGN serves as a classical entry receptor for both Ebola and Marburg viruses,[40,54] whereas MGL only increases infectivity of Ebola and not of Marburg virus.[55]

In the liver, L-SIGN and ASGP-R facilitate binding of filoviruses. L-SIGN recognizes both Ebola GP and Marburg virus GP.[54,56] In addition, ASGP-R has been identified as a receptor for Marburg virus that enhances the susceptibility to Marburg virus infection.[57] The targeting of filoviruses to ASGP-R and L-SIGN might explain the severe liver pathology in filovirus infection. DC-SIGN, L-SIGN, and ASGP-R can transmit Ebola to permissive cell types, via a similar mechanism reported for HIV transmission.[56,58]

The binding of C-type lectins to filovirus GP depends largely on the filovirus strain and which cell type was used for producing the virions. DC-SIGN and L-SIGN selectively interact with filovirus GPs that contain high-mannose glycans, consistent with their shared specificity for these types of oligosaccharides. ASGP-R recognizes terminal galactose residues present not only on the N-linked glycans of GP, but also on the mucin-type O-linked glycans.[58] The interaction of MGL with Ebola has not been studied in detail, although it probably recognizes terminal GalNAc residues in the mucin domain of GP.

Another pathogen that selectively targets the liver is the HCV. HCV was only identified in 1989 and is an enveloped flavivirus with an RNA genome. It is transmitted primarily via sexual intercourse and through infected blood. Twenty percent of the patients clear the virus, whereas the other 80% develop chronic hepatitis, with a high risk for liver failure, cirrhosis or hepatocellular carcinoma. About 170 million people are infected with HCV worldwide (World Health Organization, 2004).

The envelope of HCV contains two structural glycoproteins, E1 and E2. E1 and E2 form a non-covalent heterodimer that forms the major protein component of the HCV envelope. Both E1 and E2 are heavily N-glycosylated with 6 and 11 potential glycosylation sites, respectively. Since HCV cannot be produced under laboratory conditions, most studies have used recombinant E1/E2, which when combined form virus-like particles or pseudotyped viruses for their carbohydrate analysis. Only high-mannose-type glycans, of the Man7-9 species, were associated with E1/E2 complexes, concurring with an observed retention in the endoplasmic reticulum (ER).[59,60] The ER localization and lack of complex glycans suggests that HCV buds directly from the ER.[61] However, E1/E2 glycans might be modified during the transit of HCV through the secretory pathway. Lectin analysis of HCV virions in the serum of infected humans demonstrated strong binding of HCV to the lectins RCAI and WGA, weak binding to Con A and no detectable binding to AAL,

LCA and PNA,[62] indicating that complex glycans also are present on HCV envelope glycoproteins.

The principal site of HCV replication is not only hepatocytes but also peripheral mononuclear cells, such as B cells and DC.[63] DC-SIGN was identified as a binding receptor for HCV E1/E2 proteins or pseudotyped viruses.[64,65] The C-type lectin MGL does not recognize E1/E2 glycoproteins (S. van Vliet, unpublished results). In the liver, the major site of HCV infection, both L-SIGN and ASGP-R are expressed on LSEC and hepatocytes, respectively. L-SIGN serves as a capture receptor for HCV virions obtained from patients sera, similar to DC-SIGN.[66] Binding of DC-SIGN and L-SIGN was dependent on the presence of high-mannose-type glycans on E2.[64] In addition, ASGP-R was shown to recognize E1 and E2 glycoproteins, although it is unclear which particular N-glycan mediates this recognition.[67]

In situ binding studies of HCV virus-like particles to liver sections demonstrated that HCV interacts with LSEC instead of the hepatocytes, in an L-SIGN-dependent manner.[68] Both L-SIGN and DC-SIGN efficiently transmit pseudotyped HCV viruses to permissive hepatocytes.[69,70] Therefore, capture of circulating HCV virions by LSEC and DC may facilitate infection of hepatocytes and other mononuclear cells.

4.2 C-type Lectins in Parasitic Infections

Another type of pathogen that targets multiple C-type lectins at several sites in the body is the helminth parasite *Schistosoma mansoni*.

Schistosomiasis affects an estimated 200 million people worldwide and is endemic in sub-Saharan Africa, the Arabian Peninsula and north-eastern Brazil. Infection occurs when parasite larvae, known as cercariae, penetrate the skin during water contact. Inside the human host, the cercariae develop into schistosomula that reside in the portal veins draining the large intestine. Mature worms mate and start releasing eggs. The eggs are partially trapped in the liver, where they cause strong immune reactions such as granuloma formation and liver fibrosis. In addition, these eggs release soluble egg antigens (SEA), composed of several polysaccharides, glycoproteins and glycolipids, which are highly antigenic and can even enhance inflammatory responses.[71] Eggs that exit via the urine or faeces hatch in fresh water and can again infect the intermediate snail host.

The N-glycosylation of *S. mansoni* strongly resembles mammalian glycosylation. *S. mansoni* can synthesize both high-mannose and complex-type glycans, although these may contain specific carbohydrate structures which are rarely found in humans. Some of these structures include the LacdiNAc (LDN) motifs and its fucosylated derivate LDNF.[72] Whereas di-antennary N-glycans of *S. mansoni* lack galactose, tri- and tetra-antennary N-glycans contain high amounts of galactose and Lewis[x] or poly-Lewis[x] structures.[73] Many membrane and secreted glycoproteins derived from adult schistosomes contain simple O-glycans of the Tn or core 1 subtype.[74] In contrast, the circulating cathodic antigen (CCA) and circulating anodic antigen (CAA) have a core 2 motif and (poly-) Lewis[x] structures, similar to the complex N-glycans. *S. mansoni* glycolipids are built up from repeating fucosylated GlcNAc residues linked to a 'schiso-core' motif, $GalNAc\beta1-4Glc\beta1$-ceramide.[75]

Because of its unique life cycle and high glycan diversity, *S. mansoni* worms and eggs encounter C-type lectins not only on DC in the skin and mucosa, but also on hepatocytes and LSEC in the liver. The DC-expressed C-type lectins DC-SIGN and MGL recognize SEA.[25,26,39] DC-SIGN mediates binding via the fucosylated Lewis[x] and LDNF structures, whereas MGL recognizes carbohydrate structures terminated with β-GalNAc, such as LDN and LDNF. In the liver the C-type lectins L-SIGN and ASGP-R can potentially interact with *S. mansoni* eggs and SEA. Although L-SIGN does not recognize Lewis[x] structures, it does interact with SEA.[28] Since ASGP-R has a high affinity for terminal GalNAc residues, it can probably interact with the *S. mansoni* eggs or SEA; however, to date, no interaction between ASGP-R and *S. mansoni* has been reported.

4.3   C-type Lectins and *Helicobacter pylori*

In 1983 a new genus of spiral gram-negative bacteria, *Helicobacter pylori*, was discovered in patients with gastritis. *H. pylori* colonizes the stomach in more than half of the global human population. *H. pylori* infection can lead to gastric and duodenal ulcers, mucosa-associated lymphoid tissue lymphoma and gastric cancer.[76] The O-antigen of *H. pylori* LPS shows a strong structural similarity with host glyco-epitopes and is subjected to phase variation. The high-frequency phase variation (up to 0.5%) and molecular mimicry of host glycans is thought to provide an adaptive mechanism for survival of *H. pylori* and evasion of host immune responses.

H. pylori LPS O-antigens predominantly carry the blood group antigens Lewis[x] and Lewis[y], and to a lesser extent Lewis[a] and Lewis[b]. H type 2 structures are lacking on *H. pylori* LPS.[77] In *H. pylori* LPS, Lewis[x] can be present in a monomeric form or as repetitive copies of single Lewis[x] structures. Translational frame shifts in fucosyltransferase genes control the mechanism of phase variation in Lewis antigen expression.[78] Clinical isolates from one patient can contain several phase variants of *H. pylori* with a differential Lewis antigen expression on the LPS O-antigen.[79]

Since *H. pylori* resides in the stomach, it can only be recognized by C-type lectins on DC in the lamina propria of the stomach. DC-SIGN, expressed by the gastric mucosal DC, recognizes phase variants of *H. pylori* that carry Lewis[y] structures on their LPS O-antigens.[80] The C-type lectin MGL does not recognize Lewis antigen positive phase variants of *H. pylori* (S. van Vliet, unpublished results).

4.4   C-type Lectins and *Candida albicans*

Candida albicans is one of the most common opportunistic pathogens known. It is part of the normal flora of the mouth, gastrointestinal tract, and membranes lining the mucosa of other cavities. In immunocompromised patients, like AIDS patients, *C. albicans* can become a real pathogen and even cause life-threatening disease. It can switch from a yeast-like morphology into various filamentous forms.[81] The ability to switch between different morphologies is believed to be an important virulence factor for *C. albicans*.

The cell wall of *C. albicans* is composed of a rigid microfibrillar network of chitin, homopolymers of unbranced β1-4GlcNAc and β-glucans, branched polymers of β1-3- and β1-6-linked glucose. β-glucans are not displayed on the cell surface of the living yeast; however, during the normal mechanism of yeast budding and cell separation, scars are created that expose β-glucans to the exterior.[82] In addition, β-glucans can be made accessible by heat-killing the yeast.

C. albicans glycoproteins are attached to the network of β-glucans and chitin. Both N- and O-glycans in *C. albicans* are formed by addition of mannose polymers, commonly referred to as mannans. Mannans represent 40% of the total cell wall carbohydrates and form the surface coating. O-linked mannose residues consist of short chains of α1-2 or α1-3-linked mannose. The N-linked mannose polymer consists of an inner core of two GlcNAc residues elongated by α1-6 linear chain with branched α1-2 and α1-3 mannose.[83] Some *C. albicans* strains are capable of synthesizing β1-2-linked mannose polymers.[84] *C. albicans* glycans control the structure and plasticity of the cell wall and are involved in yeast–host interactions. Similar to *H. pylori*, *C. albicans* can actively modify its glycans to regulate and control the co-existence between pathogen and host.[85]

α-Mannan is the main immunodominant component of the *C. albicans* cell wall. DC-SIGN has a high affinity for yeast α-mannan.[6] Moreover, DC-SIGN is expressed by DC located at the mucosal sites, where *C. albicans* resides. DC-SIGN can function as uptake receptor and mediates phagocytosis of the *C. albicans* particles.[86] Although no interaction between L-SIGN and *C. albicans* has been reported, the high affinity of L-SIGN for α-mannans[11] suggests that L-SIGN also can bind *C. albicans*.

An overview of the pathogens discussed and the recognition of these pathogens by C-type lectins are given in Table 3.

5 Conclusion

C-type lectins are widely expressed on APC and in the liver on both hepatocytes and LSEC. Many pathogens exploit self-glycans or glycans that specifically target C-type lectins to direct the immune response in favour of pathogen survival.

Table 3 *Pathogen recognition of human C-type lectins. The carbohydrate structures, which mediate pathogen recognition, are indicated in the table*

Pathogen	DC-SIGN	L-SIGN	MGL	ASGP-R
HIV	+ high mannose	+ high mannose	−	−
Ebola virus	+ high mannose	+ high mannose	+ terminal GalNAc?	+ terminal Gal
HCV	+ high mannose	+ high mannose	−	+ terminal Gal?
Helicobacter pylori	+ Lewis antigens	−	−	−
Candida albicans	+ mannan	+ mannan	−	−
Schistosoma mansoni	+ Lewis[x], LDNF	+ unknown structures	+ LDN, LDNF	+?

Although C-type lectins can enhance antigen presentation, targeting of these receptors supports immune suppression via IL-10 production or induction of regulatory T cells. Several pathogens that target C-type lectins can cause chronic infections and alter the Th1/Th2 balance.

The carbohydrate specificity of C-type lectins is detrimental to understanding the biological ligands for these receptors. Binding specificity is often determined by single amino acid residues. Minor changes in amino acid sequence in the binding domain of lectins define its selectivity for carbohydrate structures. By doing this, lectins may fine-tune their binding specificity and selectivity and therefore can specifically be employed for certain biological processes.

The development of glycan arrays provides an efficient and fast tool to study the binding properties of lectins. Crystallography will give us structural information on lectin–carbohydrate interactions; however, crystal structures of lectins and in particular those complexed with a ligand are difficult to obtain. In particular, molecular modelling may give additional information and can show why some carbohydrates bind to a lectin and why others would not. Also, nuclear magnetic resonance has been successfully used for obtaining structural information on lectin–carbohydrate complexes, e.g. sialic acid binding by sialoadhesin.[87]

As we understand more about carbohydrates and C-type lectin function, an intriguing possibility arises whether carbohydrates can be of therapeutic use in treatment of Th1-dominated autoimmune diseases. Recent studies show a decreased rate of autoimmune disease mouse models of *S. mansoni* infection. Administration of a specific carbohydrate-based vaccine might prove beneficial in the redirectioning of a disturbed Th1/Th2 balance, in the induction of anti-flammatory cytokines or even in the generation of antigen-specific regulatory T cells.

Acknowledgement

CHG and SvV are supported by the Dutch Scientific Research ZonMw, Pionier grant 900-02-002.

References

1. C.A. Janeway and R. Medzhitov, *Annu. Rev. Immunol.*, 2002, **20**, 197.
2. P.A. Knolle and G. Gerken, *Immunol. Rev.*, 2000, **174**, 21.
3. A. Engering, T.B.H. Geijtenbeek, S.J. van Vliet, M. Wijers, E. van Liempt, N. Demaurex, A. Lanzavecchia, J. Fransen, C.G. Figdor, V. Piguet and Y. van Kooyk, *J. Immunol.*, 2002, **168**, 2118.
4. L. Bonifaz, D. Bonnyay, K. Mahnke, M. Rivera, M.C. Nussenzweig and R.M. Steinman, *J. Exp. Med.*, 2002, **196**, 1627.
5. T.I. Prigozy, P.A. Sieling, D. Clemens, P.L. Stewart, S.M. Behar, S.A. Porcelli, M.B. Brenner, R.L. Modlin and M. Kronenberg, *Immunity*, 1997, **6**, 187.
6. T.B.H. Geijtenbeek, R. Torensma, S.J. van Vliet, G.C.F. van Duijnhoven, G.J. Adema, Y. van Kooyk and C.G. Figdor, *Cell*, 2000, **100**, 575.

7. A. Engering, S.J. van Vliet, K. Hebeda, D.G. Jackson, R. Prevo, S.K. Singh, T.B.H. Geijtenbeek, H. van Krieken and Y. van Kooyk, *Am. J. Pathol.*, 2004, **164**, 1587.

8. A. Engering, S.J. van Vliet, T.B.H. Geijtenbeek and Y. van Kooyk, *Blood*, 2002, **100**, 1780.

9. E.J. Soilleux, L.S. Morris, G. Leslie, J. Chehimi, Q. Luo, E. Levroney, J. Trowsdale, L.J. Montaner, R.W. Doms, D. Weissman, N. Coleman and B. Lee, *J. Leukoc. Biol.*, 2002, **71**, 445.

10. E.J. Soilleux, L.S. Morris, B. Lee, S. Pohlmann, J. Trowsdale, R.W. Doms and N. Coleman, *J. Pathol.*, 2001, **195**, 586.

11. A.A. Bashirova, T.B.H. Geijtenbeek, G.C.F. van Duijnhoven, S.J. van Vliet, J.B. Eilering, M.P. Martin, L. Wu, T.D. Martin, N. Viebig, P.A. Knolle, V.N. KewalRamani, Y. van Kooyk and M. Carrington, *J. Exp. Med.*, 2001, **193**, 671.

12. G. Ashwell and J. Harford, *Annu. Rev. Biochem.*, 1982, **51**, 531.

13. P.H. Weigel, *Subcell. Biochem.*, 1993, **19**, 125.

14. F. Pacifico, L. Laviola, L. Ulianich, A. Porcellini, C. Ventra, E. Consiglio and V.E. Avvedimento, *Biochem. Biophys. Res. Commun.*, 1995, **210**, 138.

15. Y.Y.T. Seow, M.G.K. Tan and K.T. Woo, *Nephron*, 2002, **91**, 431.

16. J. Valladeau, V. Duvert-Frances, J.J. Pin, M.J. Kleijmeer, S. Ait-Yahia, O. Ravel, C. Vincent, F. Vega, A. Helms, D. Gorman, S.M. Zurawski, G. Zurawski, J. Ford and S. Saeland, *J. Immunol.*, 2001, **167**, 5767.

17. N. Suzuki, K. Yamamoto, S. Toyoshima, T. Osawa and T. Irimura, *J. Immunol.*, 1996, **156**, 128.

18. N. Higashi, A. Morikawa, K. Fujioka, Y. Fujita, Y. Sano, M. Miyata-Takeuchi, N. Suzuki and T. Irimura, *Int. Immunol.*, 2002, **14**, 545.

19. H. Lis and N. Sharon, *Chem. Rev.*, 1998, **98**, 637.

20. K. Drickamer, *Nature*, 1992, **360**, 183.

21. B.M. Curtis, S. Scharnowske and A.J. Watson, *Proc. Natl. Acad. Sci. USA*, 1992, **89**, 8356.

22. E.J. Soilleux, R. Barten and J. Trowsdale, *J. Immunol.*, 2000, **165**, 2937.

23. D.A. Mitchell, A.J. Fadden and K. Drickamer, *J. Biol. Chem.*, 2001, **276**, 28939.

24. H. Feinberg, D.A. Mitchell, K. Drickamer and W.I. Weis, *Science*, 2001, **294**, 2163.

25. B.J. Appelmelk, I. van Die, S.J. van Vliet, C.M.J.E. Vandenbroucke-Grauls, T.B.H. Geijtenbeek and Y. van Kooyk, *J. Immunol.*, 2003, **170**, 1635.

26. I. Van Die, S.J. van Vliet, A.K. Nyame, R.D. Cummings, C.M.C. Bank, B. Appelmelk, T.B.H. Geijtenbeek and Y. van Kooyk, *Glycobiology*, 2003, **13**, 471.

27. Y. Guo, H. Feinberg, E. Conroy, D.A. Mitchell, R. Alvarez, O. Blixt, M.E. Taylor, W.I. Weis and K. Drickamer, *Nat. Struct. Mol. Biol.*, 2004, **11**, 591.

28. E. van Liempt, A. Imberty, C.M.C. Bank, S.J. van Vliet, Y. van Kooyk, T.B.H. Geijtenbeek and I. van Die, *J. Biol. Chem.*, 2004, **279**, 33161.

29. T.B.H. Geijtenbeek, G.C.F. van Duijnhoven, S.J. van Vliet, E. Krieger, G. Vriend, C.G. Figdor and Y. van Kooyk, *J. Biol. Chem.*, 2002, **277**, 11314.

30. W. Liu, L. Tang, G. Zhang, H. Wei, Y. Cui, L. Guo, Z. Gou, X. Chen, D. Jiang, Y. Zhu, G. Kang and F. He, *J. Biol. Chem.*, 2004, **279**, 18748.

31. K. Drickamer, *Curr. Opin. Struct. Biol.*, 1999, **9**, 585.

32. P.H. Weigel and J.H.N. Yik, *Biochim. Biophys. Acta*, 2002, **1572**, 341.
33. M. Meier, M.D. Bider, V.N. Malashkevich, M. Spiess and P. Burkhard, *J. Mol. Biol.*, 2000, **300**, 857.
34. M. Sarkar, J. Liao, E.A. Kabat, T. Tanabe and G. Ashwell, *J. Biol. Chem.*, 1979, **254**, 3170.
35. S.T. Iobst and K. Drickamer, *J. Biol. Chem.*, 1994, **269**, 15512.
36. A.R. Kolatkar and W.I. Weis, *J. Biol. Chem.*, 1996, **271**, 6679.
37. N. Higashi, K. Fujioka, K. Denda-Nagai, S. Hashimoto, S. Nagai, T. Sato, Y. Fujita, A. Morikawa, M. Tsuiji, M. Miyata-Takeuchi, Y. Sano, N. Suzuki, K. Yamamoto, K. Matsushima and T. Irimura, *J. Biol. Chem.*, 2002, **277**, 20686.
38. M. Tsuiji, M. Fujimori, Y. Ohashi, N. Higashi, T.M. Onami, S.M. Hedrick and T. Irimura, *J. Biol. Chem.*, 2002, **277**, 28892.
39. S.J. van Vliet, E. van Liempt, E. Saeland, C.A. Aarnoudse, B. Appelmelk, T. Irimura, T.B.H. Geijtenbeek, O. Blixt, R. Alvarez, I. van Die and Y. van Kooyk, *Int. Immunol.*, 2005, **17**, 661.
40. C.P. Alvarez, F. Lasala, J. Carrillo, O. Muñiz, A.L. Corbí and R. Delgado, *J. Virol.*, 2002, **76**, 6841.
41. T.B.H. Geijtenbeek, D.S. Kwon, R. Torensma, S.J. van Vliet, G.C.F. van Duijnhoven, J. Middel, I.L. Cornelissen, H.S. Nottet, V.N. KewalRamani, D.R. Littman, C.G. Figdor and Y. van Kooyk, *Cell*, 2000, **100**, 587.
42. A. Moris, C. Nobile, F. Buseyne, F. Porrot, J.P. Abastado and O. Schwartz, *Blood*, 2004, **103**, 2648.
43. M. Stevenson, *Nat. Med.*, 2003, **9**, 853.
44. Y. Feng, C.C. Broder, P.E. Kennedy and E.A. Berger, *Science*, 1996, **272**, 872.
45. H. Deng, R. Liu, W. Ellmeier, S. Choe, D. Unutmaz, M. Burkhart, P. Di Marzio, S. Marmon, R.E. Sutton, C.M. Hill, C.B. Davis, S.C. Peiper, T.J. Schall, D.R. Littman and N.R. Landau, *Nature*, 1996, **381**, 661.
46. X. Zhu, C. Borchers, R.J. Bienstock and K.B. Tomer, *Biochemistry*, 2000, **39**, 11194.
47. T. Mizuochi, T.J. Matthews, M. Kato, J. Hamako, K. Titani, J. Solomon and T. Feizi, *J. Biol. Chem.*, 1990, **265**, 8519.
48. K.B. Gurney, J. Elliott, H. Nassanian, C. Song, E. Soilleux, I. McGowan, P.A. Anton and B. Lee, *J. Virol.*, 2005, **79**, 5762.
49. S. Mahanty and M. Bray, *Lancet Infect. Dis.*, 2004, **4**, 487.
50. C. Will, E. Muhlberger, D. Linder, W. Slenczka, H.D. Klenk and H. Feldmann, *J. Virol.*, 1993, **67**, 1203.
51. J.A. Wilson, M. Hevey, R. Bakken, S. Guest, M. Bray, A.L. Schmaljohn and M.K. Hart, *Science*, 2000, **287**, 1664.
52. H. Feldmann, C. Will, M. Schikore, W. Slenczka and H.D. Klenk, *Virology*, 1991, **182**, 353.
53. H. Geyer, C. Will, H. Feldmann, H.D. Klenk and R. Geyer, *Glycobiology*, 1992, **2**, 299.
54. A. Marzi, T. Gramberg, G. Simmons, P. Moller, A.J. Rennekamp, M. Krumbiegel, M. Geier, J. Eisemann, N. Turza, B. Saunier, A. Steinkasserer, S. Becker, P. Bates, H. Hofmann and S. Pohlmann, *J. Virol.*, 2004, **78**, 12090.

55. A. Takada, K. Fujioka, M. Tsuiji, A. Morikawa, N. Higashi, H. Ebihara, D. Kobasa, H. Feldmann, T. Irimura and Y. Kawaoka, *J. Virol.*, 2004, **78**, 2943.

56. G. Simmons, J.D. Reeves, C.C. Grogan, L.H. Vandenberghe, F. Baribaud, J.C. Whitbeck, E. Burke, M.J. Buchmeier, E.J. Soilleux, J.L. Riley, R.W. Doms, P. Bates and S. Pöhlmann, *Virology*, 2003, **305**, 115.

57. S. Becker, M. Spiess and H.D. Klenk, *J. Gen. Virol.*, 1995, **76**(Pt 2), 393.

58. G. Lin, G. Simmons, S. Pöhlmann, F. Baribaud, H. Ni, G.J. Leslie, B.S. Haggarty, P. Bates, D. Weissman, J.A. Hoxie and R.W. Doms, *J. Virol.*, 2003, **77**, 1337.

59. S. Duvet, L. Cocquerel, A. Pillez, R. Cacan, A. Verbert, D. Moradpour, C. Wychowski and J. Dubuisson, *J. Biol. Chem.*, 1998, **273**, 32088.

60. V. Deleersnyder, A. Pillez, C. Wychowski, K. Blight, J. Xu, Y.S. Hahn, C.M. Rice and J. Dubuisson, *J. Virol.*, 1997, **71**, 697.

61. A. Goffard and J. Dubuisson, *Biochimie*, 2003, **85**, 295.

62. K. Sato, H. Okamoto, S. Aihara, Y. Hoshi, T. Tanaka and S. Mishiro, *Virology*, 1993, **196**, 354.

63. M.C. Navas, A. Fuchs, E. Schvoerer, A. Bohbot, A.M. Aubertin and F. Stoll-Keller, *J. Med. Virol.*, 2002, **67**, 152.

64. P.Y. Lozach, H. Lortat-Jacob, A. De Lacroix De Lavalette, I. Staropoli, S. Foung, A. Amara, C. Houlès, F. Fieschi, O. Schwartz, J.L. Virelizier, F. Arenzana-Seisdedos and R. Altmeyer, *J. Biol. Chem.*, 2003, **278**, 20358.

65. S. Pöhlmann, J. Zhang, F. Baribaud, Z. Chen, G.J. Leslie, G. Lin, A. Granelli-Piperno, R.W. Doms, C.M. Rice and J.A. McKeating, *J. Virol.*, 2003, **77**, 4070.

66. J.P. Gardner, R.J. Durso, R.R. Arrigale, G.P. Donovan, P.J. Maddon, T. Dragic and W.C. Olson, *Proc. Natl. Acad. Sci. USA*, 2003, **100**, 4498.

67. B. Saunier, M. Triyatni, L. Ulianich, P. Maruvada, P. Yen and L.D. Kohn, *J. Virol.*, 2003, **77**, 546.

68. I.S. Ludwig, A.N. Lekkerkerker, E. Depla, F. Bosman, R.J.P. Musters, S. Depraetere, Y. van Kooyk and T.B.H. Geijtenbeek, *J. Virol.*, 2004, **78**, 8322.

69. E.G. Cormier, R.J. Durso, F. Tsamis, L. Boussemart, C. Manix, W.C. Olson, J.P. Gardner and T. Dragic, *Proc. Natl. Acad. Sci. USA*, 2004, **101**, 14067.

70. P.Y. Lozach, A. Amara, B. Bartosch, J.L. Virelizier, F. Arenzana-Seisdedos, F.L. Cosset and R. Altmeyer, *J. Biol. Chem.*, 2004, **279**, 32035.

71. R.D. Cummings and A.K. Nyame, *Biochim. Biophys. Acta*, 1999, **1455**, 363.

72. J. Srivatsan, D.F. Smith and R.D. Cummings, *Glycobiology*, 1992, **2**, 445.

73. J. Srivatsan, D.F. Smith and R.D. Cummings, *J. Biol. Chem.*, 1992, **267**, 20196.

74. K. Nyame, R.D. Cummings and R.T. Damian, *J. Biol. Chem.*, 1987, **262**, 7990.

75. C.K. Makaaru, R.T. Damian, D.F. Smith and R.D. Cummings, *J. Biol. Chem.*, 1992, **267**, 2251.

76. P.B. Ernst and B.D. Gold, *Annu. Rev. Microbiol.*, 2000, **54**, 615.

77. M.A. Monteiro, P. Zheng, B. Ho, S. Yokota, K. Amano, Z. Pan, D.E. Berg, K.H. Chan, L.L. MacLean and M.B. Perry, *Glycobiology*, 2000, **10**, 701.

78. B.J. Appelmelk, S.L. Martin, M.A. Monteiro, C.A. Clayton, A.A. McColm, P. Zheng, T. Verboom, J.J. Maaskant, D.H. Van den Eijnden, C.H. Hokke, M.B. Perry, C.M.J.E. Vandenbroucke-Grauls and J.G. Kusters, *Infect. Immun.*, 1999, **67**, 5361.

79. B.J. Appelmelk, B. Shiberu, C. Trinks, N. Tapsi, P.Y. Zheng, T. Verboom, J. Maaskant, C.H. Hokke, W.E.C.M. Schiphorst, D. Blanchard, I.M. Simoons-Smit, D.H. Van den Eijnden and C.M.J.E. Vandenbroucke-Grauls, *Infect. Immun.*, 1998, **66**, 70.

80. M.P. Bergman, A. Engering, H.H. Smits, S.J. van Vliet, A.A. van Bodegraven, H.P. Wirth, M.L. Kapsenberg, C.M.J.E. Vandenbroucke-Grauls, Y. van Kooyk and B.J. Appelmelk, *J. Exp. Med.*, 2004, **200**, 979.

81. F.C. Odds, *Crit. Rev. Microbiol.*, 1987, **15**, 1.

82. B.N. Gantner, R.M. Simmons and D.M. Underhill, *EMBO J.*, 2005, **24**, 1277.

83. N. Shibata, H. Kobayashi, Y. Okawa and S. Suzuki, *Eur. J. Biochem.*, 2003, **270**, 2565.

84. M. Nitz, C.C. Ling, A. Otter, J.E. Cutler and D.R. Bundle, *J. Biol. Chem.*, 2002, **277**, 3440.

85. D. Poulain and T. Jouault, *Curr. Opin. Microbiol.*, 2004, **7**, 342.

86. A. Cambi, K. Gijzen, I.J.M. de Vries, R. Torensma, B. Joosten, G.J. Adema, M.G. Netea, B.J. Kullberg, L. Romani and C.G. Figdor, *Eur. J. Immunol.*, 2003, **33**, 532.

87. P.R. Crocker, M. Vinson, S. Kelm and K. Drickamer, *Biochem. J.*, 1999, **341**(Pt 2), 355.

CHAPTER 8

Targeting Microbial Sialic Acid Metabolism for New Drug Development

ERIC R. VIMR AND SUSAN M. STEENBERGEN

Laboratory of Sialobiology, Department of Pathobiology, University of Illinois, Urbana, IL 61802, USA

1 Introduction

Anyone opposed philosophically or otherwise to Darwinian evolution as the fundamental explanation for all life or the biomedical scientist who may not see the need for evolutionary thought in his or her professional activities need only look at the details of an infectious disease to witness evolution's effects played out on a daily scale. As summarized by Dawkins,[1] extant species bear in their genomes the ultimately successful struggles of their ancestors. Therefore, and unlike any other pure or applied science, biology is profoundly historical and naturally (though not massively) contingent. The concept of contingency will be used several times in this chapter when describing the host–microbe interaction. For now, just consider the example of that laboratory workhorse, *Escherichia coli*, or any of its pathogenic derivatives, which may have begun its most recent evolutionary trajectory in the Permian- or Triassic-period gut of some therapsid host. If this time period seems too remote, then certainly the *E. coli* tale began as early as the great Jurassic die-off and the resulting mammalian evolutionary ascendance.[2] The point is that the sheer volume of combined fecal output from animals with recognizable alimentary canals provided means of efficient transmission between hosts and free-living environments for that first intrepid *E. coli* ancestor. From this scenario we can conclude that the presumably exclusively free-living ancestor of *E. coli* faced environmental contingency, and since the gut can be thought of as a tube connecting mouth or nasopharynx to anus, the opportunity for diversity (speciation) in this tube is, at least metaphorically, comparable to the vertical niche stratification of macroscopic life in a tropical rain forest, thus offering the *E. coli* ancestor dramatically rich evolutionary possibilities. Should there be any wonder when a comparative microbial genomic study indicated that *E. coli* and the obligate human commensal *Haemophilus influenzae* shared a common ancestor that was more similar to *E. coli* than it was to *H. influenzae*?[3] One of these descendents, *E. coli*, retained its free-living potential while exploiting the host's intestinal "rain forest," as *H. influenzae* lost its ability to live free while becoming increasingly specialized and ultimately dependent on a single host. The emergence of diarrheagenic *E. coli* strains is then easily seen as a strategy for increasing dissemination by horizontal acquisition of genes encoding exotoxins that derange host fluid retention. In contrast, persistence of host-adapted *H. influenzae* required evolution of mechanisms preventing clearance by the host. Therefore, when the host–microbe interaction is seen as an arms race between and among species,[4] understanding the details of particular microbial metabolic processes provides a variety of potential targets for new therapeutic development.[5,6] In this context we will discuss how *H. influenzae*, and closely related bacteria, gained evolutionary success through genomic divestiture coupled with prudent genetic reinvestments exploiting one of the central features of all host–microbe interactions: the nexus between host glycocalyx and the outside world.

Another example of thinking evolutionarily is central to the topics in this book. Infectious disease on a worldwide or global scale remains the leading cause of death.[7] Of course, it was not until the success of antibiotics during World War II that infectious diseases ceased being the preeminent winnowing agents of all previous generations. Today, the developing world still recognizes the cruel efficiency with which so many microbes dispatch the very young, aged, or immunocompromised, the latter increasingly common as modern medicine continues to advance in the developed world. However, it is the overuse or misuse of antibiotics, the failure to develop new antibiotics, and the economic disincentives, if not practical difficulties, with vaccine development that has made the reemergence of once treatable infections so ominous. In the worst-case scenario of a truly postantibiotic era,[7] the ensuing threats to worldwide well-being may dwarf our current preoccupation with "weaponized" versions of "select agents." Assuming that the worst-case scenario is avoided, we will be left with a dearth of standard antibiotics to treat an increasingly greater percentage of the industrialized world's population, let alone the increased burden caused by infectious diseases in developing nations.

None of this gloom should be treated as theoretical; evolution is about nothing more than getting the genetic endowment of one generation of organisms into the next. Given their potentially rapid generation times and opportunities for promiscuous lateral gene transfer (horizontal evolution), microbes can rapidly gain a significant evolutionary advantage over multicellular hosts. Although the exact Darwinian "unit" that is selected may be controversial, Dawkins's[1] simplification that it is the gene (or replicator), or more precisely the fitness of that gene's interactions with all others in the organism is the fundamental unit of evolution. The conclusion that the spectacular success of traditional antibiotic therapy may, like petroleum, be finite urges us to marshal these resources now and to develop alternatives for the future. Targeting virulence mechanisms is an attractive potential alternative for new drug development. The impetus for pursuing such targets may increase if a range of previous "chronic" diseases is convincingly shown to have a microbial basis.[8] Cell wall biosynthesis and certain aspects of microbial protein or nucleic acid synthesis are so distinct from their mammalian counterparts that drugs blocking these processes have broad action spectra with limited side effects. Unfortunately, microbes have targeted the same processes during eons of internecine competition, so that the mechanisms of resistance were already in place when humankind "discovered" antibiotics. One consequence of our rediscovery is that there has "been little *de novo* evolution giving rise to new gene functions forced by man's use of antibiotics,"[9] which may explain why some antibiotics lost their efficacy so rapidly after introduction.

All Gram-negative bacteria produce lipid A as a unique and essential component of lipopolysaccharide (LPS) and the outer membrane. Elucidation of lipid A biosynthesis by Raetz and Whitfield[10] identified the first committed step involving the deacetylation of acyl-UDP-*N*-acetylglucosamine (UDP-GlcNAc) by UDP-3-*O*-(*R*-3-hydroxymyristoyl)-GlcNAc deacetylases as the target for two new lead drugs. Unfortunately, mutation to resistance to one of these drugs was relatively high, and while the other effectively inhibited most bacterial deacetylases *in vitro*, it was ineffective *in vivo* either because it failed to enter cells or was metabolized, or because it entered but was rapidly excreted.[11] These discouraging results suggest that targeting essential metabolic processes, as in the case of traditional antibiotics, may have limited utility for effective long-term therapeutic design.

Other microbial targets receiving scrutiny include chaperone inhibitors,[6] inhibitors of type III secretion systems,[5] and potential inhibitors of the generalized bacterial protein secretory (Sec) apparatus.[12] Large-scale screening efforts are being made to identify new microbial drug targets and essential as well as nonessential gene functions by individually knocking out all the genes in a given organism.[13,14] The wide range of targets involving microbial adhesion to host cells is not discussed in this chapter. However, it is worth noting that the earlier impression that the influenza virus hemagglutinin, which is necessary for host invasion, was a prime target for drug design was diminished when simple transition state analogues of viral neuraminidases (sialidases) entered clinical practice as one of the, if not the first, rationally designed antimicrobial agents.[15,16] It is the bias of this chapter that low-molecular-weight inhibitors of specific virulence mechanisms, and not adhesion mimetics or agents blocking essential functions, offer the best chance of achieving a balance between effective prophylaxis or treatment and acceptable rates of mutation to resistance, thus preventing rapid loss of efficacy. The focus will be exclusively on microbial metabolism of sialic acids (sialometabolism). Readers are referred to earlier reviews discussing the background supporting this focus.[17,18]

2 What is Virulence?

The question "what is virulence?" cannot be separated from a definition of what it means to be a pathogen. While any microbe may be capable of causing disease in a given host under some set of conditions, pathogenic or facultatively pathogenic microbes are known exclusively because of their frequent association with disease. In other words, we recognize these organisms because they are the causative agents of certain predictable pathologies. Virulence is then defined as the relative damage done to a host during the host–microbe interaction.[19–22] Using this definition of virulence, the invasive pathogen *E. coli* K1 is of low virulence to otherwise healthy adults but is highly virulent in neonates because of the relative immaturity of their immune systems, predisposing them to the high levels of bacteremia that are necessary for diseases ranging from sepsis to meningitis.[18] Note that this definition of virulence does not rely on a concept of relative transmissibility, which though essential for understanding epidemics and other public health issues is not central to a concept of disease (or disease prevention) when defined as the outcome of a given host–microbe interaction. The following aspects of the host–microbe interaction therefore identify individual systems where therapeutic intervention may reduce or prevent disease:[23]

- attachment to hosts (adhesion),
- intoxication of host cells (toxins),
- subversion of host defenses (inhibition or evasion of innate or adaptive immune functions),
- synchronized regulation of virulence-associated gene expression (controlling the ordered expression of regulons or other physiologically interconnected gene systems), and
- preferential secretion of virulence-associated products (*e.g.*, type III secretion systems).

All of these mechanisms may be thought of as assisting in the persistence and, ultimately, transmission of microbes to other susceptible hosts. Although the genes for these virulence mechanisms are increasingly recognizable in the burgeoning genomic databases, it is our contention that understanding the processes most central to microbial persistence offers the best hope for long-term control and eventual eradication of infectious diseases.

3   Rational Drug Design

The phrase "rational drug design" has come to be associated with studies directed toward discovery of lead compounds when these efforts are (usually) assisted by a crystal structure or other three-dimensional information about the target. While structural information can certainly help in drug design (see, *e.g.*, the fascinating story behind the anti-influenza agents Relenza and Tamiflu),[15,16] it is not always or even frequently necessary for initial identification of lead compounds, which can then guide future developments with or without a three-dimensional structure. For example, the chemical intuition of Cushman and colleagues[24] regarding the likely similarities between carboxypeptidase A and dipeptidyl carboxypeptidase (angiotensin-converting enzyme) helped to design the first antihypertensive drug, Captopril (3-mercapto-2-methylpropanolyl-L-proline). Cocrystals of additional analogues with bacterial thermolysin assisted the rational design of inhibitors with fewer side effects than Captopril,[25] and analysis of these cocrystals by Hassall *et al.*,[26] using the interactions between the inhibitors and the functionally related microbial enzyme, thermolysin, facilitated an approximately 100-fold reduced screening effort that resulted in the development of even more effective new drugs. While the merits of this rational approach are obvious, the time and expertise needed for success may preclude or at least diminish efforts to identify new antimicrobial agents by such approaches, especially when the payoff may seem less than for a drug targeting a more common organic disease such as hypertension. Fortunately, advances in modern chemistry and the mandated interest for select agents in the US offers an alternative to the rational approach.

4   Irrational [*sic*] Drug Design

High-throughput (robotic) screening of combinatorial chemical libraries offers an attractive alternative approach for the identification of anti-infective lead agents. Although the details for synthesizing large chemical libraries is outside the scope of this review, once available such libraries can be rapidly screened as long as there is a robust, inexpensive assay for a defined virulence pathway. Perhaps the simplest screen to imagine involves an assay for compounds that inhibit microbial growth under a set of conditions that would be expected to result in defective propagation and subsequent host clearance. This type of assay would rely on the availability of a simple, automated screen measuring growth by turbidometric or other photometric method. The obvious drawback of such a screen is its inability to distinguish between compounds with generalized instead of specific effects. Additional work would be necessary to identify generalized toxicity, which would be likely to also affect the host, from an effect on the specifically targeted virulence pathway. All negative screens may suffer from similar problems.

In contrast to negative screening methods, a positive screen would identify compounds targeting microbial pathways that are not required for growth under *in vitro* or laboratory conditions, but that when inhibited *in vivo* would be expected or known to cause loss of virulence.[5,6] The ability to design such screens depends on detailed basic understanding of particular virulence pathways. However, even with detailed knowledge of basic virulence mechanisms, how to develop a positive screen may not always be obvious. The followings sections describe some aspects of microbial sialometabolism that have facilitated positive screening for inhibitors of sialic acid uptake.

5   The Sialic Acids

Sialic acid is the designation given to a family of nine-carbon keto-sugars with representation in all organisms of the developmental lineage (deuterostomes) that includes humans but are infrequently produced by most other organisms. Sialic acid negative organisms also appear to lack the genes for synthesizing sialic acids, suggesting these organisms either never synthesized sialic acids or lost the function during evolution. In contrast, a growing list of bacterial pathogens and commensals as well as some free-living bacteria either synthesize sialic acid by a *de novo* pathway or acquire it directly from their animal hosts (Table 1). The unusually limited phylogenetic distribution of sialic acids has attracted attention. One proposal suggests that sialic acid synthesis evolved in free-living eubacteria and spread horizontally to an ancestor of the deuterostomes.[27] Despite this, or any other study pertaining to evolution of the sialic acids, we are left with if not the origin of these sugars in the deuterostome lineage, then at least

Table 1 *Sialic acid uptake in microbial commensals or pathogens*

Organisms	Genes	System type	Function	Refs.
E. coli, enterics in general, and others	*nanT*	MFS	C, N, and energy source Cell wall precursors Signaling Colonization	17, 18, 42, 62
P. multocida, *H. influenzae,* *V. cholerae,* *Fusobacterium nucleatum*	*nanP, nanU*	TRAP	Nutritional Surface modification Persistence	17, 18, 37
Streptococcus pyogenes, HAP?	spy0233-0237	ABC	Surface modification Persistence	18, 87

their functional apotheosis in higher metazoans where sialic acids, as the frequently terminal, nonreducing sugars for a wide range of glycoproteins and glycolipids, function in diverse developmental and regulatory phenomena that are crucial to homeostasis. Therefore, for many microbes encountering the animal glycocalyx the most prevalent molecular host species are sialoglycoconjugates, a situation that has led to the wide range of microbial responses to host sialic acids including

- adhesion to host surfaces,
- molecular mimicry of host sialoglycoconjugate structures,
- enzymatic removal of terminal sialic acid residues, and
- nutritional use of host sialic acids.

Any of these processes may provide a target for new antimicrobial drug development.

5.1   Monosaccharides

Figure 1 shows several sialic acid or sialic acid-like monosaccharides. Keto-deoxy neuraminate (KDN; Figure 1C) and legionaminic acid (Leg5Ac7Am; Figure 1D) have the same stereochemistry as the most common sialic acid, *N*-acetylneuraminic acid (Neu5Ac; Figure 1A), but either lack the *N*-acetyl group at carbon-5 or the carbon-9 hydroxyl, respectively. Pseudaminic acid (Pse5,7Ac) is structurally similar to Leg5Ac7Am but has a different stereochemistry (Figure 1E). Keto-deoxy octonate (KDO; Figure 1F) is a ubiquitous eight-carbon constituent of bacterial LPS that behaves like Neu5Ac in most chemical reactions. Dehydro-Neu5Ac (Neu5Ac2en; Figure 1B) is a derivative of Neu5Ac that mimics the carbonium ion transition state produced during sialidase cleavage of terminal sialic acid residues from glycoconjugates. Neu5Ac2en served as the lead compound for the development of Relenza (zanamivir), an influenza virus sialic acid inhibitor in clinical practice. The carbon-5-hydroxylated, or *N*-glycolyl, derivative of Neu5Ac (Neu5Gc; Figure 1A), is common in higher deuterostomes except humans, where a deletion of the hydroxylase gene after the split from the last common ancestor of humans, chimpanzees, and bonobos resulted in the loss of Neu5Gc in the

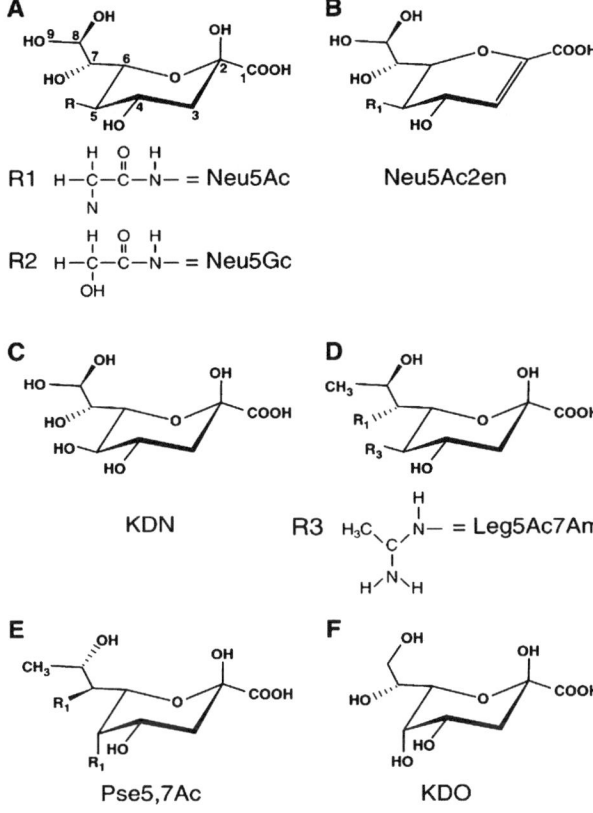

Figure 1 *Structures of sialic acids and related keto-sugar acids. (A) 2-Keto-3-deoxy-5-acetamido-7,8,9-D-glycero-D-galacto nonulosonic acids: N-acetylneuraminate (Neu5Ac) and N-glycolylneuraminate (Neu5Gc) indicated by R1 and R2, respectively. (B) Neu5Ac2en, N-acetyl-2,3-didehydro-2-deoxyneuraminic acid. (C) KDN, keto-deoxy neuraminic acid. (D) Leg5Ac7Am (legionaminic acid), 5,7-diamino-3,5,7,9-tetradeoxy-D-glycero-D-galacto-nonulosonate. (E) Pse5,7Ac (pseudaminic acid), 5,7-diamino-3,5,7,9-tetradeoxy-L-glycero-L-manno-nonulosonate. (F) KDO, keto-deoxy octonate. (Reprinted from ref.18 with permission.)*

human lineage. Varki[28] has popularized a variety of ideas surrounding this evolutionary biochemical event. However, given the importance of human carbohydrate polymorphisms for mediating resistance to a range of infectious agents,[29,30] the loss of Neu5Gc probably has similar significance for resistance to at least some infectious diseases. In any event, the widespread occurrence of microbial KDO, Leg5Ac7Am, and Pse5,7Ac suggests that the basic metabolic machinery to synthesize sialic acids was in place long before the emergence of the deuterostome lineage.

5.2 Structural Partners, Synthesis, and Catabolism

Except as short-lived precursors or the hydrolytic products of glycohydrolytic activity, sialic acids are never found free in nature. They may be glycoketosidically connected to internal sialic acids by α2,8- or α2,9-linkages, or to other sugar residues, such as galactose, *N*-acetylglucosamine (GlcNAc), or *N*-acetylgalactosamine, usually by α2,3- or α2,6-linkages. In all cases investigated to date, sialic acids are activated for transfer to acceptors by coupling with CTP to generate the activated CMP-β-glycosides. The CMP–sialic acid synthetase from *E. coli* K1 was the first activating enzyme to be characterized;[31–33] all subsequent examples of it in both eukaryotes and prokaryotes are orthologues, though important functional distinctions exist even between the bacterial enzymes.[34] The similarity between KDO and sialic acid synthetases, as noted previously,[27,29] is expected from the structural similarity between the eight- and nine-carbon keto-sugar substrates (Figure 1). The group of glycosyltransferases known as sialyltransferases binds CMP–sialic acid and transfers carbohydrate units to appropriate acceptors to complete sialoglycoconjugate biosynthesis. Despite primary structural divergence, sialyltransferases may share a limited number of three-dimensional folds.[35] Therefore, sialyl activation and transfer, as the terminal steps for synthesizing sialoglycoconjugates, are similar in all organisms so far investigated. Despite differences in the enzymatic properties of individual synthetases and sialyltransferases, the evolutionary (convergent as well as divergent) similarities suggest that these enzymes may not be good therapeutic targets.

In contrast to the similarity of sialyl activation and transfer mechanisms, synthesis of the monosaccharides differs fundamentally between prokaryotic and eukaryotic species. In prokaryotes, *N*-acetylmannosamine (ManNAc) is used in place of ManNAc-6-phosphate (ManNAc-6-P) to generate Neu5Ac directly, instead of the Neu5Ac-9-P synthesized in eukaryotes from ManNAc-6-P (Figure 2). Neu5Ac-9-P requires a specific or nonspecific phosphatase prior to activation, whereas the free Neu5Ac generated by prokaryotes is immediately ready for activation. Bravo *et al.*[27] have argued that a C-terminal domain of eukaryotic-type CMP–sialic acid synthetases acts as the Neu5Ac-9-P phosphatase, but experiments have shown that this domain in the *E. coli* K1 activating enzyme is an esterase.[36] The evolutionary value of fusing an esterase to sialic acid synthetase is not obvious, but it is interesting to note that orthologues of both the synthetase and esterase domains are found in separate though sialometabolically related operons in the obligate commensal *Pasteurella multocida*.[37] Bravo *et al.*[27] also argue that there are two prokaryotic pathways for the synthesis of precursor ManNAc: epimerization from UDP-GlcNAc yielding UDP and ManNAc (*E. coli* Model 1, Figure 2) or from GlcNAc-6-P to ManNAc-6-P, followed by dephosphorylation to free ManNAc (*Neisseria meningitidis* Model, Figure 2). However, genetic and physiological studies ruled out ManNAc-6-P as a prokaryotic sialic acid precursor,[38] a conclusion, which was subsequently confirmed by direct biochemical and structural studies,[39] indicating that there is a single prokaryotic pathway: that shown for *E. coli* Model 1 in Figure 2. The most recent structural work on bacterial Neu5Ac and ManNAc synthesis has largely confirmed the previous genetic and physiological studies.[39] In our opinion, it is the confirmatory instead of novel research-generating aspect of the so-called rational approach that limits its effectiveness for understanding previously unexplored basic biological phenomena. In any event, ManNAc derived from UDP-GlcNAc in eukaryotes is phosphorylated by a specific kinase that is fused to the epimerase domain prior to synthesis of Neu5Ac-9-P by condensation with phosphoenolpyruvate, whereas free ManNAc in prokaryotes is condensed directly to generate

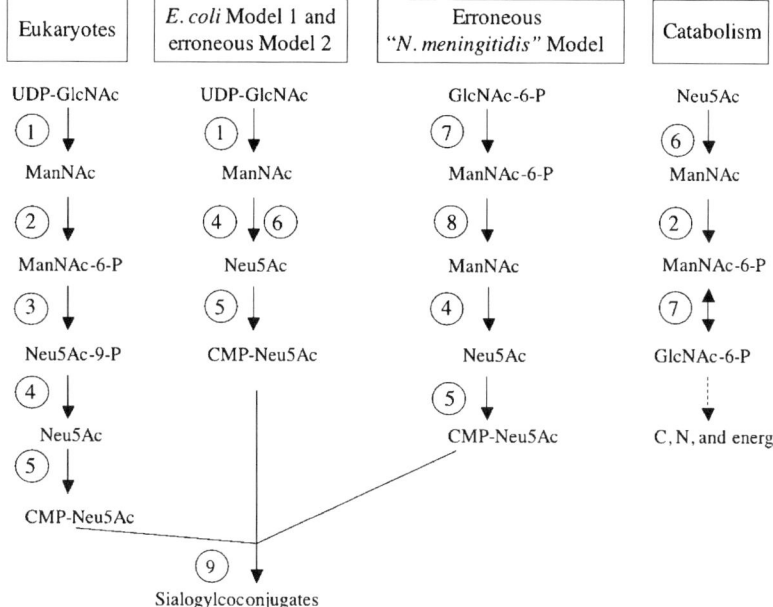

Figure 2 *Synthesis and catabolism of sialic acids in eukaryotes and bacteria.[17,18,27,32,33,39–42,82,83,86] Abbreviations of intermediates in the synthesis or catabolism of sialic acids are defined in the text. Circled numbers refer to individual metabolic steps: 1, UDP-GlcNAc 2′-epimerase; 2, ManNAc kinase; 3, sialic acid synthase (condensing enzyme); 4, Neu5Ac-9-P phosphatase; 5, CMP–sialic acid synthetase (activating enzyme); 6, sialate aldolase (lyase); 7, GlcNAc-6-P 2′-epimerase; 8, ManNAc-6-P phosphatase; 9, sialyltransferase. C and N refer to carbon and nitrogen, respectively, and the dashed line indicates that more than one metabolic step is involved in the indicated transformation. Note that steps1 and 2 in eukaryotes are carried out by a single bifunctional enzyme, whereas orthologues of the epimerase and kinase are found in bacterial synthetic and catabolic pathways, respectively. Erroneous model 2 was suggested to operate by the 1-6-5 pathway, whereas model 1 operates as 1-4-5. The N. meningitides model was based on erroneous assignment of enzyme 7 to the synthetic pathway, and the apparent absence of enzyme 8 in bacteria.[86] Note that while the eukaryotic pathway is assumed to be universal in this lineage, the process has only been confirmed in mammals*

Neu5Ac (Figure 2). Thus, the prokaryotic synthesis of sialic acids dispenses with both the kinase and phosphatase steps.

The mechanistic differences between prokaryotic and eukaryotic sialic acid synthesis most likely reflects the widespread microbial utilization of sialic acids as nutritional sources of carbon, nitrogen, energy, and cell wall precursors (Figure 2), while sialates liberated in higher eukaryotic glycoconjugates are either rapidly excreted or recycled by a lysosomal pathway involving endo- or pinocytosis. Eukaryotes thus lack efficient cytoplasmic sialic acid uptake systems for scavenging sialates from the external environment,[40] while bacteria have evolved several different transport mechanisms mediating this process.[18] Once internalized by cells with an intact *nan* system, sialic acids are cleaved by sialate aldolase (lyase) to release ManNAc and pyruvate

(Figure 2). ManNAc is usually then converted to ManNAc-6-P by an orthologue of the mammalian sialate kinase and epimerized to GlcNAc-6-P by a unique 2′-epimerase designated NanE.[38,41,42] However, *Bacteroides fragilis* epimerizes free ManNAc and then phosphorylates the GlcNAc product prior to conversion to fructose 6-phosphate by GlcNAc deacetylase and glucosamine deaminase.[43] Although the aldolase is a unique reagent for the synthesis of sialate derivatives by aldol condensation, it appears to play an exclusively catabolic role in bacteria grown under normal physiological conditions.[18] An earlier report did not detect the relatively labile condensing enzyme *in vitro*, leading to the erroneous conclusion (*E. coli* Model 2, Figure 2) that some bacteria used the aldolase for sialate biosynthesis.[44] This conclusion ignored earlier genetic and physiological data excluding aldolase as a biosynthetic enzyme *in vivo*,[45] further demonstrating the relative power of the genetic approach and the caution that must be exercised when considering (especially negative) biochemical results. The reader is referred to Ringenberg *et al.*[38] and Vimr *et al.*[18] for both the primary data and review supporting our understanding of prokaryotic UDP-GlcNAc 2′-epimerase and Neu5Ac synthase. These reports also include a summary of the historical development of our results.

5.3   Supermolecular Structures

The transfer of activated sialic acids to appropriate acceptors generates sialoglycoconjugates that are primarily localized to the prokaryotic or eukaryotic cell surface. Although our focus is the sialometabolism of prokaryotes, it is interesting to mention the remarkable molecular mimicry of the mammalian polysialic acid glycan of the neural cell adhesion molecule (NCAM) by the bacterial polysialic acid capsule. In mammals, polysialic acid is an oncofetal antigen primarily located in the developing central nervous system, where it functions as a negative regulator of cell adhesion.[46] It was generally assumed that the equivalent bacterial structure would be synthesized by homologues of the mammalian sialyltransferases, while in fact all prokaryotic sialyltransferases appear only distantly related to their mammalian counterparts.[47] We suggest that polysialylation may be an example of convergent evolution that has arisen at least twice – bacteria and the deuterostome central nervous system – in both cases arising when a system was needed to provide maximum antirecognition function resulting from charge–charge repulsion and a hydrophilic cell surface. In bacteria, this antirecognition function effectively inhibits phagocytosis and other activities of the innate immune system.[18]

Aside from *E. coli* K1 and a few meningococcal strains that also produce polysialic acid, a wide range of bacteria modify (decorate) their surfaces with Neu5Ac, Leg5Ac7Am, or Pse5,7Ac, usually attached to surface polysaccharides or O-linked to flagellar proteins. To date, no example of an N-linked prokaryotic sialoglycoconjugate has been reported. It is the increasingly common confirmation of sialic acids in prokaryotes that makes detection of these sugars in plants,[48] insects,[49] and fungi[50] so interesting. Unfortunately, most reports of nonmammalian eukaryotic sialic acids are suspect because of potential methodological defects,[51] leaving deuterostomes and (mostly) pathogenic or commensal bacteria as the only solidly confirmed examples of sialic acid positive organisms. However, a recent description of the *Drosophila* sialyltransferase (D. SiaT)[49] may be convincing, making it perhaps the first example of a nonbacterial or nondeuterostome organism that synthesizes sialic acid.[49] Note that sialic acid may be produced in *Drosophila* only during a short larval period, which may explain why so many other studies failed to detect sialic acid biosynthesis in protostomes.

6 Functions of Sialic Acid Uptake During the Host–Microbe Interaction

Whereas eukaryotes discussed above do not efficiently transport environmental sialic acids, a range of Gram-positive and Gram-negative bacteria do so by internalizing it against a concentration gradient. The first genetic analysis of this process was conducted with *E. coli* while investigating the biosynthesis of polysialic acid in the K1 serotype.[32,33] *E. coli* expressing the K1, or polysialic acid, capsule synthesizes precursor sialic acid by the *de novo* prokaryotic pathway described above, using sialic acid uptake solely for catabolism. However, if sialate lyase is inactivated by mutation, transported sialic acid may be activated and transferred to the growing polysialic acid chain, demonstrating that the site of capsule biosynthesis is intracellular. Because sialic acids are ubiquitous components of animal glycoconjugates, it can be surmised that microbes catabolizing these sugars have a competitive advantage over microbes that are unable to derive carbon or energy from host sialic acids. However, for obligate commensals such as *H. influenzae* and *P. multocida*, catabolism *per se* does not appear necessary for host colonization or disease.[37,52] In contrast, a recent study of mouse intestinal colonization by *E. coli* indicates that sialic acid utilization may be an important phenotype for living within mucus overlaying the luminal epithelium.[53]

As described by Corfield,[54] mucus is a sialic acid rich glycoconjugate representing a potentially important nutritional source for colonizing bacteria like *E. coli*. However, *E. coli* appears to lack the sialidase and other glycosylhyrolases needed for extracting monosaccharides from mucus for transport and subsequent catabolism. Presumably, host sialidases or sialidases produced by other microbes sharing overlapping niches release sufficient monosaccharides to make scavenging an evolutionarily successful strategy in this niche. The remarkable result coming from the analysis of sugar metabolism by *E. coli* is that it was not amino acid or fatty acid but carbohydrate metabolism that was found to be central to microbial colonization or persistence in the mammalian gut.[53] The approximate hierarchy gluconate > GlcNAc > Neu5Ac = glucuronate > mannose > fucose > ribose indicates that better models or more precise methods are needed to fully understand the role of sialic acid catabolism in colonization of host mucosal environments by obligate commensals or frank pathogens. This and a related study[55] of lactate utilization by *N. meningitidis* emphasizes that there may be no fundamental distinction between virulence and metabolism.

Microorganisms in the *Haemophilus–Actinobacillus–Pasteurella* (HAP) group synthesize activating and transferase enzymes but are dependent on a supply of free sialic acid that must be scavenged from the external or host environment. HAP organisms have truncated biosynthetic pathways that may reflect an obligate commensal association with hosts that constantly synthesize sialylated cell surface or systemic glycoconjugates. Some of these commensals, such as *Haemophilus ducreyi* (human chancroid) and *Haemophilus somnus* (cattle disease), also may lack sialocatabolism, indicating that the primary function of sialic acid transport in HAP organisms may be cell surface decoration instead of nutrient acquisition. This conclusion is consistent with observations that sialic acid catabolism *per se* is not required for systemic disease or host colonization.[37,52]

A critical test of the hypothesis that host sialic acid is necessary for disease came from the analysis of an *H. influenzae* mutant with defective activating enzyme. This mutant transports and catabolizes sialic acid but is unable to activate it for transfer to glycoconjugate acceptors, and thus unable to decorate its surface. Assuming that cell surface decoration through sialylation is necessary for colonization and disease, the *H. influenzae* synthetase mutant was investigated in the chinchilla model of otitis media.[56] Otitis media (middle ear infection) is common in young children, and with one-third of all cases caused by *H. influenzae*, this bacterium is responsible for a million doctor visits and nearly as many antibiotic prescriptions written in the United States alone each year.[57] Inactivation of the synthetase was found to confer an avirulent phenotype in the chinchilla model, demonstrating the required host source of sialic acid *in vivo*.[56] Despite the experimentally established connection between environmental sialic acid uptake, cell surface sialylation, and otitis media, the exact function of sialylation in this or other diseases, such as pasteurellosis or chancroid, where sialic acid uptake is also known or suspected to be an essential virulence determinant, remains to be explained.

Although the well-known antirecognition functions of sialylation offer at least a partial explanation for the prevalence of sialylation in pathogens and commensals, the full answer is likely to be more complicated. For example, it is said that chinchillas lack serum bactericidal activity in the absence of exogenous complement,[56] although another study disputes this claim.[58] It is likely that the function of sialylation in the development of otitis media involves molecular recognition events other than simple complement avoidance. These other functions most likely involve host–microbe interactions where the shifting cell surface sialylation pattern of the microbe is either passively or directly connected to persistence. An example of a direct interaction is the sensing of environmental sialic acid through the sialic acid binding protein, NanR.[18,59] In addition to controlling expression of the *nan* catabolic operon in *E. coli*, NanR indirectly regulates type I fimbriae expression.[60] Other functions of microbial sensing are being revealed through the analysis of transcriptional microarray patterns.[18] For pathogens or commensals with truncated biosynthetic systems, all of the functions mediated by sialic acid metabolism will depend on specific recognition and uptake systems.[17,18] At least some of these uptake systems are likely to be good targets for new drug development.

7   Sialic Acid Uptake

There are five recognized mechanisms of bacterial solute uptake, four of which are shown in Figure 3:[18]

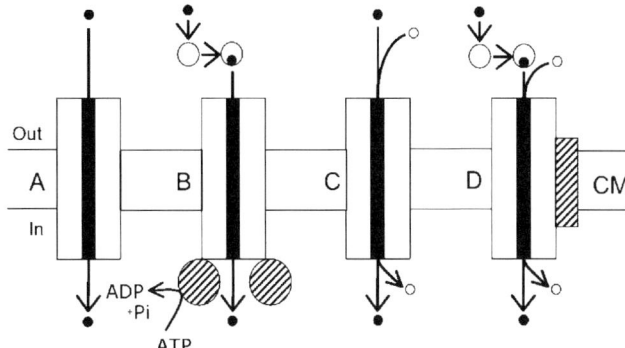

Figure 3 *Bacterial solute transporters. Solute (solid circles) channels (solid rectangles) for uptake through the cytoplasmic membrane (CM) from the extracellular environment (Out) to the cytoplasm (In) are indicated by the large rectangles representing integral membrane transport proteins. (A) Facilitated diffusion. (B) ABC transporter, where intermediate-size open circles indicate the periplasmic binding component and the hatched circles indicate the ATPase. (C) Symporter of the MFS of membrane transporters, where the small circles indicate protons or metal ions coupled to solute uptake. (D) TRAP transporter, indicating features in common with ABC and secondary transporters and the addition of a second (hatched rectangle) membrane protein of unknown function that is necessary for solute uptake. (Reprinted from ref.18 with permission.)*

- phosphoenolpyruvate:glycose phosphotransferase (PTS), or group translocation system;
- facilitated diffusion;
- antiport (countertransporters)/symport (cotransporters) of the major facilitator superfamily (MFS);
- ATP-binding cassettelike (ABC) systems; and
- tripartite ATP-independent periplasmic (TRAP) transporters.

PTS systems couple the phosphate group of phosphoenolpyruvate in a phosphorelay or kinase cascade resulting in uptake of phosphorylated and, therefore membrane-impermeable, carbohydrate ligands. Of the five systems, all except facilitated diffusion use energy to concentrate solute against its electrochemical gradient; glycerol is the only known example of facilitated diffusion (Figure 3). Permeases of the MFS, which are structurally similar to the glycerol facilitator in that they are composed of a single inner membrane protein with multiple membrane spanning domains, couple the counter- or cotransport of one solute (usually a proton or metal ion) down its electrochemical gradient to provide the energy for concentrating specific ligand. In contrast, the ATP-dependent, or ABC, systems catalyze the unidirectional uptake of ligands by coupling ATP hydrolysis with recognition of solute by a periplasmic component that donates it to an inner membrane component for uptake. These systems are usually composed of three or four gene products. ATP-independent, or TRAP transporters are also multigenic and employ a periplasmic binding protein. TRAP transporters appear to be hybrids of ancestral ABC and MFS systems. Depending on the bacterial species, sialic acid uptake may be catalyzed by an MFS, a TRAP, or an ABC system.[18]

7.1   The *E. coli* Paradigm

Although microbial sialic acid uptake was originally described as a physiological process in the Gram-positive organism *Clostridium perfringens*,[61] it was not until the early 1980s that the transporter was linked to an identified gene as part of the *E. coli* sialocatabolic operon that also included the aldolase structural gene.[32,33] Since then, the *E. coli* operon has been shown to include genes for the phosphorylation and epimerization of ManNAc (Figure 2). On the basis of the expected clustering of gene functions, computer-assisted analysis of approximately 200 bacterial stains or species identified over 10 distinct sialate transport and catabolic systems in bacteria.[18]

When the *E. coli* N-acylneuraminate transporter gene (*nanT*) was sequenced, it was predicated to encode an MFS permease with 14 instead of the usual 12 membrane spanning domains.[62] An amphipathic intracellular domain was postulated to confer selectivity for sialic acid, though this suggestion awaits verification. Although orthologues of NanT were readily identifiable in some species known to catabolize sialic acid, these genes products lack the NanT amphipathic domain and do not map near the aldolase or other sialocatabolic structural genes, suggesting that they were unlikely to function in sialic acid uptake and that the majority of organisms may use systems unrelated to NanT for sialic acid transporter.

A remarkable phenotype of *nanA* (aldolase) mutants was their sensitivity to exogenous sialic acid when cells were growing on a non-PTS substrate such as glycerol.[32] The observed toxicity to environmental sialic acid results in growth stasis (bacteriostatic) instead of a bactericidal effect, suggesting that intracellular sialic acid accumulation causes a nonlethal drop in cytoplasmic pH or directly inhibits some metabolic process necessary for net cell growth. Given the structural similarity between sialic acid and KDO (Figure 1), it is tempting to think that Neu5Ac accumulation inhibits KDO biosynthesis or transfer of KDO to LPS, which are known to be lethal or to block net cell growth, respectively.

A direct connection between NanT and KDO was detected during an investigation of KDO uptake in *Salmonella typhimurium* mutants with a conditional (temperature-sensitive) lethal defect in KDO biosynthesis. Suppressor mutants that could be rescued by the exogenous addition of KDO, which is not normally transported by *S. typhimurium*, suffered mutations that mapped in the same general position as *nanT* was known to map.[63] The suppressor mutants retained NanT function, suggesting either upregulation of *nanT* or synthesis of an altered NanT that now efficiently recognized and transported KDO as a potential analogue of sialic acid. If NanT recognizes KDO as a substrate, the permease would have no absolute discrimination between either the glycerol tail, carbon-5 acetamido group, or anomeric configuration. In unpublished experiments designed to distinguish between the overproduction and altered specificity hypotheses for explaining KDO uptake in these conditional lethal mutants, we observed that none of the *S. typhimurium* suppressor strains had rates of sialic acid uptake that were higher than wild type under a given growth condition. This result suggested that the suppressor mutations altered NanT function instead of its expression or the expression of a "cryptic" sialate transporter. Definitive evidence that NanT substrate specificity is altered in these mutants will require DNA sequencing to detect the altered amino acid residue(s).

To further investigate the substrate specificity of NanT, we took advantage of the bacteriostatic effect of sialic acid accumulation in a *nanA* mutant. This assay yields quantitative information because the diameter of the zone of inhibition is proportional to the sialic acid concentration down to about 5 µg.[32] Figure 4 shows that the typical bacteriostatic effect of Neu5Ac extends to Neu5Gc, the carbon-1 methyl ester (Neu5AcOMe), and Neu5Ac2en. Furthermore, although Neu5AcOMe and Neu5Ac2en are not aldolase substrates, all sialic acids exhibiting growth stasis also served as sole carbon source for wild-type *E. coli*, as indicated by the "+" signs shown in Figure 4. In the case of Neu5AcOMe, a wide range of esterases may convert the ester to free Neu5Ac either before (periplasmic) or after (cytoplasmic) transport. Similarly, there must be an enzymatic process to rehydrate the Neu5Ac2en double bond, generating free Neu5Ac.

Aldolase cleavage of transported Neu5Gc is expected to release *N*-glycolylman-nosamine (ManNGc) and pyruvate. Although the metabolic fate of ManNGc is

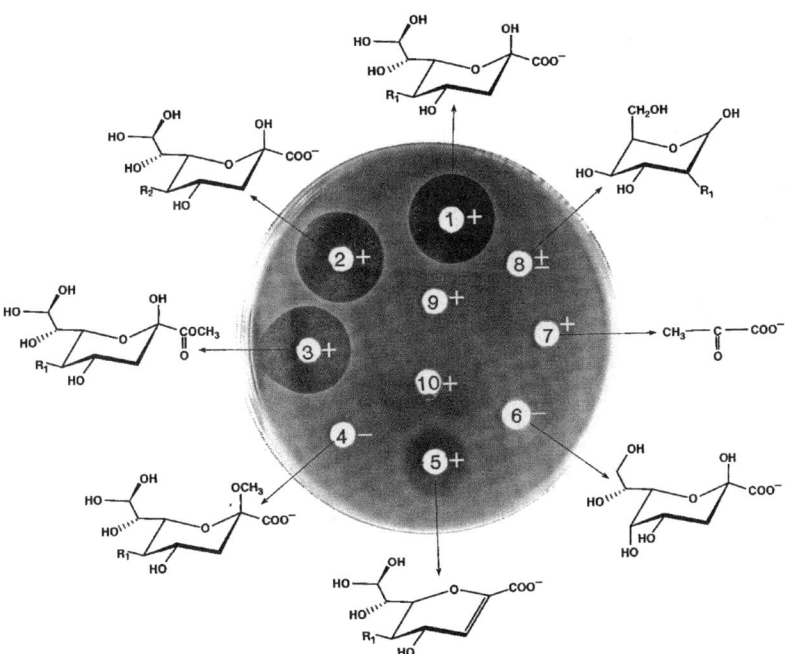

Figure 4 *Sialic acid toxicity in an E. coli nanA mutant. Bacteria are pregrown in minimal salts medium containing glycerol as sole carbon source and plated onto agar-solidified minimal media in a large Petri dish. 1, Neu5Ac; 2, Neu5Gc; 3, Neu5AcOMe; 4, Neu5AcβMe; 5, Neu5Ac2en; 6, KDO; 7, pyruvate; 8, ManNac; 9 and 10 are maltose and fucose, respectively (structures not shown). The compounds indicated were added to sterile paper disks and the culture was allowed to grow for 24 hours before photography. Growth on the indicated compounds as sole carbon sources was determined in a separate experiment using wild type: +, good growth in 48 hours; −, no growth; ±, weak growth. The zones of growth inhibition around compounds 1–3 and 5 indicate toxicity for sialic acid in the nanA mutant background.*

unclear, the common occurrence of Neu5Gc in deuterostomes other than humans means that microbes have the metabolic machinery to deal with both natural and unnatural (laboratory) sialic acids. Finally, *E. coli* does not produce sialidase and so cannot convert the β-methyl glycoside (Neu5AcβMe) to free Neu5Ac (Figure 4). Whether this glycoside is a NanT substrate will require direct competition assays, but since the two available glycosides "lock" sialic acid into either the α or β configuration, it should be straightforward to determine which form NanT "selects" (recognizes) *in vivo*. This information is expected to guide inhibitor design based on the keto-sugar backbone. Indeed, one of these glycosides should already be considered a viable lead compound for further drug development.

If the Petri dish shown in Figure 4 is incubated for an additional 24–48 h, macroscopic colonies arise within the zones of inhibition. When bacteria derived from these escape mutants are tested for sialic acid uptake, they are found to be defective, suggesting selection of spontaneous mutations in *nanT*, or in regions of the genome controlling *nanT* expression. The ability to positively select uptake mutants provides a robust system to investigate, among other issues, the structural requirements for recognition of sialic acids by NanT. In addition to the basic matters discussed above, the phenotypes of *nanA* and *nanT* mutants suggest practical approaches for identifying lead compounds that block sialic acid uptake.

The bacteriostatic effect of sialic acid added to a *nanA* mutant growing on a non-PTS substrate such as glycerol or succinate provides a positive screen for uptake inhibitors. A culture of aldolase-defective bacteria is diluted in duplicate into a microtiter dish. To one replicate, sialic acid is added and both cultures receive an aliquot of putative uptake inhibitor. Using robotics, thousands of compounds can be assayed in a few hours. Quantitative (growth curve) information is possible by taking turbidometric readings over time. Thus, compounds that are weakly recognized by the permease may still be detected from reduced growth rates. Compounds that also inhibit the control sample that did not receive sialic acid would be excluded. The subset of compounds with potential inhibitory activity is then rescreened to exclude false positives. One is then left with a collection of lead compounds that can be further refined. Although ring compounds with negative charge would be expected to be the most likely candidate inhibitors, the screen makes no assumptions about potential efficacy of a given compound. Even with no prior knowledge of sialic acid uptake systems, an inhibitor of NanT is likely to be effective against organisms with different or even unknown sialic acid uptake systems. Fortunately, a new class of sialic acid transporter has been identified in the HAP group. This group includes a variety of human and animal pathogens or commensals.

7.2   The Carboxylic Acid or TRAP Paradigm

H. influenzae has a relatively small genome (1.83 Mb) that suggests divestiture of all genetic information not needed for survival in the obligate human host.[3] Despite its limited genetic repertoire, *H. influenzae* is a successful commensal and occasional pathogen. It is responsible for about one-third of all otitis media cases, with combined costs from doctor visits and prescriptions exceeding $1 billion annually.[57] Although an effective anticapsular polysaccharide vaccine has eliminated invasive disease caused by type b strains, *H. influenzae* is still isolated from most cases of lower respiratory tract disease, the fourth leading cause of death in the US in 2001.[64] Owing to vaccination, most cases of disease in the developed world are now caused by nontypeable (unencapsulated) strains. As described above, elimination of sialylation prevents otitis media in one animal model,[56,58] suggesting approaches for treating it, as well as sinusitis, pneumonia, and chronic bronchitis.

H. influenzae is unable to use a wide range of carbohydrates for nutrition because it lacks PTS or other uptake systems for scavenging.[3] Given this situation, we were struck by the retention of a *nan* system allowing *H. influenzae* to use sialic acid as carbon source, suggesting that sialic acid metabolism was a central feature of the bacterium–human host interaction. Two interrelated and extreme hypotheses seemed tenable: sialic acid utilization was essential for colonization and persistence, or sialic acid uptake was primarily needed for surface modification. By denying *H. influenzae* of an active sialate aldolase, we were able to test the relative contributions of catabolism versus modification in the rodent model.[52] The surprising result was that the ability to catabolize sialic acid was not essential for host colonization or systemic disease, a conclusion that was confirmed, using similar methodology, for the related HAP pathogen *P. multocida*.[37] Therefore, while sialic acid catabolism is likely to provide a fitness advantage to *H. influenzae*, catabolism *per se* is not essential for short-term survival. In contrast, the essential function of sialic acid uptake for nutrition and decoration is now incontrovertible. Blocking sialic acid uptake by TRAP transporters is then, evidently, a starting point for new therapeutic development despite our current inability to completely separate the role of catabolism versus surface decoration in the disease process. The prudent conclusion is that a compound blocking sialic acid uptake will prevent both catabolism and decoration resulting in avirulence. Given the domestic and worldwide disease burden caused by HAP organisms, the projected economic benefit of such a nonantibiotic treatment exceeds $1 billion annually.

Unlike the process of sialic acid uptake by *E. coli*, which depends on the *nanT*-encoded symporter, compounds blocking TRAP transporter systems have two potential targets: the periplasmic-binding protein and the membrane permease. Although unpublished experiments attempting to crystallize NanP from *P. multocida*, overproduced as a C-terminal histidine-tagged fusion polypeptide, have not been successful, these studies allowed us to characterize the kinetics of sialic acid uptake by a TRAP system and to link the inhibitory effects of lead compounds to their effects on sialic acid binding by direct biochemical assay. This assay provides an experimental approach to characterize on–off rates and affinities of inhibitory compounds. Note that despite differences in uptake mechanisms (Table 1), some classes of inhibitory compounds should be equally effective regardless of the bacterial species. This conclusion follows from the convergent evolution of sialic acid binding proteins.[18]

7.3 ABC Systems

The cattle pathogen *H. somnus* sialylates its surface by precursor scavenging of sialic acid from its animal host, suggesting two possible mechanisms: uptake followed by activation and sialyltranfser and sialylation from host-derived CMP–sialic acid by an extracellular sialyltransferase.[65] It has been our contention that CMP–sialic acid is an unlikely candidate precursor because of its chemical instability resulting from spontaneous hydrolysis of the β-glycoside.[52] For example, in a given preparation of CMP–sialic acid, free Neu5Ac contributes as much as 10% by weight or specific activity of the total sialic acid. With the current appreciation of sialic acid uptake as the mechanism of precursor scavenging,[17,18] it is possible that CMP–sialic acid is not a physiological precursor in any system.

H. somnus is further distinguished from H. influenzae and P. multocida in that it does not appear to contain *nan* orthologues in its genome, indicating that it does not catabolize sialic acid. This observation is consistent with the dispensability of sialic acid as a nutritional source by some members of the HAP group[37,52] but suggests that other organisms, such as *H. somnus*, have the genes for activating as well as transferase enzymes and may use sialic acid uptake solely for surface decoration as an obligate feature of the host–microbe interaction. We suggest that *H. somnus* may be similar to *S. pyogenes* in using an ABC transporter instead of MFS or TRAP system for sialic acid uptake (Table 1). The prevalence of diverse sialic acid uptake systems is evidence of an intense microbial arms race in which some species have been selected, at least in part, for their use of host sialic acids for nutrition alone (*E. coli*), nutrition and decoration (*H. influenzae* and *P. multocida*), or just decoration (*H. somnus*). Therefore, blocking sialic acid uptake may be an effective therapeutic strategy aimed at a wide range of different commensals and pathogens, possibly including all other HAP species that have not previously been shown to metabolize sialic acids.

8   Polysialic Acid Acetylation

Dawkins and Krebs[4] have argued that for the fox not to catch the hare usually just means a night without dinner, whereas for the hare to get caught means, if it has not already bred, that its genes are eliminated. Faced with the apotheosis of sialic acids by the deuterostomes, we argue that the explanation for the wide range of different sialic acid uptake systems reflects the evolution of microbial survival strategies that have been selected in the context of the host interaction – that is, a reflection of the host–microbe arms race.[4] Once mechanisms of microbial cell surface sialylation are in place, additional mechanisms controlling the type or extent of modification offer the potential for further change. The classical control mechanisms for this change include transcriptional, translational, and posttranslational systems such as the sialic acid binding protein NanR, which regulates sialic acid metabolism in *E. coli*.[59] In this section, we will focus on the newly discovered mechanism of polysialic acid *O*-acetylation in *E. coli* K1.[66] Although the exact function of this modification in the host–microbe interaction is not yet clear, the modification is not required for systemic disease.[67] Similarly, *O*-acetylation of the *N. meningitidis* group A ManNAc-1-P capsule is not required for sepsis or meningitis but is reported to be essential for colonization of the nasopharynx.[68] It is likely that polysialic acid *O*-acetylation is required for intestinal colonization by *E. coli* K1 and that this process is so important (of such high selective value) that a novel mechanism controlling the modification has been selected.[66]

8.1 Contingency Loci

There is a trade-off for microbes, with their potentially rapid generation times, between increased mutation rates for short-term gain and selection against deterioration of the genome's housekeeping functions, which are necessary for their rapid growth. Moxon and colleagues[69] argued that the deleterious aspects of high mutation rates are offset by mechanisms of localized hypermutability. Loci (genes) undergoing hypermutation and whose products interact with the environment, meaning surface structures such as flagella, adhesins, and (especially) polysaccharides, are contingent because their change benefits the organism only insofar as how well the "new" phenotype increases fitness in a given environment. Therefore, the contingency phenotypes we see today reflect previous struggles to gain selective advantage, which in the context of the host–microbe interaction involves microbial surface properties.

Figure 5A shows the mechanism controlling the variation of flagella and some fimbriae contingency loci: an invertible promoter alternating between "on" and "off" transcriptional states. This mechanism was predicted purely on theoretical grounds[70] and was later the first example of a reversible contingency locus to be verified experimentally. There is no reason why iterations of this kind of recombinational switch

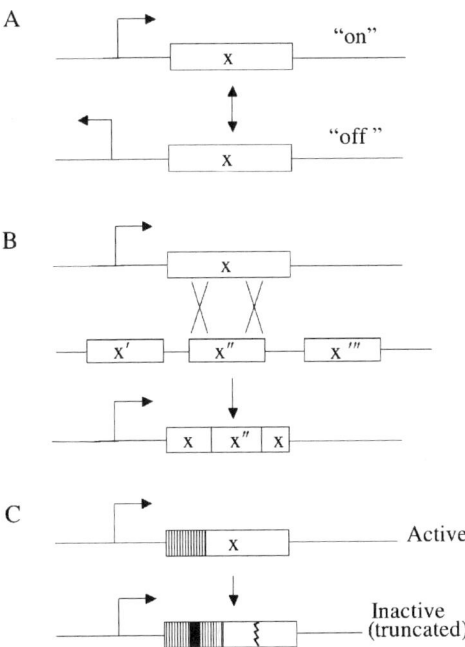

Figure 5 *Mechanism of gene x phase (form) variation. Genes are indicated by open boxes and promoters by bent arrows. (**A**) Reversible promoter switch. (**B**) A promoter-driven expression locus recombines with transcriptionally silent variant forms of gene x to generate antigenically altered gene products. (**C**) Slipped-strand mispairing of iterated repeats (small open rectangles) undergo expansion or retraction (solid rectangle), generating a frameshift mutation (jagged line) leading to polypeptide truncation (inactivation) during translation.*

should not exist, and Krinos *et al.*[71] have recently shown that expression of at least eight different *Bacteroides fragilis* surface polysaccharides is controlled by one recombinase catalyzing the reversible promoter inversion. Another type of recombinational switch is reminiscent of eukaryotic antibody selection, involving recombination between an active expression locus and inactive or silent copies encoding antigenic variants (Figure 5B), such as found in gonococcal pilin variation.[72] Note that both mechanisms involve specialized recombination machinery. In contrast, slipped-stand DNA mispairing involves standard DNA synthesis and repair mechanisms.[73] Variation caused by the translational switch between full-length and truncated (inactive) polypeptides may thus depend solely on the type of nucleotide or oligonucleotoide repeat (Figure 5C).

Homopolymeric or oligonucleotoide tracts are most often observed in genes encoding enzymes that synthesize or modify surface polysaccharides.[69] The relatively large number of contingency loci regulated by slipped-strand mispairing of (usually) di-, tri-, or tetranucleotides repeats in nasopharyngeal bacteria, such as *H. influenzae* and *N. meningitidis*, but their absence in *E. coli* may suggest potential differences between DNA synthesis or repair mechanisms by the two species. However, when independent *H. influenzae* loci were recombined into the *E. coli* genome, essentially the same mutation rates as those of the parental strain were observed in the heterologous host.[74] This observation suggested that the apparent lack of contingency loci in *E. coli* reflected differences between a free-living versus obligate commensal. In other words, in the arms race between microbes and hosts, rapid surface change may have greater selective value to an obligate parasite than it does to a facultative parasite like *E. coli*.

8.2   *E. coli* K1 Acetylation

Antipolysialic acid antibodies for the detection of the *E. coli* K1 or group b meningococcal capsules revealed two classes of *E. coli* K1. The "nonreactors" were subsequently shown to produce polysialic acid containing *O*-acetyl esters at carbon-7 or carbon-9 (Figure 6). Several subclasses of *E. coli* K1 isolates were found: those that were never nonreactive (always unacetylated), some that were always acetylated, and the largest group that cycled between reactivity and nonreactivity.[75] The authors speculated that a recombinational switch might be involved in the mechanism controlling acetylation.[75] The remarkable aspect of capsule acetylation was its frequency (approaching 10%), indicating a highly active hypermutable switch-type mechanism.

While carrying out a signature-tagged mutagenesis approach for the identification of systemic virulence genes in *E. coli* K1,[76,77] we noted an orthologue encoding endo-*N*-acetylneuraminidase (endo-N). This endo-polysialic acid depolymerase was first isolated and characterized in the K1-specific lytic bacteriophage (phage), K1F.[78] Like many phage depolymerases, endo-N binds and degrades its receptor to facilitate a biased random walk that is a necessary step in the phage infection process.[79] Phage tail proteins are under intense selective pressure because they dictate the host specificity (selectivity) needed for infection and subsequent viral propagation. Therefore, bacteria surviving K1F infection are almost always capsule negative,

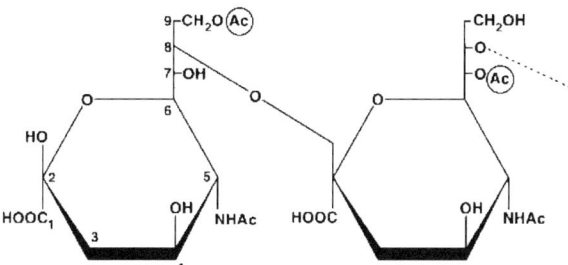

Figure 6 *Acetylation of polysialic acid. The disaccharide repeat unit of E. coli K1 polysialic acid is indicated with the carbon atoms of Neu5Ac numbered in the reducing sugar residue. Polymers may exceed 50 repeat units in length and be acetylated at either carbon position 7 or 9. The O-acetyltransferase encoded by neuO may transfer acetyl units to either or both positions, including the possibility of preferential transfer to a single position followed by nonenzymatic transesterification to the alternate position.*

providing the first genetic system for investigating capsule biosynthesis.[80] However, the remarkable feature of the chromosomal K1 endo-N structural gene was its association with a lysogenic phage designated CUS-3.[66] Contiguous with CUS-3 endo-N was a potential receptor modifying the gene encoding an *O*-acetyltransferase of the hexapeptide repeat family of acyltransferases.[66] We designated this gene *neuO* (*N*-acetylneurmainic acid *O*-acetyltransferase) to reflect its role in polysialic acid metabolism. *E. coli* K1 that were CUS-3 lysogens invariably cycled between acetylation "on" and "off" states, whereas nonlysogens were invariably acetylation negative. These observations explained two of the subclasses of strains previously isolated:[75] those with variable acetylation states and those that were always "off." Although we cannot yet explain the third subclass, that of those always "on," our results suggest that these bacteria either use a different acetylase from *neuO* or have lost the mechanism of hypervariation.

8.3 Hypervariation of *neuO*

The mechanism of *neuO* hypermutability was inferred from the presence of tandem 5'-AAGACTC-3' heptanucleotide repeats located in the 5' region of the *neuO* coding sequence. This inference was confirmed by radiometric enzyme assay of form variants, inverse PCR (transcript mapping), and direct cloning and DNA sequencing of *neuO* variants detected by antibody screening.[66] Thus, *neuO* variants arising from slipped-strand DNA synthesis or unequal recombination generated either in-frame ("on") or out-of-frame ("off") *neuO* transcripts resulted in premature truncation caused by frame shift mutations in the *neuO* catalytic domain.[66] Moxon and co-workers[81] have provided evidence that mutation frequency increases with tandem repeat length, so the heptanucleotide *neuO* iteration is consistent with the observed form variation rates that in some strains may exceed 10%.[66,75] The general principle that emerges from these studies is that because *neuO* is located on a mobile genetic element (CUS-3), the opportunity for hypervariation is increased because the polyΨ (hypervariable region of *neuO*) domain can vary under nonselective pressure and then has the opportunity of being transferred after prophage induction and reinfection of susceptible hosts. After dissemination it may then come under pressure where a very high rate of form variation, conferred by the heptanucleotide translational switch, may be of selective value. This conclusion suggests that the function of *neuO* is likely to include more than just, if at all, modifying CUS-3 lysogen susceptibility to other K1-specific phage.

8.4   Function of *neuO*

The covalent modification of surface polysaccharides has two obvious functions: antigenic variation in the context of the host interaction and resistance to phage infection. Although it is clear that capsules may block underlying structures that may serve as receptors for phage with different tail specificities, receptor modification by a resident prophage is expected to specifically reduce susceptibility to other K1-specific phages. Classical approaches such as the one-step growth curve and determination of adsorption constants will provide quantitative information addressing this hypothesis. However, preliminary evidence indicates that CUS-3 lysogens are as sensitive to lytic PK1E infection as nonlysogens and that acetylation did not affect infectivity by K1F or PK1E in a model system using cloned *neuO*. These preliminary results suggest that the evolution of capsule hypervariability was not driven by phage interactions. What then might be the adaptive value of *neuO* hypermutability?

E. *coli* K1 transiently colonizes the oropharynx, survives the gastric acidity barrier, colonizes the bowel, and traverses the mucosal barrier; it may survive and propagate systemically and invade the central nervous system.[82,83] Once established in the intestine, it may move on to invade the bladder and cause cystitis or chronic infections by existing as an intracellular biofilm in bladder cells.[84] Thus, E. *coli* K1 is a facultatively intracellular and invasive commensal and pathogen. This spectrum of host–microbe interaction is fundamentally different from that of meningococci, which also synthesize the polysialic acid capsule but where the repulsive properties (high net negative charge and low hydrophobicity) of the capsule inhibits microbe–host cell interactions and is shut off during colonization. In contrast, as far as we know from laboratory studies, the K1 capsule is always expressed. How, then, does an E. *coli* K1 biofilm progress? Of course downregulation of capsule genes by classical regulatory mechanisms induced by host cell contact is one mechanism. Another is to reduce the repulsiveness of polysialic acid by acetylation. Therefore, *neuO* hypermutability may reflect selection for a mechanism that fine-tunes the relative hydrophobicity of the cell surface.[85] Experiments addressing this hypothesis may lead to new understanding of biofilm formation in microbes with constitutive hydrophilic surfaces. These studies may also contribute to a deeper appreciation of polysialic acid function in eukaryotes. Another area for further study includes elucidating the functions (host and microbial) influencing hypervariability of contingency loci during colonization. These studies could lead to the identification of additional targets for therapeutic intervention.

9 Conclusion

Sialic acids are the external-most prevalent molecular components of the mammalian glycocalyx, and among their many functions is to act as a barrier against microbial attack. A wide range of different, but also closely related, bacterial commensals and pathogens have evolved mechanisms for disrupting or exploiting the host sialic acid barrier. These mechanisms include adhesion, molecular mimicry, removal, and scavenging of host sialic acids. A surprisingly large group of clinically and economically important bacterial species scavenges host sialic acids. The individual precursor scavenging systems include diverse uptake mechanisms, conserved activating enzyme, and only distantly related sialyltransferases. Scavenging host sialic acid for reexpression of sialoglycoconjugates at the microbial cell surface modifies bacteria properties in ways that are, mostly, still poorly defined but that are necessary for successful host–microbe interactions or disease. Therefore, chemical agents designed to block sialic acid uptake are likely to have broad therapeutic potential. In addition to scavenging environmental sialic acids, other bacteria have evolved strategies for timing the expression of cell surface sialylation or for modifying its chemical properties by covalent modification. The newly discovered mechanism of polysialic acid *O*-acetylation, described in the context of the host–microbe interaction, suggests additional targets for therapeutic intervention.

Acknowledgment

The National Institutes of Health Grant R01 AI42015 and the Illinois Governor's Individual Venture Technology Research Grants program supported original research described or summarized in this report.

References

1. R. Dawkins, *The Ancestor's Tale*, Houghton Mifflin Company, Boston, MA, 2004.
2. A.L. Koch, *Pers. Biol. Med.*, 1976, **20**, 44.
3. R.L. Tatusov, A.R. Mushegian, P. Bork, N.P. Brown, W.S. Hayes, M. Borodovsky *et al.*, *Curr. Biol.*, 1996, **6**, 279.
4. R. Dawkins and J.R. Krebs, *Proc. R. Soc. London, Ser. B*, 1979, **205**, 489.
5. A.M. Kauppi, R. Nordfelth, H. Uveil, H. Wolf-Watz and M. Elofsson, *Chem. Biol.*, 2003, **10**, 241.
6. Y.M. Lee, F. Almqvist and S.J. Hultgren, *Curr. Opin. Pharm.*, 2003, **2**, 513.
7. J.C. Gannon, *The Global Infectious Disease Threat and Its Implications for the United States*, http://www.cia.gov/cia/reports/nie/report/ie-99-17d.html, 2000.
8. K.M. Carbone, R.B. Luftig and M.R. Buckley, *Microbial Triggers of Chronic Human Illness*, http://www.asm.org, 2005.
9. A.L. Koch, *Microbiol. Rev.*, 1981, **45**, 355.
10. C.R. Raetz and C. Whitfield, *Annu. Rev. Biochem.*, 2002, **71**, 635.
11. J.A. Yethon and C. Whitfield, *Curr. Drug Targets-Infect. Dis.*, 2001, **1**, 91.

12. A. Economou, *Emerg. Ther. Targets*, 2001, **5**, 141.
13. B.J. Akerley, E.J. Rubin, V.L. Novick, K. Amaya, N. Judson and J.J. Mekalanos, *Proc. Natl. Acad. Sci. USA*, 2001, **99**, 966.
14. J.E. Shea, J.D. Santangelo and R.G. Feldman, *Emerg. Ther. Targets*, 2001, **5**, 155.
15. G. Laver and E. Garman, *Science*, 2001, **293**, 1776.
16. G. Laver and E. Garman, *Microbes Infect.*, 2002, **4**, 1309.
17. E. Vimr and C. Lichtensteiger, *Trends Microbiol.*, 2002, **10**, 254.
18. E.R. Vimr, K.A. Kalivoda, E.L. Deszo and S.M. Steenbergen, *Microbiol. Mol. Biol. Rev.*, 2004, **68**, 132.
19. A. Casadevall and L.-A. Pirofski, *Infect. Immun.*, 2000, **68**, 6511.
20. A. Casadevall and L.-A. Pirofski, *J. Infect. Dis.*, 2001, **184**, 337.
21. A. Casadevall and L.-A. Pirofski, *Ann. Med.*, 2002, **34**, 2.
22. A. Casadevall and L.-A. Pirofski, *Trends Microbiol.*, 2003, **11**, 157.
23. M.R. Buckley, *The Genomics of Disease-Causing Organism: Mapping a Strategy for Discovery and Defense*, www.asm.org/Academy/index.asp?bid=29532, 2004.
24. D.W. Cushman, H.S. Cheung, E.F. Sabo and M.A. Ondetti, *Biochemistry*, 1977, **16**, 5484.
25. A.A. Patchett, E. Harris, E.W. Tristram, M.J. Wyvratt, M.T. Wu *et al.*, *Nature*, 1980, **288**, 280.
26. C.H. Hassall, A. Kröhn, C.H. Moody and W.A. Thomas, *FEBS Lett.*, 1982, **147**, 175.
27. I.G. Bravo, S. García-Vallvé, A. Romeu and Á. Reglero, *Trends Microbiol.*, 2004, **12**, 120.
28. A. Varki, *Am. J. Phys. Anthropol. Suppl.*, 2001, **33**, 54.
29. T. Angata and A. Varki, *Chem. Rev.*, 2002, **102**, 439.
30. M. Ridley, *Genome*, Harper Collins, New York, NY, 1999.
31. G. Zapata, W.F. Vann, W. Aaronson, M.S. Lewis and M. Moos, *J. Biol. Chem.*, 1989, **264**, 14769.
32. E.R. Vimr and F.A. Troy, *J. Bacteriol.*, 1985, **164**, 845.
33. E.R. Vimr and F.A. Troy, *J. Bacteriol.*, 1985, **164**, 854.
34. H. Yu, H. Yu, R. Karpel and X. Chen, *Bioorg. Med. Chem.*, 2004, **12**, 6427.
35. P.K. Qasba, B. Ramakrishnan and E. Boeggeman, *Trends Biochem. Sci.*, 2005, **30**, 53.
36. G. Liu, C. Jin and C. Jin, *J. Biol.Chem.*, 2004, **279**, 17738.
37. S.M. Steenbergen, C.A. Lichtensteiger, R. Caughlan, J. Garfinkle, T.E. Fuller and E.R. Vimr, *Infect. Immun.*, 2005, **73**, 1284.
38. M.A. Ringenberg, S.M. Steenbergen and E.R. Vimr, *Mol. Microbiol.*, 2003, **50**, 961.
39. M.E. Tanner, *Bioorg. Chem.*, 2005, **33**, 216.
40. C.C. Wreden, M. Wlizla and R.J. Reimer, *J. Biol. Chem.*, 2005, **280**, 1408.
41. M. Ringenberg, C. Lichtensteiger and E. Vimr, *Glycobiology*, 2001, **11**, 533.
42. J. Plumbridge and E. Vimr, *J. Bacteriol.*, 1999, **181**, 47.
43. C.J. Brigham and M.H. Malamy, *J. Bacteriol.*, 2005, **187**, 890.
44. M.A. Ferrero, Á. Reglero, M. Fernandez-Lopez, R. Ordas and L.B. Rodriguez-Aparicio, *Biochem. J.*, 1996, **317**, 157.

45. E.R. Vimr, *J. Bacteriol.*, 1992, **174**, 6191.

46. J.L. Bruses and U. Rutishauser, *Biochimie*, 2001, **83**, 635.

47. S.M. Steenbergen and E.R. Vimr, *J. Biol. Chem.*, 2003, **278**, 15349.

48. M.M. Shah, K. Fujiyama, C.R. Flynn and L. Joshi, *Nature Biotech.*, 2003, **21**, 1470.

49. K. Koles, K.D. Irvine and V.M. Panin, *J. Biol. Chem.*, 2004, **279**, 4346.

50. C.S. Alviano, L.R. Travassos and R. Schauer, *Glycoconj. J.*, 1999, **16**, 545.

51. M. Séveno, M. Bardon, T. Paccalet, V. Gomord, P. Lerouge and L. Faye, *Nature Biotech.*, 2004, **22**, 1351.

52. E. Vimr, C. Lichtensteiger and S. Steenbergen, *Mol. Microbiol.*, 2000, **36**, 1113.

53. D.-E. Chang, D.J. Smalley, D.L. Tucker, M.P. Leatham, W.E. Norris *et al.*, *Proc. Natl. Acad. Sci. USA*, 2004, **101**, 7427.

54. T. Corfield, *Glycobiology*, 1993, **2**, 509.

55. R.M. Exley, J. Shaw, E. Mowe, Y. Sun, N.P. West *et al.*, *J. Exp. Med.*, 2005, **201**, 1637.

56. V. Bouchet, D.W. Hood, J. Li, J.-R. Brisson, G.A. Randle *et al.*, *Proc. Natl. Acad. Sci. USA*, 2003, **100**, 8898.

57. S.E. Stool and M.J. Field, *Pediatr. Infect. Dis. J.*, 1989, **8**, S11-S14.

58. J. Jurcisek, L. Grenier, H. Watanabe, A. Zaleski, M.A. Apicella and L.O. Bakaletz, *Infect. Immun.*, 2005, **73**, 3210.

59. K.A. Kalivoda, S.M. Steenbergen, E.R. Vimr and J. Plumbridge, *J. Bacteriol.*, 2003, **185**, 4806.

60. B.K. Sohanpal, S. El-Labany, M. Lahooti, J.A. Plumbridge and I.C. Blomfield, *Proc. Natl. Acad. Sci. USA*, 2004, **101**, 16322.

61. S. Nees and R. Schauer, *Behring. Inst. Mitt.*, 1974, **55**, 68.

62. J. Martinez, S. Steenbergen and E. Vimr, *J. Bacteriol.*, 1995, **177**, 6005.

63. R.C. Goldman and E.M. Devine, *J. Bacteriol.*, 1987, **169**, 5060.

64. S. Sethi and T.F. Murphy, *Clin. Microbiol. Rev.*, 2001, **14**, 336.

65. T.J. Inzana, G. Glindemann, A.D. Cox, W. Wakarchuk and M.D. Howard, *Infect. Immun.*, 2002, **70**, 4870.

66. E.L. Deszo, S.M. Steenbergen, D.I. Freedberg and E.R. Vimr, *Proc. Natl. Acad. Sci. USA*, 2005, **102**, 5564.

67. J. Colino and I. Outschoon, *Microb. Pathog.*, 1999, **27**, 187.

68. S.K. Gudlavalletti, A. Datta, Y.L. Tzeng, C. Noble, R.W. Carlson and D.S. Stephens, *J. Biol. Chem.*, 2004, **279**, 42765.

69. E.R. Moxon, P.B. Rainey, M.A. Nowak and R.E. Lenski, *Curr. Biol.*, 1994, **4**, 24.

70. B.A. Stocker and P.H. Makela, *Proc. R. Soc. London, Biol. Sci. B*, 1978, **202**, 5.

71. C.M. Krinos, M.J. Coyne, K.G. Weinacht, A.O. Tzianabos, D.L. Kasper and L.E. Comstock, *Nature*, 2001, **414**, 555.

72. J.R. Saunders, in *Genetics of Bacterial Diversity*, D.A. Hopwood and K.E. Chater (eds), Academic Press, New York, 1989, 267.

73. G. Levinson and G.A. Gutman, *Mol. Biol. Evol.*, 1987, **4**, 203.

74. J. Torres-Cruz and M.W. van der Woude, *J. Bacteriol.*, 2003, **185**, 6990.

75. F. Ørskov, I. Ørskov, A. Sutton, R. Schneerson, W. Lin, W. Egan, G.E. Hoff and J.B. Robbins, *J. Exp. Med.*, 1979, **149**, 669.

76. M.D. Gonzalez, C.A. Lichtensteiger and E.R. Vimr, *FEMS Microbiol. Lett.*, 2001, **198**, 125.

77. M.D. Gonzalez, C.A. Lichtensteiger, R. Caughlan and E.R. Vimr, *J. Bacteriol.*, 2002, **184**, 6050.

78. J.G. Petter and E.R. Vimr, *J. Bacteriol.*, 1993, **175**, 4354.

79. E.R. Vimr, *J. Bacteriol.*, 1992, **174**, 6191.

80. E.R. Vimr, W. Aaronson and R.P. Silver, *J. Bacteriol.*, 1989, **171**, 1106.

81. X. De Bolle, C.D. Bayliss, D. Field, T. van de Ven, N.J. Saunders, D.W. Hood and E.R. Moxon, *Mol. Microbiol.*, 2000, **35**, 211.

82. E. Vimr, S. Steenbergen and M. Cieslewicz, *J. Ind. Microbiol.*, 1995, **15**, 352.

83. E.R. Vimr, *Trends Microbiol.*, 1994, **2**, 271.

84. G.G. Anderson, J.J. Palermo, J.D. Schilling, R. Roth, J. Heuser and S.J. Hultgren, *Science*, 2003, **301**, 105.

85. C.P. Johnson, I. Fujimoto, U. Rutishauser and D.E. Leckband, *J. Biol. Chem.*, 2005, **280**, 137.

86. M. Petersen, W. Fessner, M. Frosch and E. Luneberg, *FEMS Microbiol. Lett.*, 2000, **184**, 161.

87. D.M. Post, R. Mungur, B.W. Gibson and R.S. Munson Jr., American Society for Microbiology National Meeting, Orlando, FL, 2005, No. B-227.

CHAPTER 9

Synthetic Carbohydrate-Based Antimalarial Vaccines and Glycobiology

ALEXANDRA HÖLEMANN AND PETER H. SEEBERGER

Swiss Federal Institute of Technology (ETH) Zurich, Laboratory for Organic Chemistry, ETH Hönggerberg, HCI F 315, Wolfgang-Pauli-Strasse 10, CH-8093 Zurich, Switzerland

1 Introduction

Vaccines are the most powerful and cost-efficient medical intervention in the control, prevention, and elimination of human infectious diseases. The principle of vaccination was established in 1796 by Edward Jenner, who successfully immunized people against smallpox by treatment with the related, but weakened, cowpox virus. About 80 years later, Louis Pasteur discovered that infectious diseases are caused by microorganisms and introduced a method to attenuate microbe virulence. By applying this approach, he developed the first live-attenuated vaccines against cholera, anthrax, and rabies. Since then,[1] remarkable progress in the development of vaccines against many different human pathogens including polio, influenza, measles, diphtheria, tetanus, pertussis, varicella, mumps, rubella, hepatitis B, *Pneumococci,* and *Haemophilus influenza* type B has been made.[2] Vaccination has enabled the eradication of smallpox worldwide and has decreased the incidence of once common childhood diseases.

Despite these successful developments, bacterial and viral infections in humans still represent one of the major health problems, killing at least 15 million people worldwide annually.[3] The search for new prophylactic and therapeutic vaccines to combat these infections has attracted considerable attention in the past years,[4] albeit with only limited success. No effective vaccines against most human parasites including malaria, leishmaniasis, and schistosomiasis have been developed so far.[3,5] Moreover, existing vaccines, *e.g.,* the *Bacillus Calmette Guerin* (BCG) vaccine against tuberculosis, are often of limited efficacy.[3,6] Thus, it is crucial to improve our understanding of the relevant glycan and protein antigens. Detailed investigations of the molecular interactions between host immune cells and parasites as well as parasitic interactions with their intermediate hosts will allow for the development of new vaccine candidates in order to control infectious diseases.

2 Immunization

The function of our immune system is to protect the body from infectious organisms and other potentially harmful invaders and thereby prevent infection. The immune system can be divided into natural (innate) and adaptive (acquired) immunity.[7] Innate immunity is a type of general, non-specific protection against infectious agents and pre-exists in all individuals. Several defense mechanisms and external barriers (skin, stomach acid, blood brain barrier, *etc.*) are present in the body, which are ready to stop invading microorganisms without prior activation or induction. In contrast, acquired immunity develops upon exposure of the body to invading agents, which express specific structural elements (antigens), and is specific for a particular antigen. The immune response starts with the recognition of an invader as non-self by the immune system. B- and T-cell clones that specifically recognize the antigen are subsequently generated and combat the invader by either neutralizing it with antibodies or by killing the cells that have been infected. These specific B- and T-cells are stored as memory cells, capable of recognizing the antigen and inducing a heightened immune response in case of re-exposure to the infectious organism. This process of protection against an infectious invader is known as immunization.

Vaccination is a way of inducing resistance to a foreign microorganism by specially training the immune system. The body is exposed to innocuous biological material that mimics the infectious agent, but does not lead to infection or serious disease. The immune system is stimulated to generate antigen-specific antibodies and to neutralize the antigens. This way, the immune system achieves resistance to the pathogen and acquires memory to it. In case of later contact with the intact pathogen, a strong and rapid antibody response (immunological memory) is immediately elicited preventing serious disease. The resulting immunity after vaccination persists for many years or even a lifetime. Therefore, vaccines offer an effective and relatively inexpensive protection against many diseases.

Vaccines can be generally divided into three different groups:[2b] *Live-attenuated vaccines* contain viruses or bacteria that have been modified to be less virulent, but without reducing their capability to elicit a strong immune response. The attenuation is performed by modifying or deleting specific genes that are crucial for the *in vivo* growth of the pathogen. This type of vaccine is widely used today, *e.g.*, in the treatment of tuberculosis, polio, measles, and mumps. *Inactivated whole vaccines* contain complete viruses or bacteria that have been killed by heat or exposure to a chemical and cannot cause an infection any more. This type of vaccine is no longer widely used today, as some vaccine components may be toxic and responsible for undesirable side effects. *Subunit vaccines* contain only parts of the microorganism, *e.g.*, antigenic protein or carbohydrate fragments derived from the pathogen. The antigens have either been purified from natural sources or produced synthetically. Several subunit vaccines are currently available on the market, *e.g.*, against *Haemophilus influenza* type B, Hepatitis A and B, diphtheria, and tetanus. The preparation of this type of vaccines has become possible due to improved synthetic methodologies, improved purification protocols, advances in analytical methods, and an improved molecular and biochemical understanding of the pathogens. The development of these future-generation vaccines represents an important chance in immunization. Fully or semisynthetic vaccines possess a number of advantages including a homogeneous and highly defined composition, highly reproducible biological properties, easy modification to produce more potent analogs, less or no toxic side effects, and extended shelf-lives. One important class of subunit vaccines are conjugate vaccines that consist of (fully synthetic) carbohydrate antigens covalently linked to an immunogenic carrier protein. Thus, the immune system is armed better to generate the response necessary for the protection. Preliminary immunological studies with conjugate vaccines have already shown promising results in the treatment of various diseases including cancer, HIV, leishmaniasis, and malaria.

3 Carbohydrate-Based Conjugate Vaccines

Carbohydrates – the most abundant group of natural biopolymers – are major components of all eukaryotic cell walls. They are found in the form of glycoconjugates, covalently linked to either proteins (glycoproteins) or lipids (glycolipids) that are embedded in the lipid bilayer of the cell. The carbohydrates of these glycoconjugates are exposed on the cell surface like molecular antennae, and this surrounding carbohydrate coat (glycocalix) is characteristic for a particular species, the cell type, and its developmental status. Due to the enormous structural diversity, carbohydrates constitute the most important group of biomolecules in the cell and are used to store biological information that is essential for cell adhesion processes. Glycoconjugates are mediators of many biological events including cell–cell recognition, inflammation, immunological response, metastasis, fertilization, viral replication, parasitic infection, and cell growth.[8] The carbohydrates form binding sites for many substances such as lectins, selectins, enzymes, hormones, pathogens, toxins, or antibodies. Moreover, unique cell surface carbohydrates act as biological markers of specific cells including tumor cells, viruses, and bacteria. These carbohydrates can be used for medicinal purposes[9] such as disease diagnosis and as vaccine components, since glycans are often dominant antigens.

To develop carbohydrate-based therapeutics, there is an increasing need for pure and structurally well-defined complex carbohydrates and glycoconjugates. These compounds are often only available in low concentration from natural sources. Glycoconjugate isolation, purification, and identification are difficult, as the natural isolates are highly heterogeneous and may contain contamination products. Detailed biological, biophysical, biochemical, and medicinal studies require sufficient quantities of defined pure oligosaccharides. Access to these glycans relies on chemical[10] and enzymatic[11] syntheses. The preparation of oligosaccharides presents a major challenge to synthetic chemists. Unlike oligonucleotides and oligopeptides – the two other major classes of naturally occurring biopolymers – oligosaccharides

are often not just linear, but branched molecules. A variety of available monosaccharide building blocks, the possibility of different stereochemical linkages, and
different chain lengths renders this class of biopolymers highly complex and structurally diverse. Remarkable progress[12] has been made in the area of carbohydrate
and glycoconjugate synthesis; however, traditional chemistry still remains time
consuming, technically difficult, and is only carried out by a few specialized laboratories. Further innovations are required to handle the structural complexity of glycoconjugates. In analogy to the highly efficient automated oligopeptide[13] and
oligonucleotide[14] synthesis, an automated solid-phase oligosaccharide synthesizer
has recently been introduced.[15] The development of this machine significantly
accelerates the carbohydrate assembly: Excess of reagents can be used to ensure
high coupling yields, but can be removed easily after each step by washing procedures, thus greatly reducing the number of purification steps. Together with easily
accessible glycosylating agents (such as glycosyl phosphates[10a,16] and trichloroace-
timidates[17]) and suitable protection and deprotection strategies, the automated synthesizer has become a powerful tool for the rapid assembly of linear and branched
oligosaccharides. Complex, naturally occurring carbohydrates as well as chemically modified oligosaccharide structures are now accessible, and a number of biologically relevant oligosaccharides including the blood group antigens Lewis X **1**,
Lewis Y **2**, and Le^y–Le^x nonasaccharide **3** (Figure 1)[18] have already been prepared
using automated strategies.[15]

The application of synthetic viral and bacterial oligosaccharides as immunopro-
phylactic and immunotherapeutic agents has been widely investigated. The first synthetic conjugate polysaccharide vaccine against *Haemophilus influenza* type B has
recently been approved and is now part of Cuba's vaccination program.[19] Many other
carbohydrate-based vaccine constructs are currently being developed and tested or
are undergoing clinical or preclinical trials.[20] Progress has been made in the development of synthetic carbohydrate-based conjugate vaccines against cancer[21] and
HIV[9,22] as well as in the area of parasitic infections.[23] Synthetic carbohydrates will
soon be a versatile basis for novel vaccines.

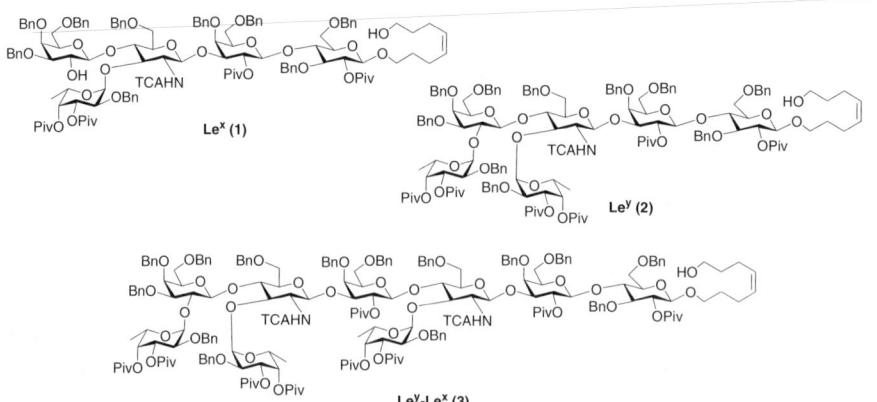

Figure 1 *Blood group determinants and tumor-associated antigens*

4 Development of a Synthetic Carbohydrate-Based Vaccine Against Malaria

4.1 Introduction

4.1.1 Disease Burden

Malaria is the most devastating tropical parasitic disease in the world with the global number of cases continuing to rise. Malaria was once spread worldwide, but was successfully eradicated in many countries during the middle of the 20th century. Together with tuberculosis and AIDS, malaria represents one of the major public health problems in more than 90 countries, mainly in tropical and subtropical regions of the world. Approximately 40% of the world's population (about 2400 million people) live with the risk of contracting malaria. Each year, malaria infects 5–10% of humanity and causes more than 300 million clinical cases.[24] Mortality due to malaria is estimated to be between one and two million deaths per year.

More than 90% of the malarial infections and deaths occur in sub-Saharan Africa, mostly affecting young children under the age of five. Malaria represents one major cause of death in children worldwide and kills one African child every 20 seconds. In absolute numbers, 3000 children die of malaria each day, mainly in rural areas with poor access to health services. Pregnant women and their unborn children are particularly exposed to malaria, a major cause of perinatal mortality, low birth rates, and maternal anemia. Infected children typically suffer fever, severe anemia, vomiting, headache, and flu-like symptoms. If untreated, malaria rapidly progresses to convulsions, coma, and death. Many children surviving malaria are chronic victims and suffer an average of six malarial bouts a year. Repeated episodes affect the child's growth and may lead to permanent neurological damages including learning impairments or brain damage.

Malaria is considered a disease of poverty as well as a cause of poverty and produces immense direct and indirect costs. Recently, it has been shown to have a major influence on the economical development of endemic areas, reducing the annual gross domestic product in sub-Saharan African countries by 1–4%.[25]

4.1.2 Malaria Infection

Until the late 19th century, it was generally believed that malaria (from Italian mal aria, "bad air") was caused by the poisonous vapor or miasma released from the swamps. In 1880, scientists finally discovered the real source of malaria, a unicellular (protozoal) parasite of the *Plasmodium* genus that is transmitted by mosquito bites. Four different *Plasmodium* species that infect humans and cause distinct disease patterns are known today: *P. falciparum* (malaria tropica) and *P. vivax* (malaria tertiana) are the most frequent types of human malaria, whereas *P. malariae* (malaria tertiana) and *P. ovale* (malaria quartana) are less common. Most lethal malaria infections are caused by *P. falciparum*.

Plasmodium parasites reproduce and mature in host tissues and have a highly complex lifecycle. When an infected *Anopheles* mosquito takes its blood meal, the insect-stage parasites (sporozoites) are transferred into the human blood. Inside the human host, the sporozoites invade the liver cells (hepatocytes) and establish an initial asymptomatic infection. The sporozoites reproduce asexually and mature into merozoites, which are then released from the hepatocytes and infect the red blood cells (erythrocytes). Within the erythrocytes, the parasites develop through different stages (trophozoites, schizonts) and then multiply, eventually resulting in lysis of the erythrocytes. New parasites are released into the blood stream immediately invading new erythrocytes. Clinical symptoms of malaria appear about 9–14 days after the infectious mosquito bite and arise from cycles of parasites bursting from infected cells and reinfecting new cells. Malaria kills humans by destroying red blood cells (anemia) and/or by clogging the prevenous capillaries that carry blood to the brain (cerebral malaria) or other vital organs. Parasites of the species *P. vivax* and *P. ovale* are able to persist in dormant stages (hypnozoites) in the liver for years and result in clinical relapses in regular intervals.

A small amount of schizonts develops into sexual stages (gametocytes) that are transferred to the mosquito when it bites an infected individual. Inside the mosquito the male and female gametocytes undergo a series of changes, until they reach the sexual stage where they can again infect a human host.

4.1.3 Drug Treatment of Malaria

The most common treatments that are used against malaria even today have their roots in a natural plant called cinchona. Exploitation of the bark of cinchona trees in the treatment of malarial fevers started in the beginning of the 17th century. During the 18th century, this bark became the main anti-malarial agent showing remarkable results.[26] In 1820, two French pharmacists, Pelletier and Caventou, isolated the alkaloid quinine (**4**, Figure 2) as the active ingredient from the cinchona bark. Because of its therapeutic value, this compound successively substituted the crude bark for the treatment of malaria. In 1891, Guttmann and Ehrlich began to search for a synthetic drug and reported the first use of a synthetic compound in humans.[27] They observed that methylene blue (**5**, Figure 2) was selectively taken up

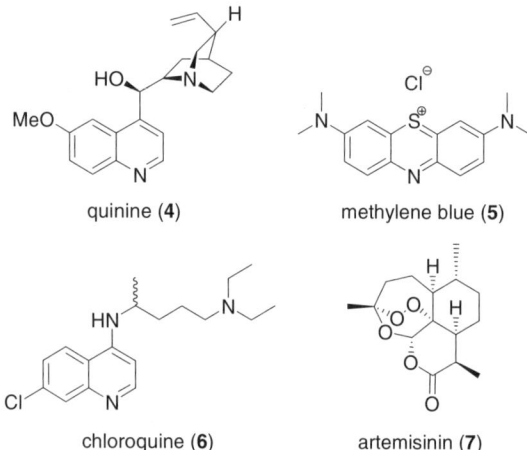

quinine (**4**)

methylene blue (**5**)

chloroquine (**6**)

artemisinin (**7**)

Figure 2 *Examples for anti-malarial drugs*

by the parasites in microscopic specimens and successfully cured two malaria patients with this compound.

During World War I, many different substitutes for quinine were prepared and tested, but no replacement was found. In 1934, chloroquine (**6**, Figure 2) was synthesized and was shown to be a highly active, prophylactic anti-malarial agent. Chloroquine was initially considered to be too toxic, but eventually became the standard and most effective anti-malarial drug. At the end of the 1950s, first cases of resistance to chloroquine occurred and chloroquine-resistant strains of *P. falciparum* are now common in all endemic areas. Since then, numerous other drugs with anti-malarial properties have been developed.[28] A limited number of drugs are currently available on the market, but variants of *P. falciparum* have developed resistance to some of these drugs. Quinine, in combination with other anti-malarial agents, has regained attention as the drug of choice. Artemisinin (**7**, Figure 2) and its derivatives, derived from traditional Chinese medicine, have exhibited potent anti-malarial activity and no resistance has been observed yet.[29] These sesquiterpenes were isolated as an active ingredient from the sweet wormhood *Artemisia annua* that has been used for a long time in China for the treatment of hemorrhoids and fever.

Due to the increasing prevalence of malaria and the emerging resistance against conventional drugs, the search for new anti-malarial agents remains important. Additional factors contributing to the rapid spread of malaria are the collapse of control programs, the resistance of the *Anopheles* mosquito against the insecticide dichlorodiphenyltrichloroethane (DDT), the migration of refugee populations, and climatic and environmental changes. Alternative medications are almost not affordable for people in endemic countries. Therefore, an effective and cheap vaccine could help to solve the malaria problem and could be integrated into existing immunization programs of children in endemic areas against common childhood diseases.

4.2 Anti-Malarial Carbohydrate-Based Vaccines

In contrast to other diseases, humans are able to develop only partial immunity to malaria even after one or two infections.[30,31] Therefore, the development of anti-malarial vaccines is extremely challenging. Impeding factors include the complex biological lifecycle of the parasites, the ability of the parasite to alter itself, and the immunological non-responsiveness of certain individuals to proteins, just to name a few.[32a] Four general types of vaccine candidates, all aiming at the prevention or reduction in morbidity and mortality, are currently being explored.[3] Instead of inducing sterile immunity where no parasites are present in the vaccinated human, all candidates are designed to induce only clinical immunization: The vaccinated individuals do not suffer malarial bouts, but still have parasitic levels in their blood. Almost all candidates contain cell surface antigens present during one of the three developmental stages of the *Plasmodium* parasites. *Sporozoite vaccines* are focused on the sporozoite and liver stages and prevent infection by either blocking the invasion of the liver cells or by destroying infected liver cells. *Erythrocytic vaccines* are designed to decrease the parasite density in the blood (merozoite stages) in order to reduce severity and duration of the disease. *Transmission-blocking vaccines* are directed at raising antibodies against the gametocyte stages of the parasite to reduce the infectivity of the mosquitoes and block further parasite development. *Anti-toxin vaccines* target the main cause of host pathology, the malarial toxins. Instead of killing the malarial parasites, the immune system is primed to attack the toxin causing lethality, and thereby reduces the malarial morbidity and mortality.[31] To develop an anti-toxin vaccine, the understanding of the molecular basis of malarial pathophysiology as well as the identification of parasitic toxins that contribute to the disease are essential. Examples for anti-toxin vaccines are the highly effective toxoid vaccines against tetanus and diphtheria, which protect humans against the most injurious consequences by targeting the toxins causing the disease.[33] The different types of vaccine candidates – including recombinant protein or DNA vaccine constructs, synthetic peptides, dimeric proteins, and viral-vectored constructs – have been tested in clinical or preclinical trials; however, to date no effective anti-malarial vaccine is available on the market.[32]

4.2.1   Identification of the Toxin

Already in 1886, Golgi[34] showed that malarial bouts are synchronous to the developmental cycle of the blood stage parasites and proposed the existence of a malarial toxin as the causative agent of malarial pathogenesis. About 90 years later,[35] the main mediator of malaria, the tumor necrosis factor-α (TNF-α), has been identified and is now widely used as a biochemical marker of malarial endotoxin activity. It was shown that macrophages from malaria infected mice produce high levels of TNF-α by stimulation with different agonists.[35] Crude extracts of rodent malaria parasites can also induce macrophages to secrete TNF-α *in vitro*.[36] It is now appreciated that malarial toxins are responsible for many severe pathological consequences of malaria and play an important role in the initiation and maintenance of the inflammatory cascade.[37] They activate macrophages and vascular endothelial cells, which leads to the production of mediators such as nitrous oxide, TNF-α, and intercellular adhesion molecule-1 (ICAM-1).[38] A detailed discussion of the mechanisms of malarial pathogenesis and of malarial immunity has been published recently.[31]

In 1992, Playfair[39] proposed that the toxin should be a phospholipid, and Schofield extracted and tentatively assigned a structure to the putative toxin from the malarial parasite *P. falciparum*.[40] After injection of the isolated material into mice, the rodents showed typical malaria symptoms. The structural assignment[41] of the toxin (**8**, Figure 3) has finally been confirmed by chemical synthesis.[42] The malarial parasite *P. falciparum* expresses a large amount of glycosylphosphatidylinositols (GPIs) anchored to proteins on its cell surface. The GPIs constitute over 95% of the total carbohydrate modification of *P. falciparum*,[43] reflecting the low levels of *N*- and *O*-glycosylation in malarial parasites.[44] The lipid chains of **8** (R^1, R^2, R^3) show high structural diversity, leading to remarkable differences in the biological activity of the GPIs.[45] Deacylation using chemical or enzymatic hydrolysis makes the carbohydrate moiety non-toxic, demonstrating that the combination of lipid and carbohydrate domains is necessary for the biological activity in host tissues.[38,40]

Malarial GPIs play a major role in the initiation and maintenance of the malarial inflammatory cascade. Initial studies, revealing that the malarial GPI can elicit an immune response in both rodents and humans, suggest that this compound has excellent properties for the development of an anti-toxin vaccine.[46] Intense efforts

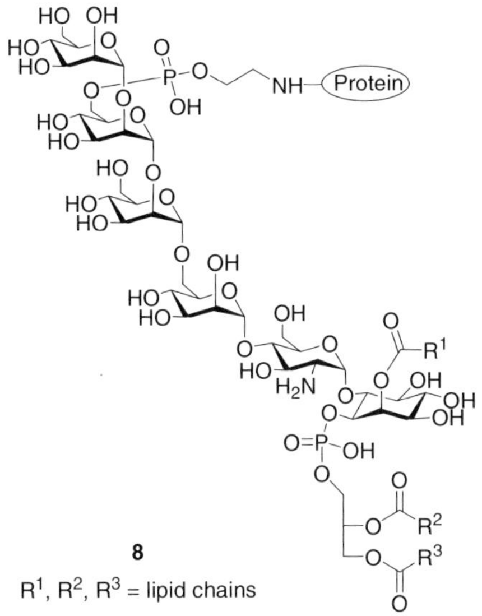

8

R^1, R^2, R^3 = lipid chains

Figure 3 *Structure of the toxin isolated from P. falciparum*

are now underway to determine the minimal immunogenic structure and the structure–activity relationship by testing defined GPI structures in animals.[47]

4.2.2   Synthesis and Testing of Carbohydrate-Based Vaccine Constructs

Glycosylphosphatidylinositol (GPI) anchored proteins are one of the most complex classes of biopolymers in nature, combining lipids, phosphates, carbohydrates, and proteins. GPIs serve to attach proteins or glycoproteins onto eukaryotic cell membranes and are involved in many biological and physiological processes. Due to the structural complexity of the GPI anchors, GPIs have been targets of intense studies, and various elegant synthetic strategies have recently been developed.[48]

The first chemical synthesis of the non-toxic[38,40] malarial GPI glycan **9** without the lipid residues was accomplished using a linear solution-phase approach with six different building blocks.[42] A more rapid assembly of **9** was achieved by combining automated solid-phase synthesis and solution-phase fragment coupling (Scheme 1).[49] GPI **9** was derived from tetramannoside **10**, readily accessible via automated solid-phase synthesis, and pseudodisaccharide **11** containing the challenging α-linkage between the inositol and the glucosamine unit that currently prevents a fully automated approach.

The automated solid-phase synthesis of tetrasaccharide **10** was performed with octenediol-functionalized Merrifield resin **12** and four readily available monosaccharide building blocks **13** to **16** (Scheme 2). Using double glycosylations with catalytic amounts of trimethylsilyl triflate (TMSOTf) to guarantee high coupling yields and a single deprotection step, the tetramannoside was obtained after cleavage from the solid support and purification by HPLC. Conversion of the anomeric pentenyl derivative into glycosyl donor **10** was smoothly achieved in two steps using standard conditions.

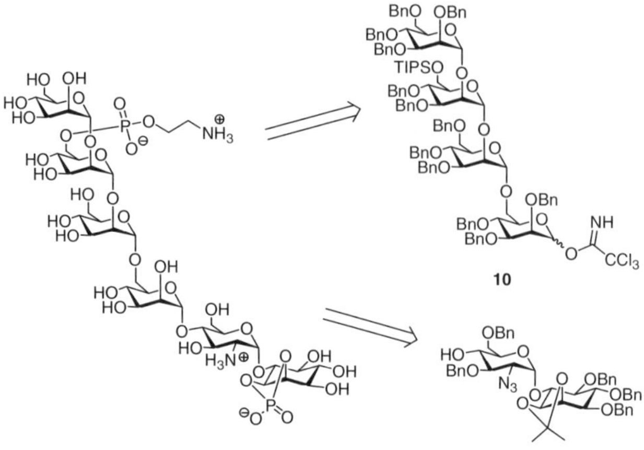

Scheme 1

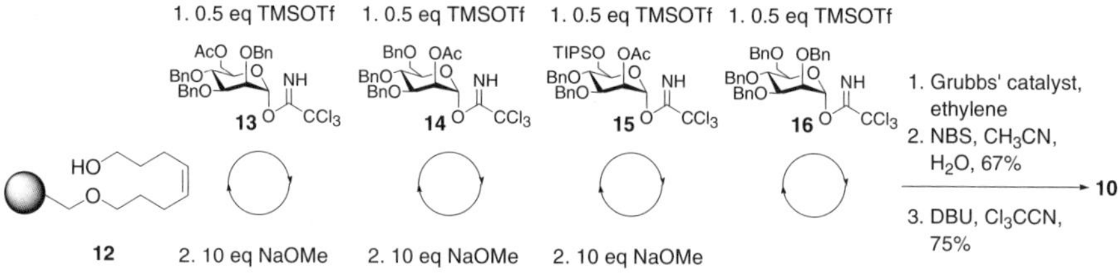

Scheme 2

Solution-phase coupling of trichloroacetimidate **10** with pseudodisaccharide **11**[49] in the presence of TMSOTf gave fully-protected pseudohexasaccharide **17** in modest yield (Scheme 3). Both acid-labile groups were removed, and the primary hydroxyl group was selectively silylated with *tert*-butyldimethylsilyl chloride (TBSCl). The cyclic phosphate was installed by treatment with methyldichlorophosphate in pyridine. Removal of the *tert*-butyldimethylsilyl (TBS) group followed by reaction with phosphoramidite **18** and oxidation gave the corresponding bisphosphate, which was treated with DBU to remove the β-cyanoethoxy group. Subsequent global deprotection using Birch conditions afforded the desired malarial GPI glycan **9**.

To prepare an immunogen, the synthetic pseudohexasaccharide **9** was reacted with 2-iminothiolane (**19**) to introduce a sulfhydryl linker at the ethanolamine unit (Scheme 4). The resulting glycan was then conjugated to maleimide-activated keyhole limpet haemocyanin (KLH) as a carrier protein in a molar ratio of 191:1, and this conjugate was used to immunize mice.

Initial studies[42] showed that the synthetic malarial GPI conjugate **20** was immunogenic in rodents producing exactly the same immune response as the natural product. Immunized mice generated high titer immunoglobulin-γ (IgG) to the

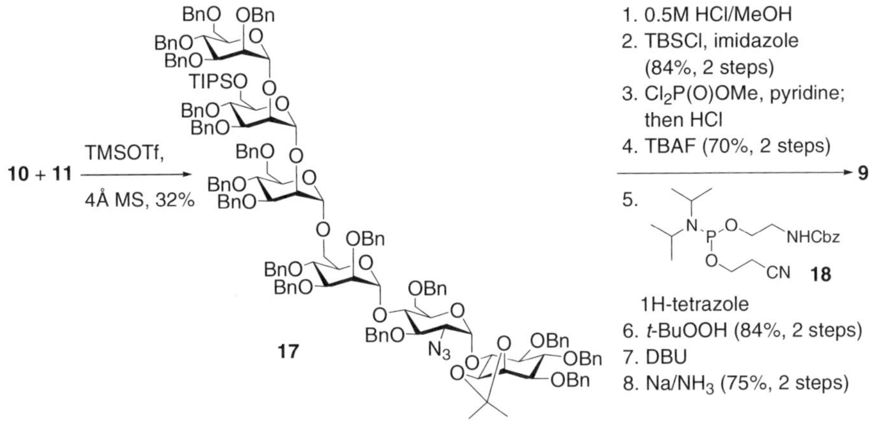

Scheme 3

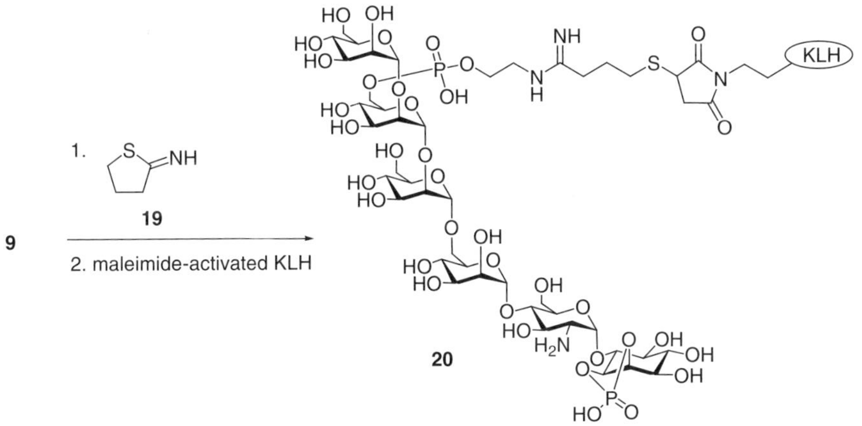

Scheme 4

synthetic GPI conjugate, whereas no reactivity was found in pre-immune sera or in animals receiving sham-conjugated KLH. The anti-glycan IgG antibodies crossreacted with *P. falciparum* trophozoites and schizonts, as detected by an immunofluorescence assay, but failed to bind to uninfected erythrocytes. In contrast to malarial GPI, the core glycan of mammalian GPI is significantly modified,[50] thus explaining the lack of reactivity. Antibodies from mice immunized with KLH-glycan were also able to neutralize the TNF-α level from macrophages induced by crude *Plasmodium* extracts.

The murine *Plasmodium berghei* ANKA malaria model is the best available model for certain aspects of lethal pathogenesis and corresponds well to several aspects of human severe and cerebral malarial syndromes. C57BL6/J mice, treated either with KLH-glycan **20** or with KLH-cysteine in Freund's adjuvant, were challenged with *P. berghei* ANKA. All infected sham-immunized and naive control mice died within five to eight days with cerebral syndromes and exhibited severe neurological dysfunctions. In contrast, mice immunized with the synthetic KLH-glycan were significantly protected against severe malaria showing a reduced death rate. While between 58 and 75% of the vaccinated mice survived until day 12, the survival rate for sham-immunized mice was only 0–9%. The level of parasites, however, does not differ significantly in immunized and control mice, indicating that prevention of fatality occurs without causing death of the parasites. These data revealed that synthetic, non-toxic GPIs conjugated to a protein serve indeed as anti-toxin vaccine candidates against malaria providing significant protection against malarial fatality and pathogenesis.

Based on these initial studies, second-generation vaccine candidate **21** (Figure 4) has been synthesized using a convergent and modular solution-phase approach with minimal late-stage modifications and robust chemistry.[51] This approach is less rapid compared to the automated assembly, but allows for the scale-up to procure sufficient quantities of GPI for preclinical and clinical trials. A [4+2] coupling between tetrasaccharide **22** and pseudodisaccharide **23**, both readily prepared on large scale in solution, has been performed as the key step. The obtained hexasaccharide was converted to the phosphate diester as described above. Global deprotection followed by conjugation to maleimide-functionalized bovine serum albumin (BSA) gave the new model vaccine, which is a useful substrate for ELISA tests for anti-GPI immunoglubolins in naturally immune and vaccinated humans.[52]

The preparation of a variety of different GPI oligosaccharides (Figure 5) that differ in the number of mannose units has been published recently using the same flexible synthetic strategy as described above.[53] These compounds will serve as molecular tools for the examination of the biosynthesis, antigenicity, and serology of GPIs. The direct incorporation of a thiol group into the inositol moiety allows for the

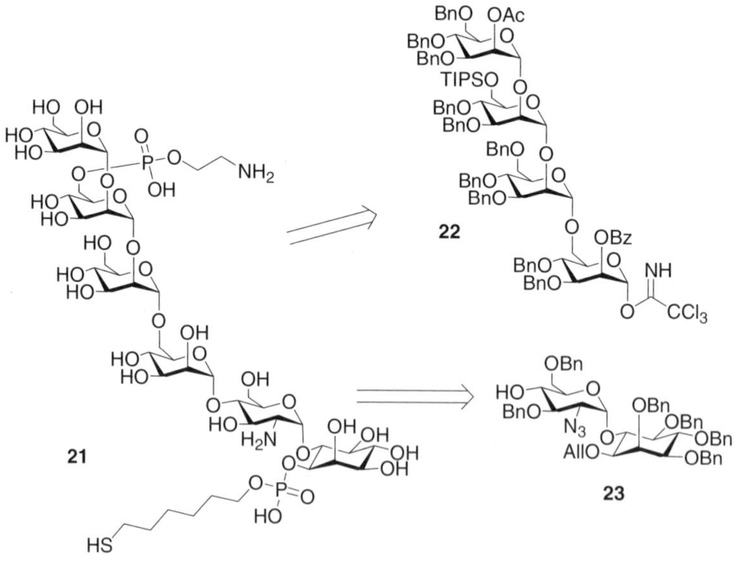

Figure 4 *Second generation anti-malarial vaccine candidate*

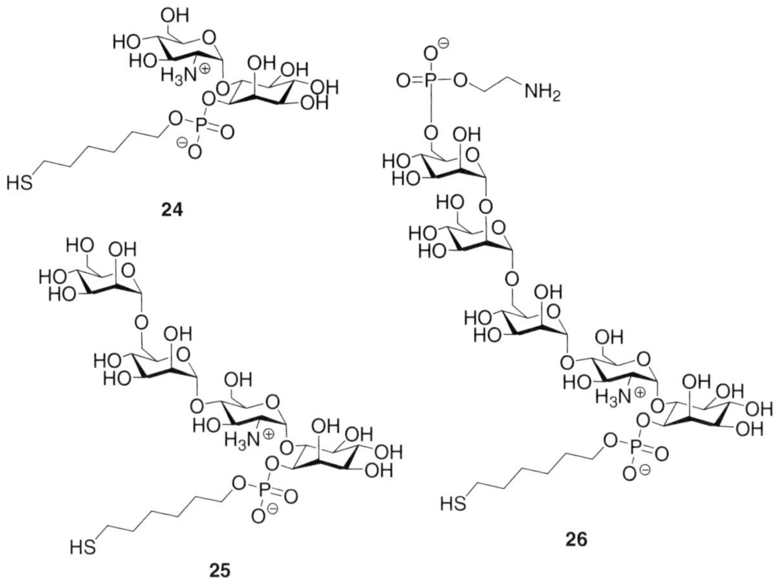

Figure 5 *Examples for synthetic GPI oligosaccharides*

rapid conjugation of these glycans to carrier proteins and for convenient screening on carbohydrate chips.[54] The synthetic molecules are useful candidates to investigate the substrate specificity of GPI biosynthetic enzymes. They will be used as anti-toxin vaccine candidates and will serve to reveal the minimal structural requirements necessary to elicit a protective immune response. Furthermore, they will find use as molecular probes to map epitopes of human anti-malarial antibodies and for other biological studies and will provide the basis for first detailed structure–activity relationship studies between GPI toxins and anti-malarial antibodies.

A lipidated and phosphorylated malarial GPI model pentasaccharide **27**[55] (Scheme 5) has been synthesized using *n*-pentenyl orthoesters[56] (NPOEs) **28**, prepared from D-mannose,[57] and α-D-glucopyranoside **29**. The *myo*-inositol building block **30** has been generated from **29** following earlier routes.[58]

Coupling of **30** with orthoester **28a** by treatment with ytterbium triflate [Yb(OTf)$_3$] and *N*-iodosuccinimide (NIS) gave pseudodisaccharide **31** that was converted to glucosamine **32** by replacement of the benzoate with triflic anhydride and subsequent azide displacement (Scheme 6).[59] Consecutive reaction of **32** with different NPOEs **28** was followed by several protecting group manipulations to yield the desired pseudopentasaccharide **33**.

Attachment of the phosphoethanolamine followed by removal of the cyclohexylidene group afforded diol **34** (Scheme 7). **34** was converted to a cyclic orthoester, which was subsequently opened by Yb(OTf)$_3$ to give the two regioisomeric acylated glycans **35a** and **35b** in a disappointing ratio of 1:2. **35b** was reacted with diacteylated glycerylphosphoamidite **36** to give the desired lipidated and phosphorylated malarial model glycan **27**.

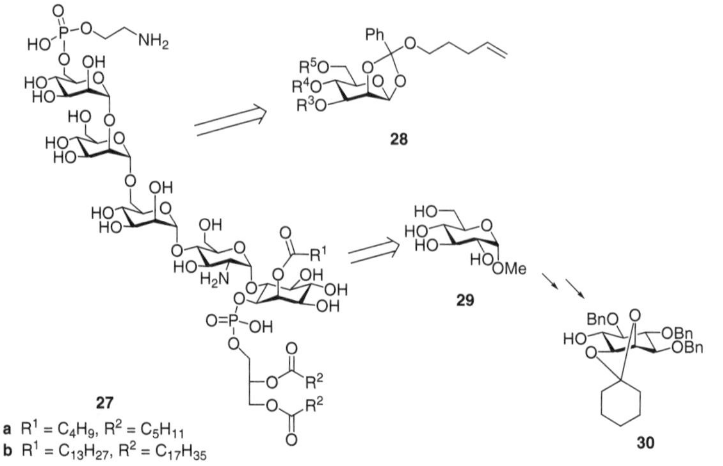

Scheme 5

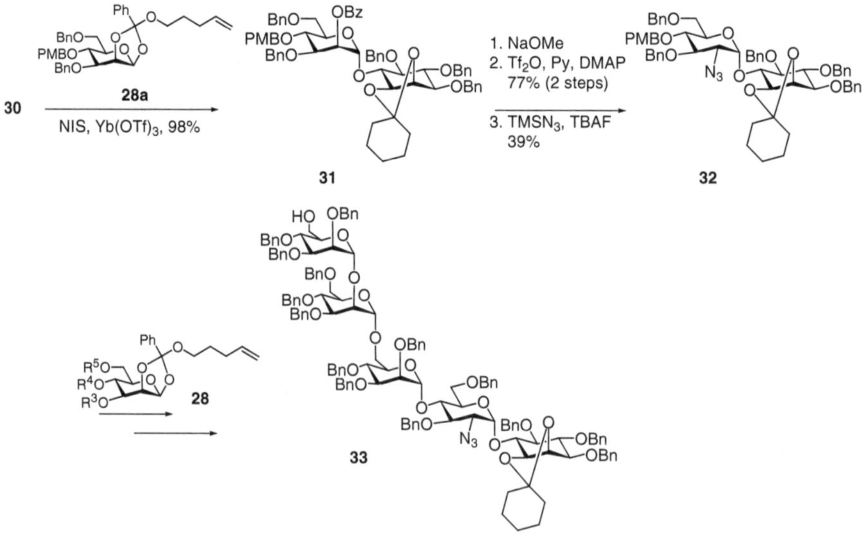

Scheme 6

Using the [4+2] glycosylation approach described above, the fully lipidated malarial GPI **40** has been synthesized for the first time by combining tetramannoside **22**
and pseudodisaccharide **37** (Scheme 8).[60] Replacement of the ester functionalities by
benzyl ethers furnished the key hexasaccharide that was subsequently decorated with
the lipid, the phosphate ethanolamine and the phospholipid. Careful evaluation of the
order of functionalizations showed that lipidation with palmitic acid followed by

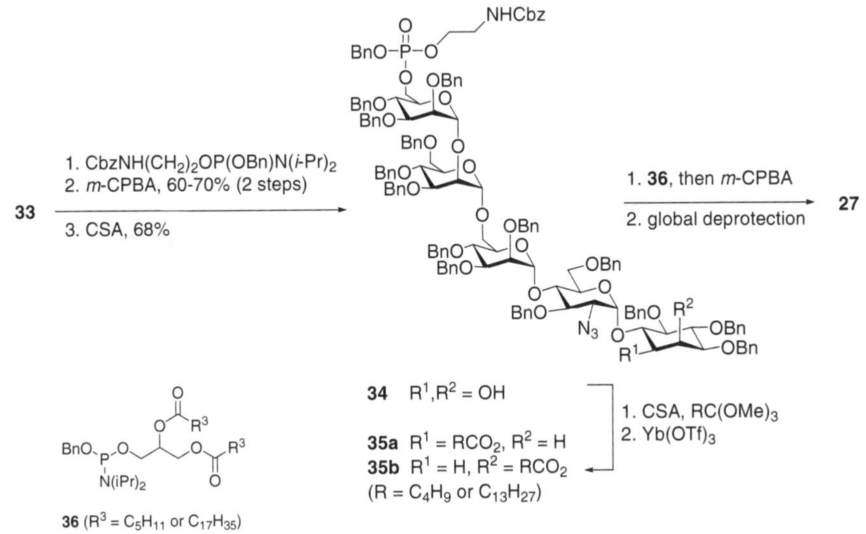

Scheme 7

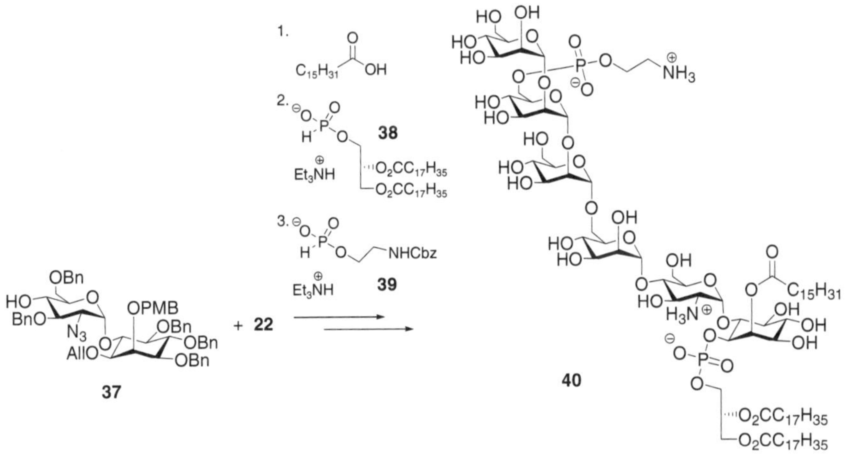

Scheme 8

installation of the phospholipid **38** prior to introduction of the phosphate ethanolamine **39** gave the best results. Finally, global deprotection by hydrogenolysis with Pearlman's catalyst and purification yielded GPI glycan **40**, which is currently used to study malarial pathogenesis and aspects of fundamental immunology.

In summary, a variety of efficient synthetic methods for the preparation of malarial GPI glycans has been developed recently. Using these strategies, sufficient quantities of pure oligosaccharides can now be produced to support biochemical, biological, immunological, and medicinal investigations. A better understanding of malarial pathogenesis will eventually help to discover an effective anti-toxin vaccine.

5 Development of a Carbohydrate-Based Vaccine Against Leishmaniasis

5.1 Disease Burden

Leishmaniasis, another widespread tropical disease, is currently endemic in 88 countries on four continents.[3] About 350 million people live at risk of leishmaniasis with a worldwide prevalence of 12 million clinical cases annually. The impact of leishmaniasis on public health has been greatly underestimated so far. It is estimated that about 1.5 to 2 million new infections and about 60,000 deaths occur each year.[61] Leishmaniasis is on the edge of becoming endemic in North America.[62]

The exact disease burden of leishmaniasis remains unknown. However, its economic and social impact is immense. The current spread of leishmaniasis corresponds to environmental changes such as new irrigation schemes, deforestation, building of dams and wells, and to the massive migration of non-immune people to endemic areas. Moreover, HIV/leishmaniasis coinfections are emerging as an extremely serious, new disease[63] producing cumulative deficiency of the immune response, as the Leishmania parasites and HIV destroy the same cells.

Leishmaniasis is caused by several different species of protozoan parasites including *Leishmania donovani*, *L. tropica*, *L. infantum*, *L. major*, and *L. mexicana* and is transmitted by the bite of an infected female *phlebotomine* sand fly. Various forms of human leishmaniasis with a wide range of clinical symptoms and devastating consequences exist: *Visceral leishmaniasis* (kala azar) is the most lethal and severe form of leishmaniasis. The disease is characterized by regular bouts of fever, substantial loss of weight, anemia, swelling of the spleen and the liver, and results in death if untreated. *Mucocutaneous leishmaniasis* (espundia) produces lesions in the face that lead to partial or total destruction of the mucous membranes of nose, mouth, and throat cavities as well as of the surrounding tissues. *Cutaneous leishmaniasis* is the most common form of the disease with 50–75% of the new cases, and is characterized by a large number of skin ulcers on the exposed part of the body. It causes serious disability leaving the patient permanently scarred.

Leishmaniasis patients are usually treated with expensive antimony drugs, which require a lengthy therapy and show toxic side effects. Moreover, some Leishmania parasites have recently become resistant to some of these drugs. Leishmaniasis has been extensively studied in murine models, but low parasite levels render the diagnosis difficult. The parasite inhabits the macrophages, the part of the immune response that is designed to kill invading organisms. Therefore, the mechanism of infection represents a challenge for therapeutic strategies. Several vaccine candidates are currently explored including whole killed antigens as well as surface antigens; however, no effective vaccine has been developed to date.[3,64]

5.2 Carbohydrate-Based Vaccine Constructs Against Leishmaniasis

With respect to glycoconjugate biosynthesis, Leishmania parasites are one of the best-studied protozoa. The parasites express large amounts of a complex oligosaccharide – a lipophosphoglycan (LPG) – on their cell surface. The LPG is composed of three parts, a GPI anchor, a repeating phosphorylated disaccharide fragment, and different cap oligosaccharides (Figure 6). Different *Leishmania* species can be distinguished by minor modifications in their backbone and the number of repeating units. Preliminary structure-activity relationship studies have been performed leading to the following assignments:[65] the GPI moiety exhibits several biological functions. Its most important role is to attach the oligosaccharide to the plasma membrane. The repeating phosphorylated disaccharide unit likely forms a macromolecular diffusion barrier preventing the binding of the host's antibodies to the LPG epitopes. The cap glycan is presumed to attach the parasite to the digestive tract of the sand fly. Additionally, the cap is thought to contain the epitope responsible for the recognition by the mammalian host macrophage.

It was shown that LPGs are disease-promoting antigens.[66] Mice vaccinated with preparations of parasitic LPGs were able to develop protective immunity[67] and to generate antibodies to LPG.[68] LPG-conjugates might be useful transmission-blocking vaccines against leishmaniasis, and therefore, a number of synthetic strategies to prepare different sections of the *Leishmania* LPG have been introduced. The GPI heptasaccharyl *myo*-inositol part has been prepared using a convergent synthetic strategy.[69] Other synthetic LPG oligosaccharides[70] have been identified as substrates for *Leishmania* transferases.[71] The branched tetrasaccharide cap, which is exclusively found on the cell surface of *Leishmania* parasites, has recently been accomplished as an ideal target for the development of an anti-leishmaniasis vaccine candidate. The first chemical synthesis[72] of this compound was accomplished using the *n*-pentenyl glycoside (NPG) protocol in linear and convergent approaches. With respect to product recovery, the convergent synthesis (Scheme 9) is the method of choice, giving the fully protected cap saccharide **44** in 69% yield (compared to 35% for the linear approach). A similar approach has been used incorporating a disaccharide trichloroacetimidate building block.[73] This strategy allows for isotope labeling of the glycan, which can be used in macrophage acceptor binding and immunological studies.

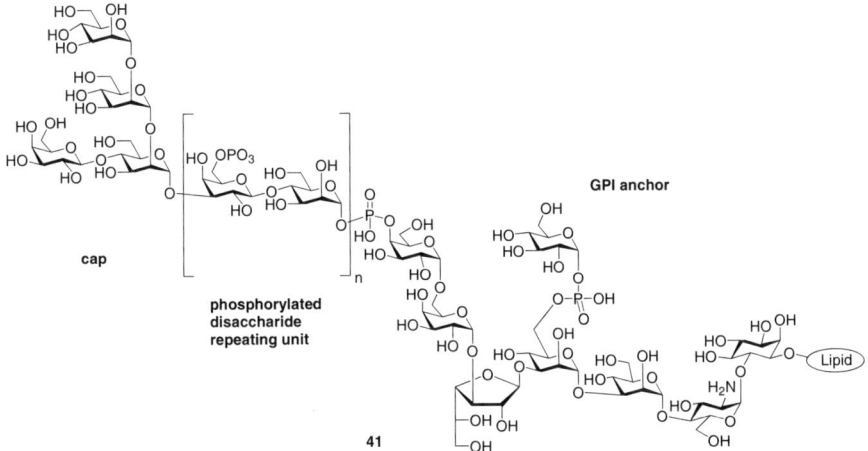

Figure 6 *Structure of lipophosphoglycans (LPGs) ubiquitous found on Leishmania parasite cell surfaces*

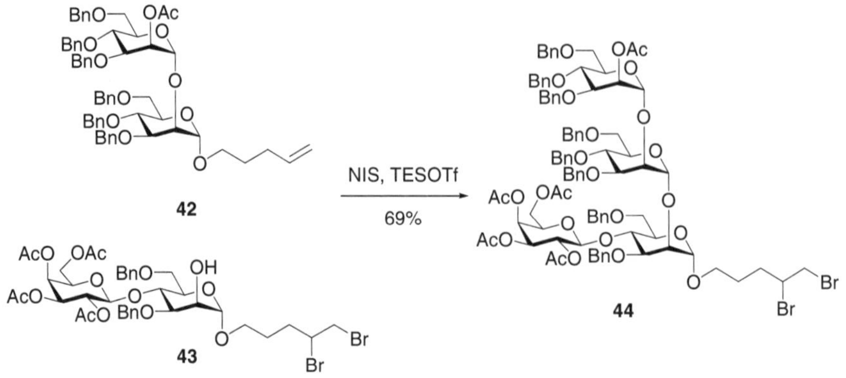

Scheme 9

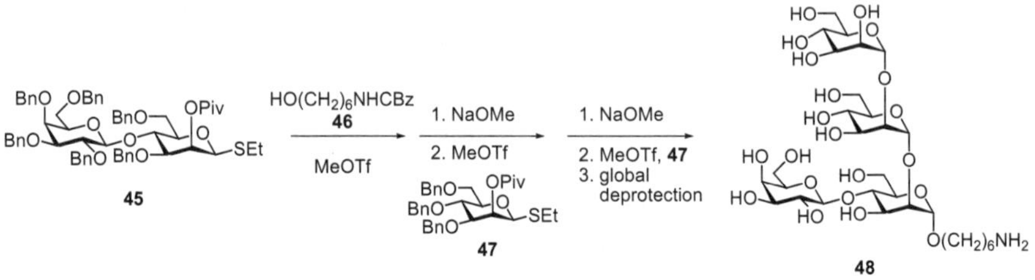

Scheme 10

A linear, efficient solution-phase synthesis of the tetrasaccharide has been accomplished involving thioglycoside building blocks (Scheme 10). Key disaccharide **45**, prepared in several steps from hexabenzyl lactal,[74] was coupled to N-protected aminohexanol **46** in the presence of methyl triflate (MeOTf). Removal of the pivaloyl group followed by reaction with thioglycoside donor **47** gave the trisaccharide that was further elongated by repetition of the deprotection/coupling steps. Global removal of the protecting groups led to tetrasaccharide **48**.[75]

The first automated solid-phase synthesis[76] of the LPG tetrasaccharide utilized the stepwise incorporation of three building blocks. The central mannose building block **49** was equipped with different, orthogonal acyl protecting groups to achieve branching. Trichloroacetimidate **49** was attached to the octenediol-functionalized Merrifield resin **12** under TMSOTf activation (Scheme 11). Selective removal of the levulinoyl group with hydrazine, followed by glycosylation with phosphate **50** enabled the installation of the galactose unit. Removal of the acetate and coupling to mannose trichloroacetimidate **51** was performed twice to furnish the resin-bound tetrasaccharide. Cleavage from the solid-support by Grubbs' catalyst in an ethylene atmosphere and purification by HPLC gave fully protected tetrasaccharide **52**.

The preparation of synthetic and semisynthetic immunogens was achieved based on the linker-equipped tetrasaccharide **48** (Scheme 12).[75] The primary amine serves to attach the glycan to the immunostimulator Pam₃Cys to furnish the desired vaccine construct **53**. Semisynthetic vaccine **54** was prepared by condensation of **48** with

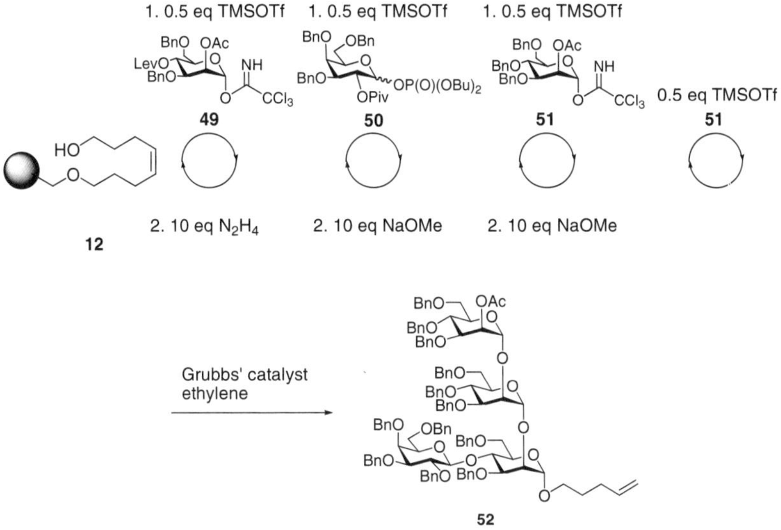

Scheme 11

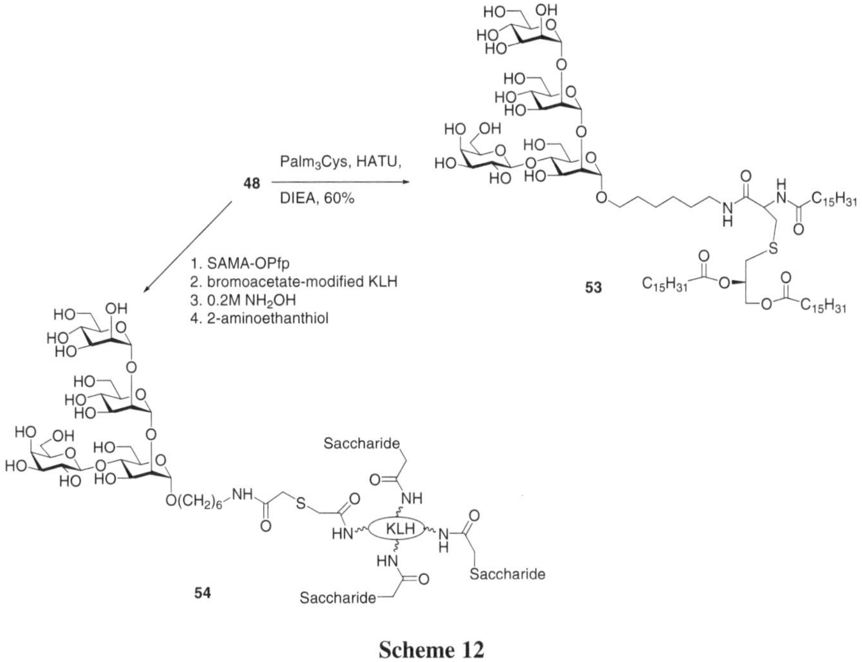

Scheme 12

S-acetylmercaptoacetate pentafluorophenyl ester (SAMA-OPfp) and subsequent conjugation to bromoacetate-modified KLH. Both constructs have entered initial immunological experiments in bulb-C mice demonstrating their ability to elicit strong B- and T-cell responses.[77] Further studies are currently performed to evaluate their role as potential vaccine candidates against leishmaniasis.

6   Conclusions and Outlook

The fundamental contributions of complex oligosaccharides and glycoconjugates to important biological processes are now understood in significant molecular detail. Specific types of carbohydrates are expressed on the cell surface of microorganisms and can be used as the basis for therapeutics, pharmaceuticals, diagnostics, and vaccines. It is anticipated that glycans are often dominant antigens, and carbohydrate-based vaccine constructs have already been shown to successfully protect humans against various diseases.

The isolation, purification, and identification as well as the preparation of oligosaccharides represented a major challenge to biologists and chemists. Recent advances in carbohydrate synthesis have led to innovative and efficient strategies facilitating the chemical and enzymatic preparation of diverse glycoconjugates. The introduction of an automated solid-phase synthesizer has greatly accelerated the carbohydrate assembly. Linear and highly branched molecules are now accessible on this machine, and its versatility has been demonstrated by preparing a series of biologically relevant oligosaccharides. These chemical approaches help to generate sufficient quantities of well-defined, homogeneous carbohydrate antigens for biological and immunological studies.

Parasitic infections, including malaria and leishmaniasis, are major public health problems, mainly afflicting humans in developing countries. The increasing prevalence of parasitic infections and the emergence of resistance to conventional drugs have led to a concerted commitment of scientific and economic communities to establish efficient health care structures. Protection generated by vaccination is a great success of medicine and has prevented more deaths than any other active medical intervention so far. Vaccines are the most effective and cost-efficient tools for the eradication of devastating infectious diseases, thus improving the quality of life. The role of vaccines will become increasingly important in the future, as new targets such as new infectious diseases, immunotherapy of tumors, chronic infections, autoimmunity and allergies emerge.

Vaccine candidates against malaria and leishmaniasis are currently being investigated. Advances in organic chemistry have led to completely synthetic carbohydrate-based antigens that have been used for the preparation of vaccine conjugates. These constructs are currently evaluated in immunological tests to demonstrate their ability to elicit an immune response. Immunization of mice with a GPI malarial-KLH vaccine construct has shown very promising results in initial experiments leading to an increased survival rate of vaccinated mice compared to sham-immunized mice. The new malaria vaccine candidate is currently progressing through preclinical studies towards clinical evaluation.

Despite these new technologies, the challenges in the development of vaccines are immense and much remains to be explored. The progress of a new vaccine through development, testing, and registration is slow and takes on average between 10 to 15 years with costs between 200 and 500 million dollars. The long-term goal is to offer better safety, efficiency, and delivery methods with lower production costs, finally resulting in an efficient distribution and better availability of vaccines especially in developing countries. Improved synthetic techniques support the preparation of analoges of various carbohydrate antigens that may facilitate the construction of "ideal" glycoconjugate-based vaccines. It is anticipated that these new breakthroughs may eventually lead to the development of effective vaccines or to the improvement of existing ones in coming years. The creative interplay of target-oriented synthesis and immunology will hopefully benefit in controlling or eradicating devastating parasitic and other infectious diseases.

Acknowledgments

We are grateful to all present and past members of the Seeberger group as well as our collaborators contributing to the results reported herein. We thank the ETH, the Swiss National Science Foundation (Grant No. 200021-101593, "Automated Solid-Phase Synthesis of Oligosaccharides") and the Deutsche Forschungsgemeinschaft (DFG, Emmy Noether Postdoc Fellowship to A.H.).

References

1. A.M. Silverstein, *A History of Immunology*, Academic Press, San Diego, 1989.
2. (a) M.F. Bachmann and M.R. Dyer, *Nat. Rev. Drug Discov.*, 2004, **3**, 81. (b) R. Rappuoli, Vaccination of Humans, *Nature Encyclopedia of Life Sciences*, Nature Publishing Group, London, 1999.
3. State of the Art of New Vaccines Research & Development, Initiative for Vaccine Research, World Health Organization, Geneva, 2003.
4. (a) P. Moingeon, J. Almond and M. de Wilde, *Curr. Opin. Microbiol.*, 2003, **6**, 462. (b) M.P. Kieny, J.-L. Excler and M. Girard, *Am. J. Public Health*, 2004, **94**, 1931.
5. R.M. Zinkernagel and *Annu. Rev. Immunol.*, 2003, **21**, 515.
6. H. Kumar, D. Malhotra, S. Goswami and R.N.K. Bamezai, *Crit. Rev. Microbiol.*, 2003, **29**, 297.
7. S.H.E. Kaufmann and D. Kabelitz, *Methods Microbiol.*, 2002, **32**, 1.
8. (a) R.A. Dwek, *Chem. Rev.*, 1996, **96**, 683. (b) Y.C. Lee and R.T. Lee, *Acc. Chem. Res.*, 1995, **28**, 322. c) A. Varki, *Glycobiology*, 1993, **3**, 97.
9. K.J. Yarema and C.R. Bertozzi, *Curr. Opin. Chem. Biol.*, 1998, **2**, 49.
10. Selected reviews: (a) E.R. Palmacci, O.J. Plante and P.H. Seeberger, *Eur. J. Org. Chem.*, 2002, 595. (b) R.R. Schmidt, J.C. Castro-Palomino and O. Retz, *Pure Appl. Chem.*, 1999, **71**, 729. (c) P.H. Seeberger and S.J. Danishefsky, *Acc. Chem. Res.*, 1998, **31**, 685. d) S.J. Danishefsky and M.T. Bilodeau, *Angew. Chem. Int. Ed. Engl.*, 1996, **35**, 1380.
11. (a) K.M. Koeller and C.-H. Wong, *Chem. Rev.*, 2000, **100**, 4465. (b) C.-H. Wong, *Pure Appl. Chem.*, 1995, **67**, 1609.
12. (a) A. Hölemann and P.H. Seeberger, *Curr. Opin. Biotechnol.*, 2004, **15**, 615. (b) D. Macmillan and A.M. Daines, *Curr. Med. Chem.*, 2003, **10**, 2733. (c) B.G. Davis, *Chem. Rev.*, 2002, **102**, 579.
13. (a) E. Atherton and R.C. Sheppard, *Solid-Phase Peptide Synthesis: A Practical Approach*, Oxford University Press, Oxford, 1989. (b) R.B. Merrifield, *Angew. Chem. Int. Ed. Engl.*, 1985, **24**, 799.
14. (a) M.H. Caruthers, *Acc. Chem. Res.*, 1991, **24**, 278. (b) M.H. Caruthers, *Science*, 1985, **230**, 281.

15. (a) O.J. Plante, E.R. Palmacci and P.H. Seeberger, *Science*, 2001, **291**, 1523. (b) P.H. Seeberger, *Chem. Comm.*, 2003, 1115. (c) E.R. Palmacci, O.J. Plante, M.C. Hewitt and P.H. Seeberger, *Helv. Chim. Acta*, 2003, **86**, 3975.

16. O.J. Plante, R.B. Andrade and P.H. Seeberger, *Org. Lett.*, 1999, **1**, 211.

17. R.R. Schmidt and W. Kinzy, *Adv. Carbohydr. Chem. Biochem.*, 1994, **50**, 21.

18. (a) K.R. Love and P.H. Seeberger, *J. Org. Chem.*, 2005, **70**, 3168. (b) K.R. Love and P.H. Seeberger, *Angew. Chem. Int. Ed.*, 2004, **43**, 602. These compounds act as tumor markers and are currently explored in cancer therapy: G. Ragupathi, P.P. Deshpande, D.M. Coltart, H.M. Kim, L.J. Williams, S.J. Danishefsky and P.O. Livingston, *Int. J. Cancer*, 2002, **99**, 207.

19. (a) V. Verez-Bencomo, V. Fernández-Santana, E. Hardy, M.E. Toledo, M.C. Rodríguez, L. Heynngnezz, A. Rodriguez, A. Baly, L. Herrera, M. Izquierdo, A. Villar, Y. Valdés, K. Cosme, M.L. Deler, M. Montane, E. Garcia, A. Ramos, A. Aguilar, E. Medina, G. Toraño, I. Sosa, I. Hernandez, R. Martínez, A. Muzachio, A. Carmenates, L. Costa, F. Cardoso, C. Campa, M. Diaz and R. Roy, *Science*, 2004, **305**, 522. (b) J. Kaiser, *Science*, 2004, **305**, 460.

20. (a) B. Kuberan and R.J. Linhardt, *Curr. Org. Chem.*, 2000, **4**, 653. (b) S. Borman, *Chemical & Engineering News*, 2004, **82**, 31.

21. (a) S.J. Keding and S.J. Danishefsky, in *Carbohydrate-Based Drug Discovery*, C.-H. Wong (ed), 1st ed., Wiley-VCH, Weinheim, 2003, 381. (b) S.J. Danishefsky and J.R. Allen, *Angew. Chem. Int. Ed.*, 2000, **39**, 836.

22. Selected papers: (a) V.Y. Dudkin, M. Orlova, X. Geng, M. Mandal, W.C. Olson and S.J. Danishefsky, *J. Am. Chem. Soc.*, 2004, **126**, 9560. (b) M. Mandal, V.Y. Dudkin, X. Geng, and S.J. Danishefsky, *Angew. Chem. Int. Ed.*, 2004, **43**, 2557. (c) X. Geng, V.Y. Dudkin, M. Mandal and S.J. Danishefsky, *Angew. Chem. Int. Ed.*, 2004, **43**, 2562. (d) D.M. Ratner, O.J. Plante and P.H. Seeberger, *Eur. J. Org. Chem.*, 2002, 826. (e) E.W. Adams, D.M. Ratner, H.R. Bokesch, J.B. McMahon, B.R. O'Keefe and P.H. Seeberger, *Chem. Biol.*, 2004, **11**, 875. (f) M.C. Bryan, F. Fazio, H.-K. Lee, C.-Y. Huang, A. Chang, M.D. Best, D.A. Calarese, O. Blixt, J.C. Paulson, D. Burton, I.A. Wilson and C.-H. Wong, *J. Am. Chem. Soc.*, 2004, **126**, 8640.

23. A.K. Nyame, Z.S. Kawar and R.D. Cummings, *Arch. Biochem. Biophys.*, 2004, **426**, 182.

24. The World Health Report, World Health Organization, 2004. http://www.who.int/whr/2004/en/report04_en.pdf.

25. J.L. Gallup and J.D. Sachs, *Am. J. Trop. Med. Hyg.*, 2001, **64**, 85.

26. S.R. Meshnick and M.J. Dobson, in *Antimalarial Chemotherapy*, P.J. Rosenthal (ed), Humana Press, Totowa, NJ, 2001, 15.

27. P. Guttmann and P. Ehrlich, *Berl. Klein. Wochenschr.*, 1891, **28**, 953.

28. J. Wiesner, R. Ortmann, H. Jomaa and M. Schlitzer, *Angew. Chem. Int. Ed.*, 2003, **42**, 5274.

29. D.L. Klaymann, *Science*, 1985, **228**, 1049.

30. S. Gupta, R.W. Snow, C.A. Donnelly, K. Marsh and C. Newbold, *Nat. Med.*, 1999, **5**, 340.

31. L. Schofield, *Chem. Immunol. (Malarial Immunology)*, 2002, **80**, 322.

32. (a) M.F. Good, *Nat. Rev. Immunol.*, 2001, **1**, 117. (b) P.L. Alonso *et al., Lancet*, 2004, **364**, 1411. (c) P.L. Alonso *et al., Lancet*, 2005, **366**, 2012.

33. F. Schofield, *Rev. Infect. Dis.*, 1986, **8**, 144.
34. (a) C. Golgi, *Arch. Sci. Med. (Torino)*, 1886, **10**, 109. (b) C. Golgi, *Arch. Sci. Med. (Torino)*, 1889, **13**, 173.
35. (a) I.A. Clark, *Lancet*, 1978, **2**, 75. (b) I.A. Clark, J.-L. Virelizier, E.A. Carswell and P.R. Wood, *Infect. Immun.*, 1991, **32**, 1058.
36. (a) C.A. Bate, J. Taverne and J.H. Playfair, *Immunology*, 1988, **64**, 227. (b) C.A. Bate, J. Taverne and J.H. Playfair, *Immunology*, 1989, **66**, 600.
37. (a) L. Schofield, M.J. McConville, D. Hansen, A.S. Campbell, B. Fraser-Reid, M.J. Grusby and S.D. Tachado, *Science*, 1999, **283**, 225. (b) I.A. Clark and L. Schofield, *Parasitol. Today*, 2000, **16**, 451.
38. (a) S.D. Tachado, P. Gerold, M.J. McConville, T. Baldwin, D. Quilici, R.T. Schwarz and L. Schofield, *J. Immunol.*, 1996, **156**, 1897. (b) L. Schofield, S. Novakovic, P. Gerold, R.T. Schwarz, M.J. McConville and S.D. Tachado, *J. Immunol.*, 1996, **156**, 1886.
39. C.A.W. Bate, J. Taverne and J.H.L. Playfair, *Infect. Immun.*, 1992, **60**, 1894.
40. (a) L. Schofield and F. Hackett, *J. Exp. Med.*, 1993, **177**, 145. (b) S.D. Tachado, P. Gerold, R. Schwarz, S. Novakovic, M. McConville and L. Schofield, *Proc. Natl. Acad. Sci. USA*, 1997, **94**, 4022. (c) L. Schofield, L. Vivas, F. Hackett, P. Gerold, R.T. Schwarz and S.D. Tachado, *Ann. Trop. Med. Parasitol.*, 1993, **87**, 617. (d) S.D. Tachado and L. Schofield, *Biochem. Biophys. Res. Commun.*, 1994, **205**, 984. (e) L. Schofield and S.D. Tachado, *Immunol. Cell Bio.*, 1996, **74**, 555.
41. (a) P. Gerold, A. Dieckmann-Schuppert and R.T. Schwarz, *J. Biol. Chem.*, 1994, **269**, 2597. (b) P. Gerold, L. Schofield, M. Blackman, A.A. Holder and R.T. Schwarz, *Mol. Biochem. Parasitol.*, 1996, **75**, 131. (c) P. Gerold, L. Vivas, S.A. Ogun, N. Azzouz, K.N. Brown, A.A. Holder and R.T. Schwarz, *Biochem. J.*, 1997, **328**, 905.
42. L. Schofield, M.C. Hewitt, K. Evans, M.-A. Slomos and P.H. Seeberger, *Nature*, 2002, **418**, 785.
43. D.C. Gowda, P. Gupta and E.A. Davidson, *J. Biol. Chem.*, 1997, **272**, 6428.
44. A. Dieckmann-Schuppert, S. Bender, M. Odenthal-Schnittler, E. Bause and R.T. Schwarz, *Eur. J. Biochem.*, 1992, **205**, 815.
45. D.C. Gowda, *Microbes. Infect.*, 2002, **4**, 983.
46. R.S. Naik, O.H. Branch, A.S. Woods, M. Vijaykumar, D.J. Perkins, B.L. Nahlen, A.A. Lal, R.J. Cotter, C.E. Costello, C.F. Ockenhouse, E.A. Davidson and D.C. Gowda, *J. Med. Exp.*, 2000, **192**, 1563.
47. L. Schofield and P.H. Seeberger, unpublished results.
48. Z.W. Guo and L. Bishop, *Eur. J. Org. Chem.*, 2004, 3585.
49. M.C. Hewitt, D.A. Snyder and P.H. Seeberger, *J. Am Chem. Soc.*, 2002, **124**, 13434.
50. M.J. McConville and M.A. Ferguson, *Biochem. J.*, 1993, **294**, 305.
51. P.H. Seeberger, R.L. Soucy, Y.-U. Kwon, D.A. Snyder and T. Kanemitsu, *Chem. Commun.*, 2004, 1706.
52. C. Evans, P.H. Seeberger and L. Schofield, unpublished results.
53. Y.-U. Kwon, R.L. Soucy, D.A. Snyder and P.H. Seeberger, *Chem. Eur. J.*, 2005, **11**, 2493.
54. M.D. Disney and P.H. Seeberger, *Drug Discov. Today: Targets*, 2004, **3**, 151.

55. a) J. Lu, K.N. Jayaprakash, U. Schlueter and B. Fraser-Reid, *J. Am. Chem. Soc.*, 2004, **126**, 7540. (b) J. Lu, K.N. Jayaprakash and B. Fraser-Reid, *Tetrahedron Lett.*, 2004, **45**, 879.

56. (a) K.N. Jayaprakash, K.V. Radhakrishnan and B. Fraser-Reid, *Tetrahedron Lett.*, 2002, **43**, 6953. (b) K.N. Jayaprakash and B. Fraser-Reid, *Synlett*, 2004, 301.

57. M. Mach, U. Schlueter, F. Mathew, B. Fraser-Reid and K.C. Hazen, *Tetrahedron*, 2002, **58**, 7345.

58. (a) S.L. Bender and R.J. Budhu, *J. Am. Chem. Soc.*, 1991, **113**, 9883. (b) Z.J. Jia, L. Olsson and B. Fraser-Reid, *J. Chem. Soc. Perkin Trans. 1*, 1998, 631.

59. E.D. Soli, A.E. Manoso, M.C. Satterson, P. DeShong, D.A. Favor, R. Hirschmann and A.B. Smith III, *J. Org. Chem.*, 1999, **64**, 3171.

60. X. Liu, Y.-U. Kwon and P.H. Seeberger, *J. Am. Chem. Soc.*, 2005, **127**, 5004.

61. B.L. Herwaldt, *Lancet*, 1999, **354**, 1191.

62. M. Enserink, *Science*, 2000, **290**, 1881.

63. J. Alvar, C. Canavate and B. Gutierrez-Solar *et al.*, *Clin. Microbiol. Rev.*, 1997, **10**, 298.

64. F. Modabber, *Ann. Trop. Med. Parasitol*, 1995, **89**(suppl 1), 83.

65. (a) P.F.P. Pimento, E.M.B. Saraiva and D.L. Sacks, *Exp. Parasitol.*, 1991, **72**, 191. (b) D.L. Tolsen, S.J. Turco, R.P. Beecroft and T.W. Pearson, *Mol. Biochem. Parasitol.*, 1989, **35**, 109. (c) B.L. Chan, M.V. Chao and A.R. Saltiel, *Proc. Natl. Acad. Sci. USA*, 1989, **86**, 1756. (d) D.D. Eardley and M.E. Koshland, *Science*, 1991, **251**, 78. (e) A.R. Saltiel, J.A. Fox, P. Sherline and P. Cuatrecasas, *Science*, 1986, **233**, 967.

66. G.F. Mitchell and E. Handman, *Parisite Immun.*, 1986, **8**, 255.

67. (a) H. Moll, G.F. Mitchell, M.J. McConville and E. Handman, *Infect. Immun.*, 1989, **57**, 3349. (b) M.J. McConville, A. Bacic, G.F. Mitchell and E. Handman, *Proc. Natl. Acad. Sci. USA*, 1987, **84**, 8941. (c) D.G. Russell and J. Alexander, *J. Immunol.*, 1988, **140**, 1274.

68. (a) W.K. Tonui, P.A. Mbati, C.O. Anjili, A.S. Orago, S.J. Turco, J.I. Githure and D.K. Koech, *East Afr. Med. J.*, 2001, **78**, 84. (b) W.K. Tonui, P.A. Mbati, C.O. Anjili, A.S. Orago, S.J. Turco, J.I. Githure and D.K. Koech, *East Afr. Med. J.*, 2001, **78**, 90.

69. K. Ruda, J. Lindberg, P.J. Garegg, S. Oscarson and P. Konradsson, *J. Am. Chem. Soc.*, 2000, **122**, 11067.

70. (a) V.A. Nikolaev, T.J. Rutherford, M.A.J. Ferguson and J.S. Brimacombe, *J. Chem. Soc. Perkin Trans. 1*, 1996, 1559. (b) V.A. Nikolaev, T.J. Rutherford, M.A.J. Ferguson and J.S. Brimacombe, *J. Chem. Soc. Perkin Trans. 1*, 1995, 1977. (c) V.A. Nikolaev, J.A. Chudek and M.A.J. Ferguson, *Carbohydr. Res.*, 1995, **272**, 179.

71. G.M. Brown, A.R. Millar, C. Masterson, J.S. Brimacombe, A.V. Nikolaev and M.A.J. Ferguson, *Eur. J. Biochem.*, 1996, **242**, 410.

72. A. Arasappan and B. Fraser-Reid, *J. Org. Chem.*, 1996, **61**, 2401.

73. M. Upreti, D. Ruhela and R.A. Vishwakarma, *Tetrahedron*, 2000, **56**, 6577.

74. W. Kinzy and R.R. Schmidt, *Carbohydr. Res.*, 1987, **164**, 265.

75. M.C. Hewitt and P.H. Seeberger, *J. Org. Chem.*, 2001, **66**, 4233.

76. M.C. Hewitt and P.H. Seeberger, *Org. Lett.*, 2001, **3**, 3699.

77. L. Schofield and P.H. Seeberger, unpublished results.

Studies Toward a Rationally Designed Conjugate Vaccine for Cholera Using Synthetic Carbohydrate Antigens

PAVOL KOVÁČ

National Institutes of Health, NIDDK, LMC, Carbohydrates, Bethesda, MD 20892-0815, USA

1 Introduction

Cholera is a very serious, both endemic and epidemic disease. It is a diarrheal illness caused by infection of the intestine with the bacterium *Vibrio cholerae*. Based on the structure of their O-specific polysaccharide, vibrios are classified into some 200 serogroups,[1] of which only a few cause disease in humans. Until the early 1990s *V. cholerae* serogroup O1 was the only known cause of cholera.[2] Later in that decade, however, large outbreaks of cholera occurred in India and Bangladesh, which were caused by a previously unrecognized serogroup of *V. cholerae*, designated O139. People living in industrialized countries acquire diarrheal diseases not only through travel to endemic areas, but also through consumption of undercooked domestic or imported seafood. From the beginning of 1995 through the end of 2000, 61 cases of cholera, including one death, were reported in the USA.[3]

The main symptom of cholera is severe watery diarrhea. The daily weight loss due to diarrhea can amount to one-third of the body weight. Rapid loss of body fluids leads to dehydration and hypotensic shock. Without treatment, death can occur within hours of the first symptoms.[4] Notwithstanding the fact that cholera has been very rare in industrialized nations for the last 100 years, continuous occurrence of cholera in the Indian subcontinent and sub-Saharan Africa augments the need for an effective and safe vaccine.

Lipopolysaccharides (LPS, Figure 1) are present on the outer membrane of Gram-negative bacteria as secondary gene products. The LPS consists essentially of three parts: lipid A; the core, sometimes subclassified into inner and outer core;[5] and the O-specific polysaccharide (O-SP, O-PS, or O-specific antigen). All three parts are immunologically active, but the serotype specificity is provided by the chemical structure of the O-PS, which is almost always made up of oligosaccharide, sometimes monosaccharide-repeating units. The O-SP protrudes into the environment, and interacts with the receptors/binding sites of elements of the immune system when the organism is challenged by a bacterial pathogen. These O-SPs are both essential virulence factors *and* protective antigens of these pathogens.[2,6,7] Thus, immunity is often provided by antibodies that recognize the O-SP part of the LPS. It should, therefore, be possible to use O-SPs or synthetic oligosaccharides that mimic their structure as antigenic components of vaccines. Carbohydrates, however, are poor immunogens. They are classified as T-independent (TI) class 2 (TI-2) antigens. Such antigens are only capable of stimulating antibody production by specialized anti-carbohydrate antibody producing B cells, which are absent in infants. This explains the inefficacy of vaccines based on capsular polysaccharides alone in children about 2 years old or younger. The inability of TI-2 antigens to induce boostable memory immune responses has been well documented.[8] As a result, the level and the spectrum of antibodies produced following immunization with carbohydrate antigens are insufficient to render protection. Early work, pioneered by Avery and Goebel,[9,10] with haptens of low molecular mass showed that small, nonimmunogenic molecules can become immunogenic by covalent attachment to large proteins. Carbohydrates are no exception, and by immunizing vertebrates with carbohydrate–protein conjugates it is possible to elicit carbohydrate-specific antibodies in a T-dependent fashion. As a result, multiple injections of such neoglycoconjugates, *i.e.*, synthetic carbohydrate-containing substances, can sharply boost antibody titers and promote antibody class switching way beyond what is normally observed as a result of priming with TI antigens. This rationale

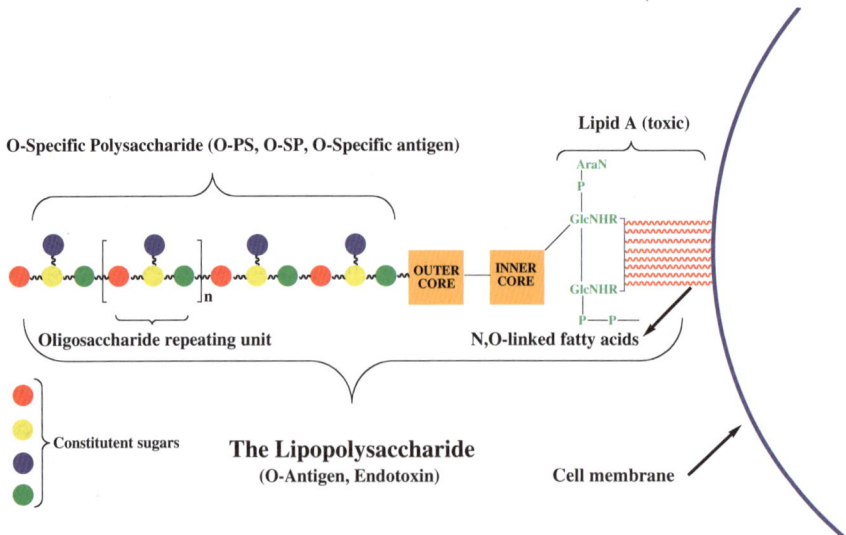

Figure 1 *General structure of the lipopolysaccharide of Gram-negative bacteria*

behind the concept of conjugate vaccines justifies hopes that further research, aimed at finding the right configuration of carbohydrate–protein constructs, would afford vaccines free from drawbacks cellular vaccines often have.

In contrast to approaches to a conjugate vaccine for cholera from base-detoxified LPS of *V. cholerae* O1[11,12] or those utilizing molecular biology,[13] or bacteriology,[14] our efforts, as well as those of others,[15] have been focused on the use of chemically synthesized fragments of the O-SP as antigenic components of the conjugate vaccine. Based on the recent success with the conjugate vaccine from *H. influenzae* type b oligosaccharide,[16] we are hopeful that our use of synthetic oligosaccharides that mimic O-SPs as components of neoglycoconjugates would yield a vaccine for cholera effective in a wide spectrum of population, including children below the age of 2 years. This chapter offers a summary of efforts aimed at developing a vaccine for cholera from synthetic fragments of the O-SP of *V. cholerae* O1.

Synthetic vaccine development is a very young, multidisciplinary endeavor. Advancement in this field is unthinkable without consolidated and collaborative effort of scientists in many fields including, but not limited to, all branches of the life sciences. Although detailed structural requirements for a potent, medically acceptable synthetic vaccine are not known, a large body of information has accumulated regarding possible chemistries of conjugation and fundamentals of the immune response to natural and man-made immunogens.[17–21] Results in the literature on immunogenicity of conjugate vaccines often appear contradictory. For example, there is considerable disagreement about the benefit of having a larger or a shorter oligosaccharide as the antigen. Similarly, some works find high-density hapten neoglycoconjugates more potent than low-density hapten ones and *vice versa*. It is the opinion of this author that universal requirements for a potent conjugate vaccine will be virtually impossible to define. The optimum size and density of the oligosaccharide in the neoglycoconjugate will vary from case to case, if for no other reason than because the constitution of the repeating unit in the individual O-SPs varies from strain to strain. With the understanding that other factors[22] may have to be considered later, our approach to developing a vaccine for cholera has thus far involved the following stages:

1. Synthesis of carbohydrate antigens
2. Determination of the carbohydrate size and carbohydrate–protein ratio in the neoglyco conjugates for optimal anti-carbohydrate response
3. Selection of the coupling chemistry and optimization of the reaction conditions
4. Development of protocols for efficient and controlled preparation of series of neoglyco conjugates with predetermined carbohydrate–protein ratio
5. Serological evaluation of responses in neonatal mice following immunization with synthetic immunogens

A schematic representation of the general structure of carbohydrate–protein constructs from oligosaccharide fragments of the O-SP of *V. cholerae* O1, serotype Ogawa, similar to those described in this chapter, is shown in Figure 2. Syntheses that led to experimental immunogens we have tested thus far will be described in detail. Independent, related, alternative work from other laboratories will also be included.

Synthesis of higher oligosaccharides that mimic partial structures of the O-SPs of *V. cholerae* O1, serotype Inaba, and Ogawa is a considerable challenge for organic

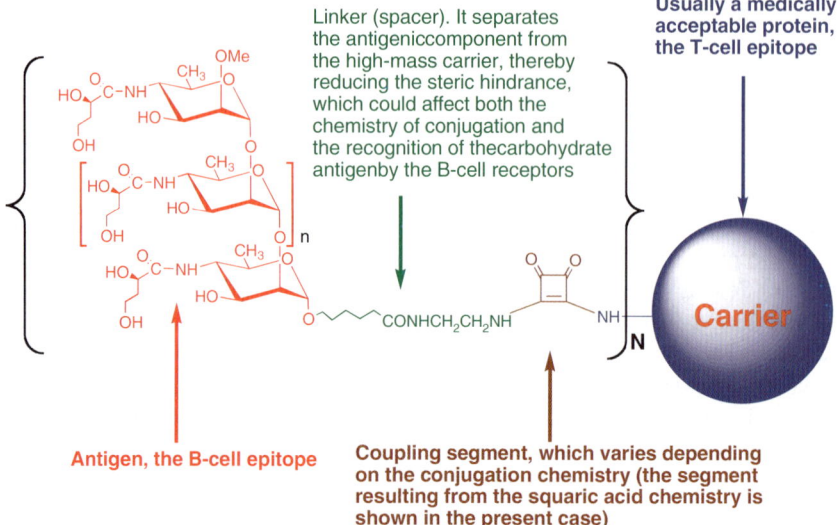

Figure 2 *Schematic representation of the general structure of carbohydrate–protein constructs from oligosaccharide fragments of the O-PS of V. cholerae O1, serotype Ogawa*

chemists, and the most practical synthetic pathway, of many that one could envision, is not immediately obvious. During the years we have been involved in the work aimed at developing a vaccine for cholera we have synthesized many saccharides relevant to the goals of the project using a number of approaches. While new synthetic schemes may have appeared to be improvements to previous pathways at the time of design and execution, some of them have later proved flawed. The individual approaches and the reasoning that led to them, as well as their merit, or lack thereof, will be explained from the historical perspective, as they have evolved to the present state.

Stein[23] has pointed out, quite rightly, that there is virtually an infinite number of choices for glycoconjugate formation. The number of variables that can affect the immunogenicity and protective capacity of a conjugate vaccine is large. Such variables are, for example, the size of the oligosaccharide, the mode of conjugation (direct attachment of the antigen to a carrier or conjugation through a linker/spacer), the nature of the spacer and the carrier, and the conjugation model, *i.e.*, single-point attachment of the antigen (to give neoglycoproteins,[24] which mimic in some way the situation on the outer membrane of the pathogenic bacteria), or multiple-point attachment (to afford cross-linked, lattice-type, high-molecular-mass material). Another attractive conjugation model can be realized by single-point attachment of several carbohydrate units to a subcarrier, to form a cluster, which is then linked to a carrier. It has been demonstrated that the efficacy of multivalent ligands increases many fold,[25–28] compared to a single ligand. In this context, another variable that must not be overlooked is the density of the hapten on the carrier protein. There is no general consensus regarding this variable. The minimum sufficient density is not known, and despite some reports of a high-carbohydrate-density glycoconjugate being more potent than its low-density analog,[28] it is likely that overderivatizing the carrier may mask certain immunologically important epitopes on it, resulting in diminished contribution of the T-cell component of the conjugate and, eventually, in decreased overall immunogenicity. In view of the above, and our limited resources, our work necessarily had to exclude studies of some of the variables. In the initial stage of our endeavor toward a conjugate vaccine for cholera we decided to first test our synthetic strategy and analytical methods by making the first set of glycoconjugates from inexpensive bovine serum albumin (BSA) as a carrier, and vary the size and the density of the haptenic molecule, which would be attached through a spacer in a single-point attachment mode. Once we identify the minimum size of the oligosaccharide and favorable loading of the carrier, we plan to make similar glycoconjugates using medically acceptable carrier proteins, and compare the immunogenicity and protective capacity of such glycoconjugates with those having the same hapten attached in form of a cluster.

2 Synthesis of Carbohydrate Antigens in the Form of Methyl α-Glycosides

The O-SPs of the two main strains of *V. cholerae* O1, Ogawa and Inaba, consist of a chain of about 15–20[2,22] (1→2)-α-linked moieties of 4-amino-4,6-dideoxy-D-mannopyranose (D-perosamine), the amino group of which is acylated with 3-deoxy-L-*glycero*-tetronic acid.[29] Only the Ogawa strain has the *O*-2 of its upstream,[30] terminal perosamine residue methylated.[31,32] The structure of the O-SP of the two serotypes is compared in Figure 3.

It is believed that detailed information about the mode of binding of the O-SP with its homologous antibodies is important for rational choice of the fragment that is to become part of the first generation of synthetic experimental immunogens. Studies toward that end have to include identification of the immunologically dominant epitope(s) in the O-SP. Within our efforts to obtain such information, we have synthesized

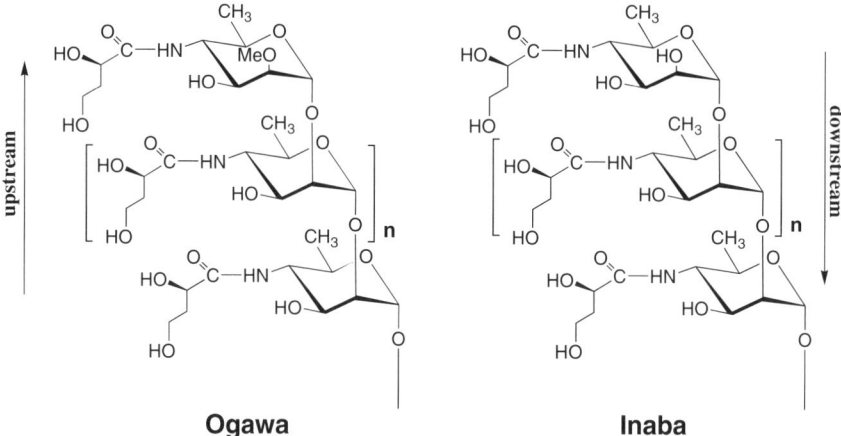

Figure 3 *Structures of the O-PSs of V. cholerae O1, serotypes Inaba and Ogawa*

a large number of saccharides, from the mono- through the hexasaccharide of the O-SP of both serotypes of *V. cholerae* O1, and measured their affinity constants. These haptens were synthesized in the form of methyl α-glycosides. In this way, the stereochemistry at the anomeric center of each perosamine moiety, including the one at the downstream[30] end (Figure 3), mimics the stereochemical situation in the native O-SP.

It is obvious from a brief inspection of formulae in Figure 3 that the monomeric unit in the O-SP consists of two unrelated fragments, D-perosamine and the 3-deoxy-L-*glycero*-tetronic acid side chain. These are rare chiral molecules, and the prerequisite for synthesizing a series of haptens related to structures shown in Figure 3 is a dependable access to a suitable derivative of D-perosamine and a powerful acylating reagent derived from 3-deoxy-L-*glycero*-tetronic acid.

2.1 Synthesis of Methyl α-D-Perosaminide and its 2-*O*-Methyl Analog

The reducing sugar D-perosamine in the free base form is unstable.[33] Stevens *et al.*[34] were unable to isolate D-perosamine hydrochloride from the acid hydrolyzate of its methyl α-glycoside and have shown that when D-perosamine is generated by hydrogenation of 4-azido-4,6-dideoxy-D-mannopyranose, *in situ* reduction of the Schiff's base, which is formed from the 4-amine generated and the aldehyde at C-1, gives only rearrangement products. Synthetic carbohydrate chemists became particularly interested in synthesizing D-perosamine when it was established that it occurs in a number of natural antimicrobial agents, *e.g.*, perimycine.[35]

One of the key steps in making the precursor of methyl α-D-perosaminide (methyl 4-amino-4,6-dideoxy-α-D-mannopyranoside, **4**, Scheme 1[36]), methyl 4-azido-4, 6-dideoxy-α-D-mannopyranoside (**3**) requires nucleophilic displacement of a leaving group in position 4 of a suitable talopyranoside. Early attempts[37,38] at this conversion were met with little success due to formation of products of rearrangement. Successful syntheses of methyl α-L- and α-D-perosaminide were later reported by Brimacombe *et al.*[39] and Stevens *et al.*[34] The above work is of great importance as it laid the foundation for the development of all subsequent, preparatively more important protocols.

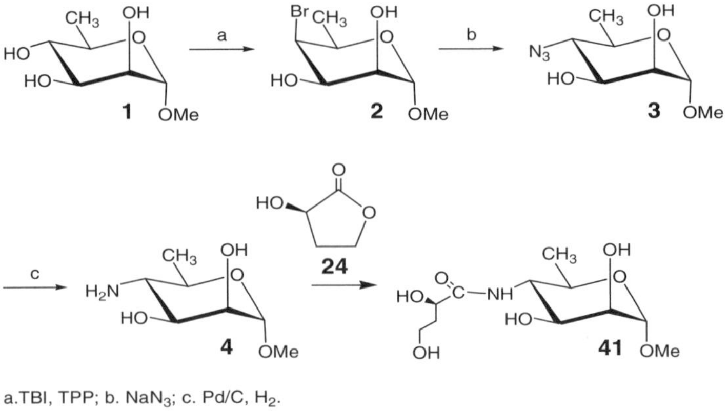

a. TBI, TPP; b. NaN₃; c. Pd/C, H₂.

Scheme 1

2.1.1   The Kenne Approach

Credit for the first successful synthesis of the two fundamental building elements of the O-SP of *V. cholerae* O1, glycoside **4** (Scheme 1) and an acylation reagent derived from 3-deoxy-L-*glycero*-tetronic acid, lactone **24** (Scheme 4), belongs to Kenne and co-workers.[36] One has to add in fairness that two other laboratories[40,41] published their independent syntheses of **4** virtually at the same time. In all these approaches, the starting material is the commercially available methyl α-D-mannopyranoside (**5**, Scheme 2). Positioning the azido group at C-4 in the *manno* configuration must, therefore, involve a double inversion at that chiral center. Kenne's synthesis (Scheme 1) is based on the regioselective bromination at position 4 of methyl 6-deoxy-α-D-mannopyranoside (**1**), which gives the corresponding 4-bromo talopyranoside (**2**). Following S_N2 displacement with the azide ion gives the crystalline azide **3** in the moderate yield of 42% (~22% from **1**). Simple catalytic reduction of the azido group then affords the target amine **4**.

2.1.2   The Bundle and the Ganem Approaches

The fundamental difference between the approach by Kenne's and those by Bundle and Ganem's laboratories is that in the latter two the first of the two required inversions of the *manno* configuration at C-4 (*manno→talo* conversion) was affected by a two-step, oxidation/reduction process. With Kenne's source of **1** undisclosed,[29] the salient points in Bundle's and Ganem's approaches to the synthesis of **4** (Scheme 2) can be summarized as follows:

1. Deoxygenation at C-6 in methyl 2,3-*O*-isopropylidene-α-D-mannopyranoside (**7**) was accomplished either by reduction of a 6-sulfonate ester (**11**) with $LiAlH_4$ (Ganem), or by hydrogenolysis over palladium-on-carbon catalyst of the 6-iodo derivative (**8**) (Bundle).
2. The azido function at the C-4 position was introduced either by treatment of methyl 6-deoxy-2,3-*O*-isopropylidene-4-*O*-trifluoromethanesulfonyl-α-D-mannopyranoside (**15**) with NaN_3 in the presence of a crown ether (Bundle), or by treatment of methyl 6-deoxy-4-*O*-methanesulfonyl(mesyl)-α-D-mannopyranoside (**14**) with NaN_3 alone (Ganem). The two laboratories were able to avoid the rearrangement previously observed during attempted similar displacement reaction[37,38] by either removing the fused 2,3-*O*-isopropylidene ring before the displacement (Ganem), or by using a better leaving group and addition of a crown ether (Bundle). It may be noted that the relatively smooth conversion **2→3** in the approach by Kenne (Scheme 1) can also be explained by the absence of the fused ring in the mannopyranoside **2**.

2.1.3   Our Approach

We examined[42] the above cited procedures leading to **4** and have found a combination of the procedures by Ganem and Bundle readily amenable to a considerable scale-up. The reaction sequence[41] **5→12** (Scheme 2) can be readily performed on 0.5 molar scale in a common academic laboratory setting. With such amount of material, it is more

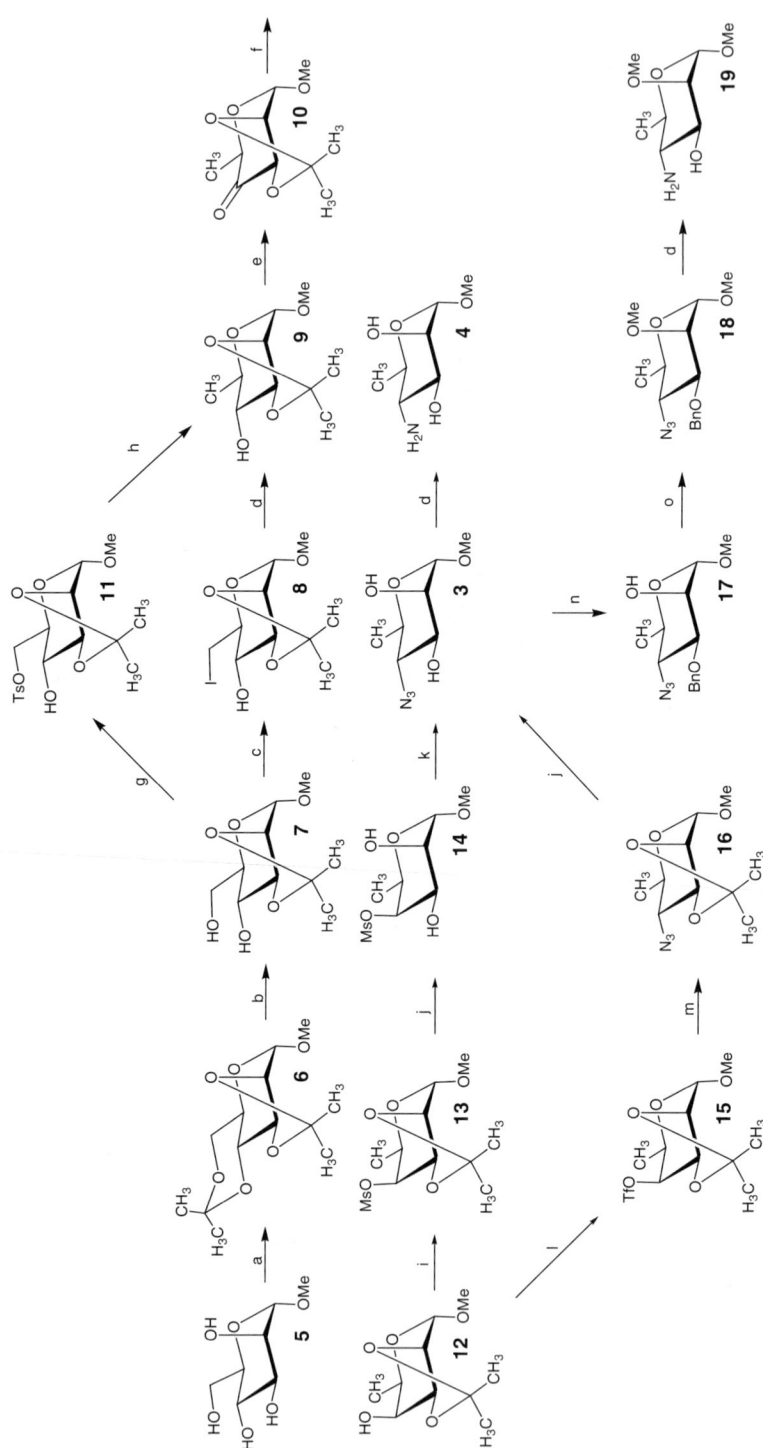

Scheme 2

a. DMP; b. H₂O, H⁺; c. TPP, TlI; d. H₂, Pd/C; e. COCl₂, DMSO; f. NaBH₄; g. TsCl; h. LiAlH₄; i. MsCl, Py (TEA, see text); j. H₂O, H⁺; k. NaN₃; l. Tf₂O; m. NaN₃, crown; n. Bu₂SnO, BnBr; o. Me, Ag₂O.

convenient to continue through the mesylate **13**[40] rather than the triflate **15**,[41] and proceed to **4** as described by Eis and Ganem.[40] It is noteworthy that we could completely avoid[43] the formation (up to 30%) of a by-product observed by Eis and Ganem[40] during conversion **12→13**, when we used pyridine instead of TEA[40] as the acid scavenger.

To obtain the 2-*O*-methyl derivative of **4**, amine **19** (Scheme 2), azide **3** (Scheme 1) was selectively benzylated at *O*-3 and then methylated (**3→17→18**). Treatment of the fully protected compound **18** with hydrogen in the presence of palladium-on-carbon catalyst then gave the target amine **19**.

2.2 Synthesis of Acylating Reagents Derived from 3-Deoxy-L-glycero-Tetronic Acid

Glattfeld and Sander[44] have reinvestigated earlier work toward α-hydroxy-γ-butyrolactone. Following the classical cyanohydrine synthesis (Scheme 3), they converted glycerol (**20**) through acrolein (**21**) and 3-hydroxy-propionaldehyde (**22**), into intermediate **23**. By way of the corresponding brucine salt, they resolved the enantiomeric mixture (**23→24**), and showed that the free acid spontaneously lactonizes upon formation. The latter information is important to take into consideration when the free acid is to be used as 3-deoxy-L-*glycero*-tetronylating reagent. In such a situation, to prevent lactonization, at least position O-4 must be protected. Often (cf. below), lactone **24** itself is, in fact, a reasonably good reagent for acylation. The rather strained ring in **24** opens readily when the compound is treated with a suitable nucleophile, *e.g.*, an amine.

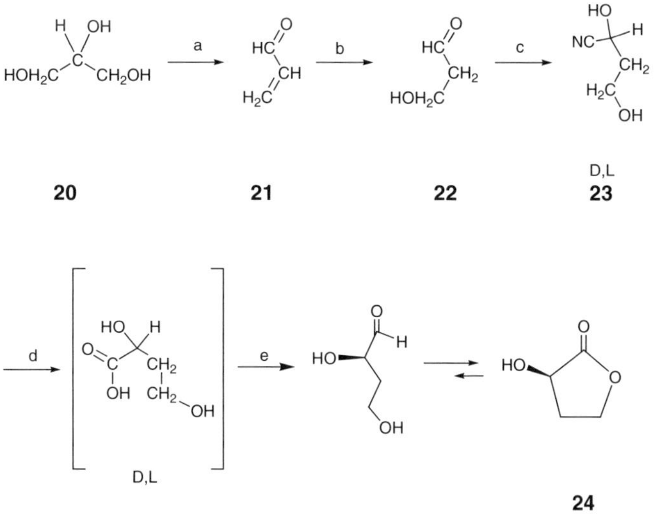

a. MgSO$_4$; b. H$_2$O; c. HCN; d. H$_2$O, HCl; e. brucine.

Scheme 3

2.2.1 The Kenne Approach

During their structural studies on the O-SP of *V. cholerae* O1, serotype Inaba, Kenne and co-workers[36] developed an elegant route to 3-deoxy-L-*glycero*-tetronolactone

(**24**), and used it to synthesize the monosaccharide repeating unit. Their strategy toward **24** (Scheme 4) is based on deamination of the corresponding amino acid L-homoserine, a reaction known[45] to proceed with retention of configuration. Thus, the commercially available L-homoserine (**25**) was deaminated and the salt **26**, formed as an intermediate, was passed through a cation-exchange resin (H⁺-form), whereupon a 4:1 equilibrium mixture of lactone **24** and the corresponding acid was obtained. Optical purity of the material was established by conversion to the acetylated (–)-2-octanyl ester derivative, whose homogeneity was verified by g.l.c.

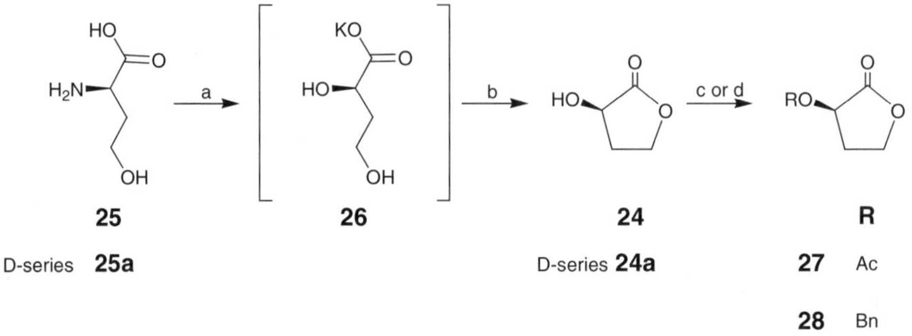

a. KNO₃; b. H⁺; c. Ac₂O; d. BnB, Ag₂O.

Scheme 4

2.2.2 The Verez-Bencomo Approach

Synthesis of a different acylating reagent derived from 3-deoxy-L-*glycero*-tetronic acid, the fully acetylated compound **34** (Scheme 5), was developed by the Cuban authors.[46] It started from the readily available L-malic acid (**29**), whose treatment with AcCl effected simultaneous dehydration and acetylation, to afford the corresponding acetylated anhydride **30**. Acylation of benzyl alcohol with virtually equimolar amount of **30** gave selectively the benzyl ester **31** whose free carboxyl group was reduced. The resulting alcohol **32** was acetylated to give **33**, and subsequent hydrogenolysis produced the targeted acylating reagent **34**.

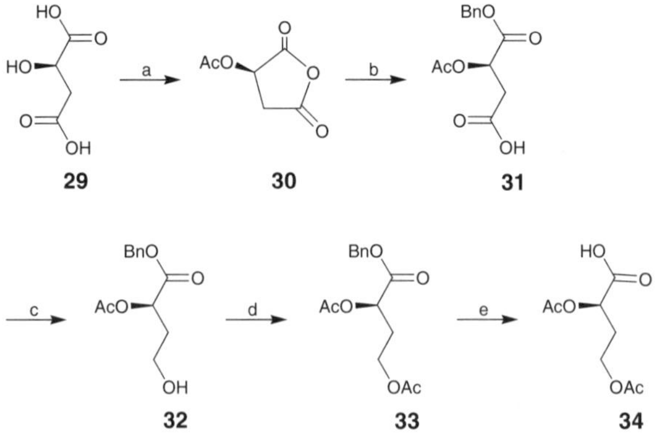

a. AcCl; b. BnOH; c. B₂H₆; d. Ac₂O; e. H₂, Pd/C.

Scheme 5

2.2.3   Our Approach

During the early stages of our synthetic work toward a conjugate vaccine for cholera, when we only had to introduce the 3-deoxy-L-*glycero*-tetronic acid side chain into a mono-[42,43,47] or a disaccharide,[43] we used lactone **24** (Scheme 4) or its 2-*O*-acetate **27** as the acylating reagent. To be able to use pure **24**, rather than a lactone-free acid mixture[36] (see above), we modified[42] Kenne's protocol. We acetylated the crude product of deamination of **25**, and obtained pure acetate **27**, after distillation. Compound **27** can itself be used for *N*-acylation,[42] or it can be deacetylated[43,47] to give **24**. We used the same protocol to prepare the D-enantiomer of **24**, compound **24a**, from commercially available D-homoserine (**25a**). The D-compound was needed in our binding studies (Section 3). We have later observed[42,48] that the diastereomeric pair of isomers, obtained by addition reaction of **24** or **24a** and methyl α-perosaminide **4** can be resolved by thin-layer chromatography, and that some NMR resonances of corresponding nuclei in these diastereomers showed distinctly different chemical shifts. We have routinely used these techniques, as well as HPLC in conjunction with chiral columns, as alternatives to the method by Kenne and co-workers,[29] to verify optical purity of different batches of **24** or its derivatives. In conjunction with later work[43] aimed at analogs of the monosaccharide determinant of the O-SP of *V. cholerae* O1, we have also prepared the 2-*O*-benzylated lactone **28** (Scheme 4).

The need for a different, more powerful acylating reagent arose during our syntheses of hexasaccharide fragments of the O-SP of *V. cholerae* O1, when we had to *N*-acylate compounds containing six amino groups[49] (compare Section 3.4). The established procedure,[36] which worked well for *N*-acylations of mono- or disaccharides,[42,43,47] failed to afford the desired *hexakis-N*-acylated product. As an alternative to the *N*-acylation with lactone **24**,[36] we chose to use a carboxylic acid as the donor of the required acyl group in conjunction with a carbodiimide-type coupling reagent. For this purpose, since the acetylated acid **34** was yet to be prepared,[46] we have developed two new acylating reagents derived from 3-deoxy-L-*glycero*-tetronic acid.[48]

As mentioned above, when free 3-deoxy-L-*glycero*-tetronic acid is to be used as an acylating reagent it must be protected at position 4, to prevent spontaneous lactonization. With a reliable method of deamination of L-homoserine in hand,[36] the commercially available, albeit costly, *N*-Boc-4-*O*-benzyl-L-homoserine (**35**) was a suitable starting material for the synthesis of such a synthon. *N*-deprotection was readily carried out with TFA, and subsequent deamination, as for **25**, afforded[48] the desired, crystalline 4-*O*-benzyl-3-deoxy-L-*glycero*-tetronic acid (**36**) (Scheme 6).

Another useful acylating reagent derived from 3-deoxy-L-*glycero*-tetronic acid is the 2,4-*O*-benzylidene derivative **40**, which we synthesized[48] from L-malic acid **29** (Scheme 6). Carboxyl groups in **29** were first reduced with BMS to give triol **37** in virtually theoretical yield, thereby considerably improving the original preparation[50] employing LiAlH$_4$ as the reducing agent. Compound **37** was converted to the crystalline[51] 2,4-*O*-benzylidene derivative **38**, which was oxidized under optimized[48] conditions using the CrO$_3$/Py/Ac$_2$O/*t*-BuOH reagent.[52,53] The *t*-butyl ester **39** thus obtained was saponified to give, after treatment of the intermediate potassium salt with cation-exchange resin (H$^+$-form), the target acid **40**. Optical purity of this material was verified when reductive opening of the 2,4-*O*-benzylidene ring in **40** gave a material that was identical with the independently synthesized 4-benzyl ether **36**.

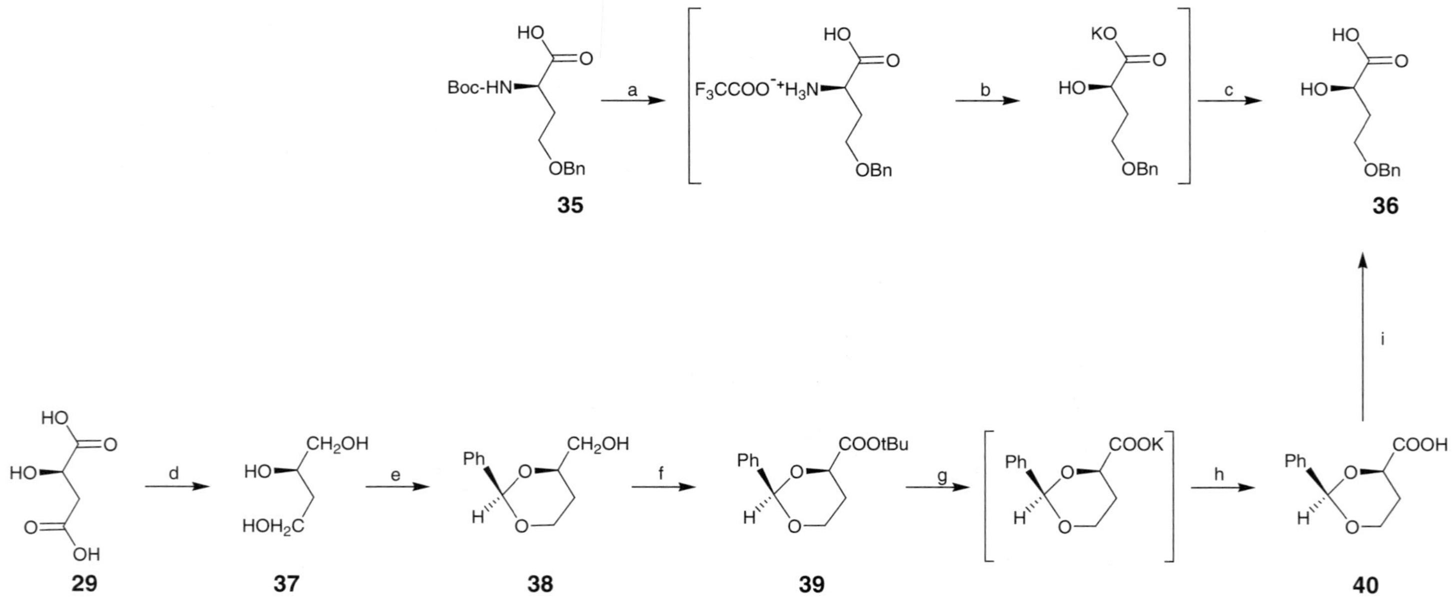

Scheme 6

a. TFA; b. KNO$_3$; c. H$^+$; d. BMS; e. PhCH(OMe)$_2$; f. CrO$_3$•Py/Ac2O/*t*-BuOH; g. KOH; h. H$^+$; i. BTMA/AlCl$_3$.

2.3 Synthesis of Methyl α-Glycosides of the Terminal, Upstream Monosaccharide Determinants of *Vibrio cholerae* O1, Serotypes Inaba and Ogawa

Two fundamentally different methods have been applied for the synthesis of monosaccharides that mimic the upstream[30] (Figure 3) terminus of the O-SP of *V. cholerae* O1, serotypes Inaba and Ogawa: addition reaction of methyl α-D-perosaminide with lactone **24**, and condensation of the same carbohydrate with 4-*O*-protected derivatives of 3-deoxy-L-*glycero*-tetronic acid using various coupling reagents.

In their original assembly of the monosaccharide-repeating unit of the O-SP of *V. cholerae* O1, serotype Inaba, Kenne and co-workers[36] treated (Scheme 1) amine **4** with crude (see above) lactone **24** in pyridine at elevated temperature. The amorphous addition product **41** was obtained in 45% yield.

Our isolation of the product of deamination of L-homoserine in the form of the 2-*O*-acetyl derivative **27** (Scheme 4) led us to use it for *N*-acylations of **4**, and we obtained our first batch of **41** in this way[42,54] (Scheme 7). It is worth mentioning, however, that *N*-acylation of **4** involving **27** is not a one-product reaction. Although a high yield of the target substance **41** can be obtained in this way, this reaction is accompanied by a cascade of *trans*-acetylations and acetyl group migrations, yielding the expected product, the 2'-*O*-acetyl derivative **46**, only in a very low yield. The main reaction product is the 4'-*O*-acetyl derivative **45** (~70%) resulting from acetyl group migration. Other minor products are acetyl derivatives **42–44**, and the *N*-acetamido compound **47**. Once we identified all the products formed in this conversion,[42] we could readily increase the yield of **41** to > 90% by subjecting the crude product of the reaction **4 + 27** to Zemplén deacetylation. This resulted in a mixture containing almost exclusively the desired compound **41**, the *N*-acetyl derivative **47** being the only other product present.

To prepare the terminal monosaccharide of the O-SP of *V. cholerae* O1, serotype Inaba, **41**, Verez-Bencomo's group utilized[46] (Scheme 8) Bundle's[41] azide **16** (Scheme 2). It was converted to amine **48**, whose EEDQ-mediated coupling with the acetylated acid **34** (Scheme 5) gave the fully protected product **49**. The deprotected, desired compound **41** was then obtained through intermediate **50**. Compound **49** is identical with the substance we obtained[42] as a minor by product, resulting from *trans*-acetylation, when we followed, essentially, the same route but used the acetylated lactone **27** as the *N*-acylating reagent.

To obtain the analogous terminal monosaccharide in the Ogawa series, **54** (Scheme 9), we used the same *N*-acylation chemistry as that involved in the synthesis of the Inaba compound. First, methyl 2-*O*-methyl-α-D-perosaminide[47] (**19**, Scheme 2) had to be prepared from the intermediate 3-benzyl ether **18** (Scheme 2). Initially, we prepared the required, intermediate 3-benzyl ether **17** (Scheme 2) following the high-yielding protocol by Eis and Ganem[40] based on selective benzylation at position 3 of the 2,3-*O*-stannylidene derivative of **3** (Scheme 1) with BnBr as both solvent and reagent (compare the protocol by Bundle and co-workers[41] involving benzylation with benzyl trichloroacetimidate). Although we were able to reproduce the high yield reported,[40] removal of the large amount of BnBr, required by the protocol adopted from earlier work,[55] proved to be cumbersome when working on >10-mmol scale. For large-scale work (100–500 mmol), we now routinely carry out

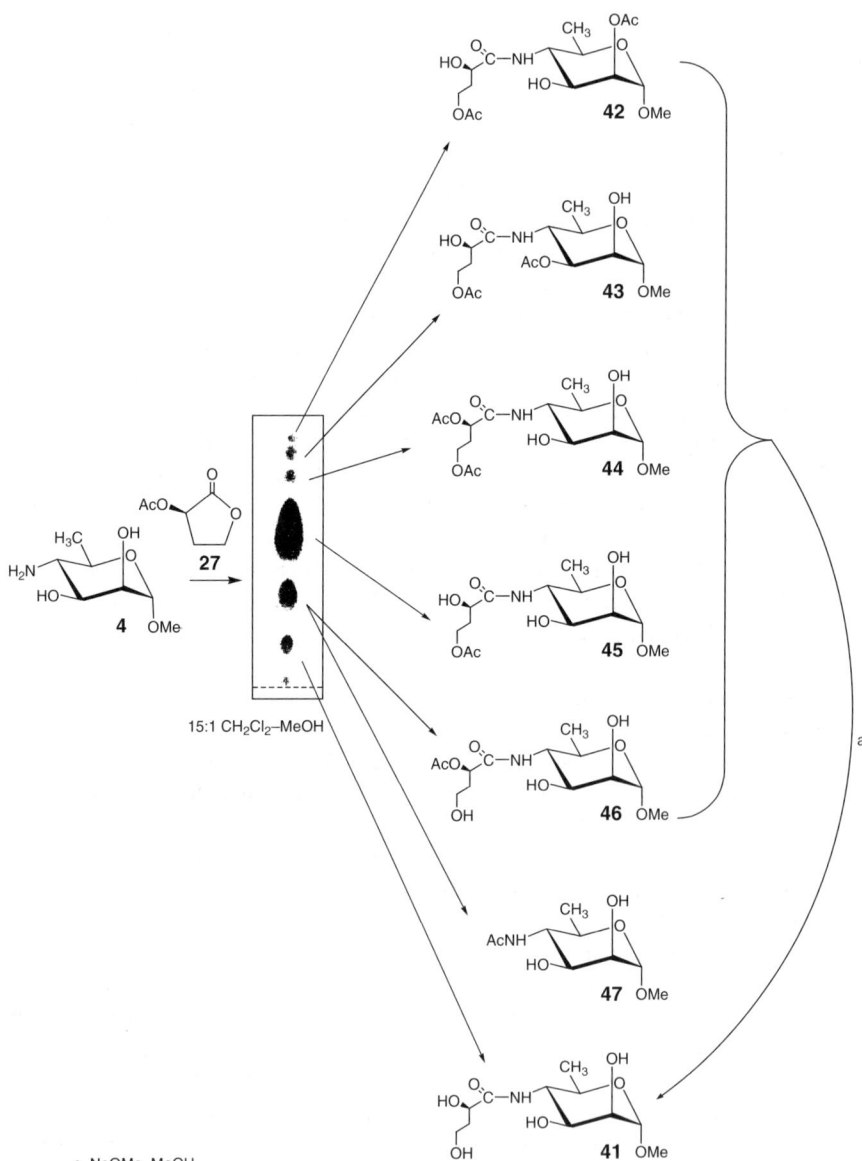

a. NaOMe, MeOH.

Scheme 7

the conversion 3→17 following the benzylation protocol developed by Nashed,[56] with addition of a little TBABr (~0.02 molar proportion). Methylation[57,58] of **17** then readily gives the fully protected azide **18**. Contrary to the findings by others,[46,59] we experienced no difficulties with the conversion **18→19** (Scheme 2, ~90%),[47] despite an amine and a product of debenzylation being formed at the same time. (The efficiency of such conversions appears to depend largely on the quality of the catalyst and absence of materials that can poison it.)

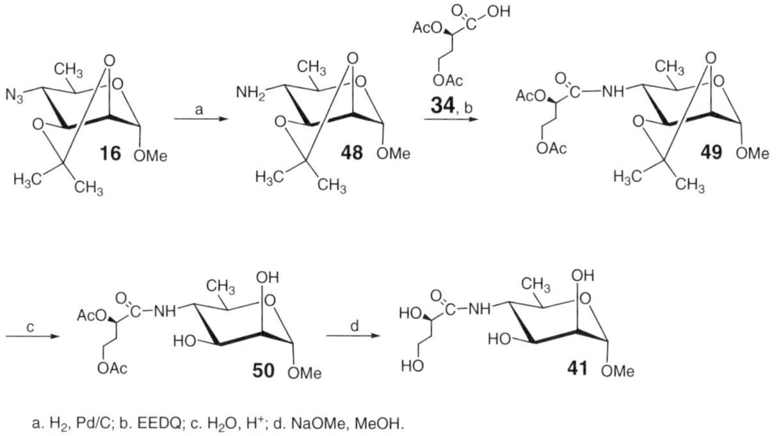
a
34, b
c
d
16
48
49
50
41
a. H_2, Pd/C; b. EEDQ; c. H_2O, H^+; d. NaOMe, MeOH.

Scheme 8

Having the derivative **19** in hand, its conversion to the target **54**, through *N*-acylation, was straightforward.[47] The only other synthesis of **54** reported to date is that from Cuba.[46] It involves condensation of the acetylated tetronic acid derivative **34** with **51** (obtained by selective azido→amino conversion in the presence of the benzyl protecting group effected by hydrogenolysis over palladium-on-charcoal catalyst) followed by stepwise deprotection (Scheme 9, **52→53→54**).

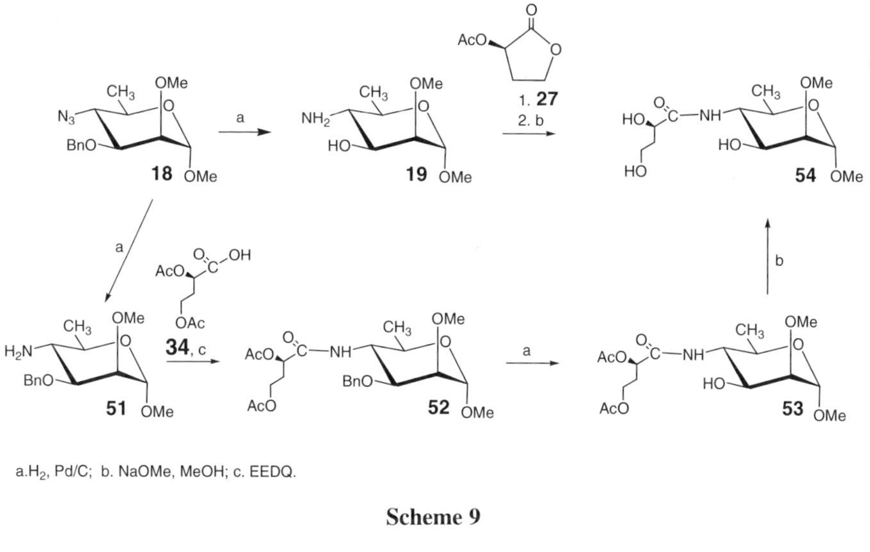
a
1. 27
2. b
34, c
18
19
54
51
52
53
a.H_2, Pd/C; b. NaOMe, MeOH; c. EEDQ.

Scheme 9

The crystal structures of monosaccharides that mimic terminal epitopes of the O-SP of *V. cholerae* O1, serotypes Inaba and Ogawa, have been elucidated.[42,47]

2.4   Synthesis of the Terminal, Upstream Oligosaccharide Determinants of *Vibrio cholerae* O1, Serotypes Inaba and Ogawa

Because of structural similarity of the O-SP of *V. cholerae* O1 and some *Brucella* polysaccharides, many syntheses of oligosaccharides that mimic the O-SP of *V. cholerae* O1 were based on the original strategy by Bundle and co-workers,[41,60–63] who synthesized a large number of D-perosamine-containing oligosaccharides. It involves construction of oligosaccharides from intermediates containing azido groups, and performing the rest of chemical manipulations – reduction of the azido functions and *N*-acylation – when the requisite-sized oligosaccharide had been obtained. A different assembly strategy can use intermediate building blocks, glycosyl donor(s) and glycosyl acceptor(s), having the *N*-acyl group already in place. We tested both strategies when we constructed[43] the methyl α-glycoside of the disaccharide that mimics the upstream terminus of the O-SP of *V. cholerae* O1, serotype Inaba (Scheme 10). When using intermediates having the tetronic acid side chain already in place, the required glycosyl acceptor **59** was obtained from alcohol **17**, which was *p*-methoxybenzylated (**17**→**55**), to place a selectively cleavable group at position of the future extension of the oligosaccharide chain. The fully protected azide **55** was then selectively reduced to the corresponding amine **56**. Subsequent acylation with lactone **24**, followed by benzoylation of the formed *N*-tetronate **57** gave the fully protected glycoside **58**. Deprotection at *O*-2 then gave the target glycosyl acceptor **59**. The glycosyl donor, chloride **61**, was readily obtained by treatment[64] with DCMME of the benzoylated compound **60**, which was prepared from previously synthesized glycoside **41** (Scheme 7). Silver trifluoromethanesulfonate (triflate)-mediated glycosylation (→**62**), followed by two-step deprotection afforded the target compound **64**, *via* alcohol **63**. Our construction of the Inaba disaccharide (**64**) from intermediates containing azido groups[43] (details not shown in Scheme 10), and Ogawa disaccharide[65] (discussed later, and shown in Scheme 12) utilized lactone **24** for *N*-acylations.

A strategy similar to ours for the synthesis of the disaccharide **64** was reported by Verez-Bencomo's group.[46] The starting point for the Cuban team was diamine **66** (Scheme 10), originally obtained by Bundle and co-workers[41] from the disaccharide **65**. Condensation of **66** with acid **34** afforded the coupling product **67**, whose two-step deprotection gave **64**, through tetraacetate **68**. The same chemists also described[46] preparation of the analogous disaccharide in the Ogawa series (Scheme 10). The key step was methylation of **65** (→**69**), followed by the same sequence of transformations (**70**→**71**→**72**→**73**) as those that led to the Inaba disaccharide **64**.

Our first approach to oligosaccharides higher than disaccharides that mimic structures of the O-SPs of *V. cholerae* O1, serotype Inaba and Ogawa, was based on the strategy applied in the synthesis of the disaccharide **64** (Scheme 10). As our glycosyl donor **61** did not have a selectively removable protecting group at *O*-2, to allow extension of the oligosaccharide chain, we had to design a different strategy. The idea,[66] which we applied first in the synthesis of the trisaccharide in the Inaba series, was to

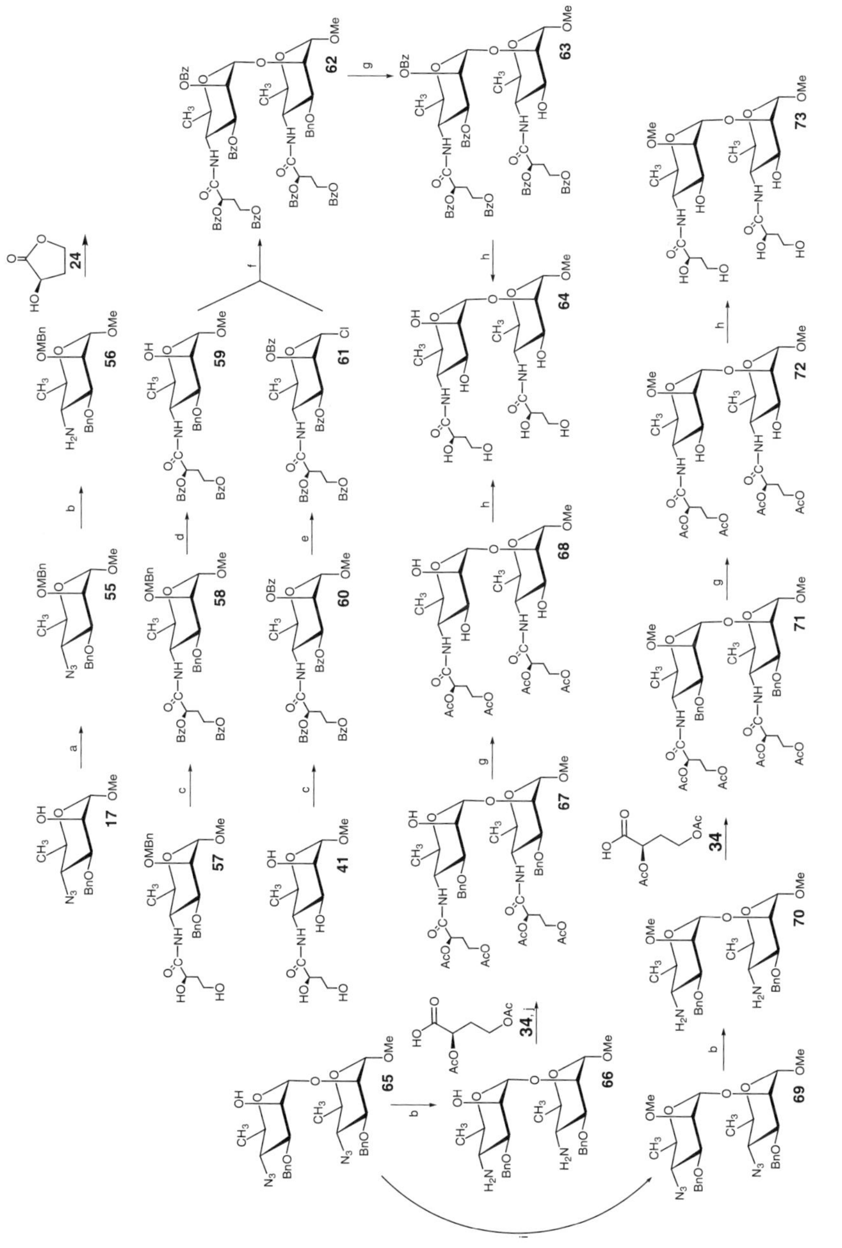

Scheme 10

a. MBnCl; b.. H$_2$S; c. BzCl; d. CAN; e. DCMME, ZnCl$_2$; f. AgOTf; g. H$_2$, Pd/C; h. NaOMe, MeOH; i. MeI, Ag$_2$O; j. EEDQ.

use the *O*-acetyl group as a permanent protecting group at *O*-3, instead of the benzyl group,[41] and the *p*-methoxybenzyl and bromoacetyl groups as temporary protecting groups for *O*-2. Our concerns regarding acetyl group migration[66] during functional group manipulations in compounds belonging to the *manno* series prompted us to use the bromoacetyl group instead of the more commonly used chloroacetyl protecting group, because the former can be removed under milder conditions.[67,68] Migration of *O*-acetyl protecting groups during de-*O*-haloacetylation would, thus, be less likely to occur. Also, a hydroxyl group protected by the 4-methoxybenzyl group, instead of the simple benzyl group, can be regenerated rather quickly under nonhydrogenolytic conditions making, again, the migration of *O*-acetyl protecting groups during deprotection at *O*-2 less likely. Experiments[66] using monosaccharide models verified that, indeed, the extent of side reactions during the required chemical transformations was preparatively unimportant. A further advantage we expected from the use of glycosyl donors and acceptors that were protected with acetyl/bromoacetyl groups only was that couplings of such synthons would give rise to fully acylated products. This was expected to simplify the final deprotection protocol.

Accordingly (Scheme 11), azide **3** (Scheme 1) was selectively *p*-methoxybenzylated under phase-transfer catalysis[47] to give predominantly the 2-ether **74**. Treatment of the latter with H_2S afforded amine **75**, whose addition reaction with lactone **24** produced the *N*-tetronate **76**. Acetylation of **76** ($\rightarrow$**77**) followed by de-*p*-methoxybenzylation then gave the glycosyl acceptor **78**. Due to considerable deactivation of the anomeric center by the neighboring bromoacetyl group,[64,69] the conversion of **79** (prepared from **78** by simple bromoacetylation) to **81** was more practical *via* the 1-*O*-acetyl intermediate **80** than directly.

Condensation of the glycosyl acceptor and glycosyl donor, **78** and **81**, respectively, then gave the fully protected disaccharide **82**, bearing a selectively removable protecting group at position 2^{II}.[70] Debromoacetylation ($\rightarrow$**83**), followed by condensation with the glycosyl donor **81**, gave the fully protected trisaccharide **84**, which was deprotected (Zemplén) to give the target trisaccharide **85**.

The methyl α-glycosides of the di- through the tetrasaccharide in the Ogawa series were synthesized[65] in a similar way (Scheme 12). It required an appropriately functionalized, 2-*O*-methylated perosaminyl glycosyl donor, to reflect the fundamental structural feature of the O-SP of *V. cholerae* O1, serotype Ogawa. To obtain such a monosaccharide (**87**), compound **54** (Scheme 9) was acetylated, and the fully protected glycoside **86** was cleaved with DCMME to give chloride **87**. Reaction of the latter with alcohol **78** (Scheme 11) gave the fully protected disaccharide **88**, which was deacetylated to give the lowest oligosaccharide in the Ogawa series, the disaccharide **89**. The glycosyl donor analogous to **87** but derived from a disaccharide, chloride **92**, was obtained (Scheme 12) in a similar way from the 1-*O*-acetylated disaccharide **91**. The latter was formed by reaction of chloride **87** with alcohol **90**, which was obtained from the bromoacetyl derivative **80** (Scheme 11). Condensation of the disaccharide chloride **92** with glycosyl acceptors **78** and **83** (Scheme 11) gave, after deacetylation of the intermediate products **93** and **95**, the target tri- (**94**) and tetrasaccharide (**96**), respectively.

Despite the fact that the target oligosaccharides had been obtained, the syntheses shown in Schemes 11 and 12 proved to be unpractical. Formation of by-products was observed, mainly due to acetyl group migration during silver-triflate-promoted

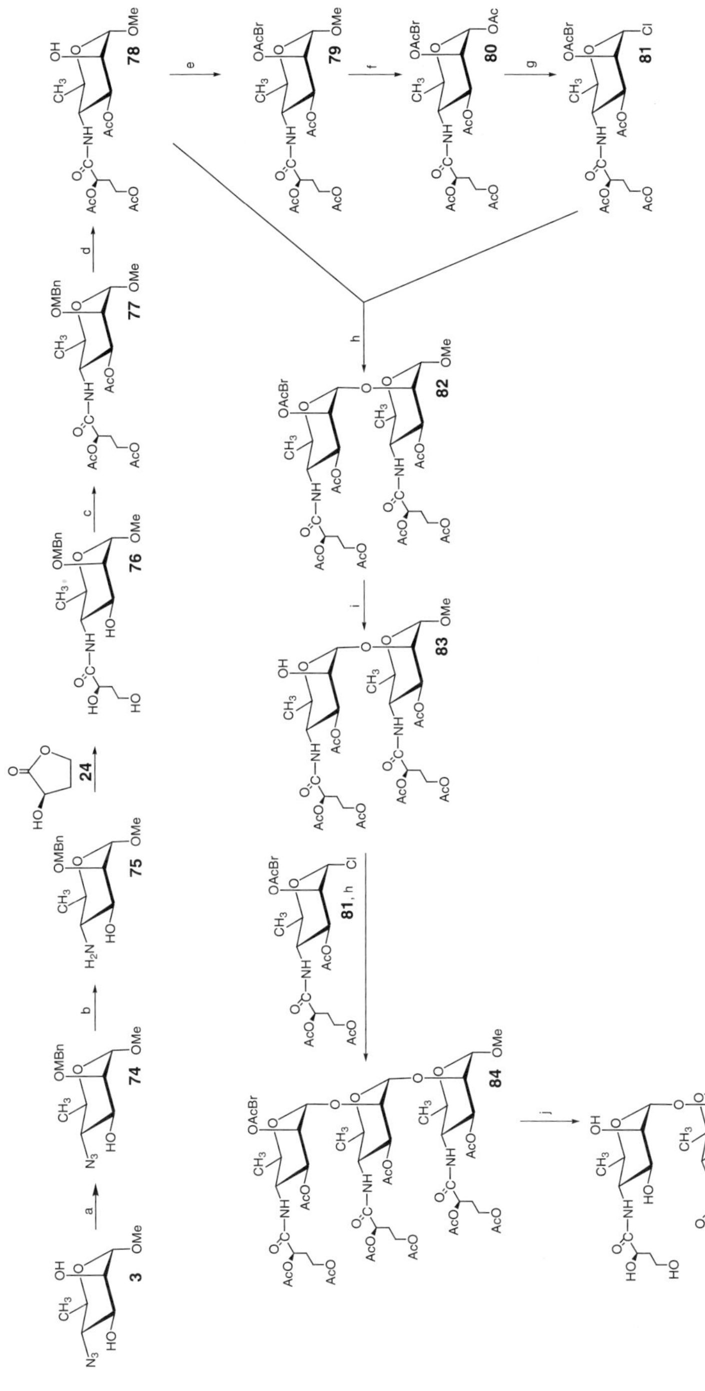

Scheme 11

a. MBnCl; b. H₂S; c. Ac₂O; d. DDQ; e. BrAcCl; f. Ac₂O/H₂SO₄; g. DCMME, ZnCl₂; h. AgOTf; i. Thiourea; j. NaOMe, MeOH.

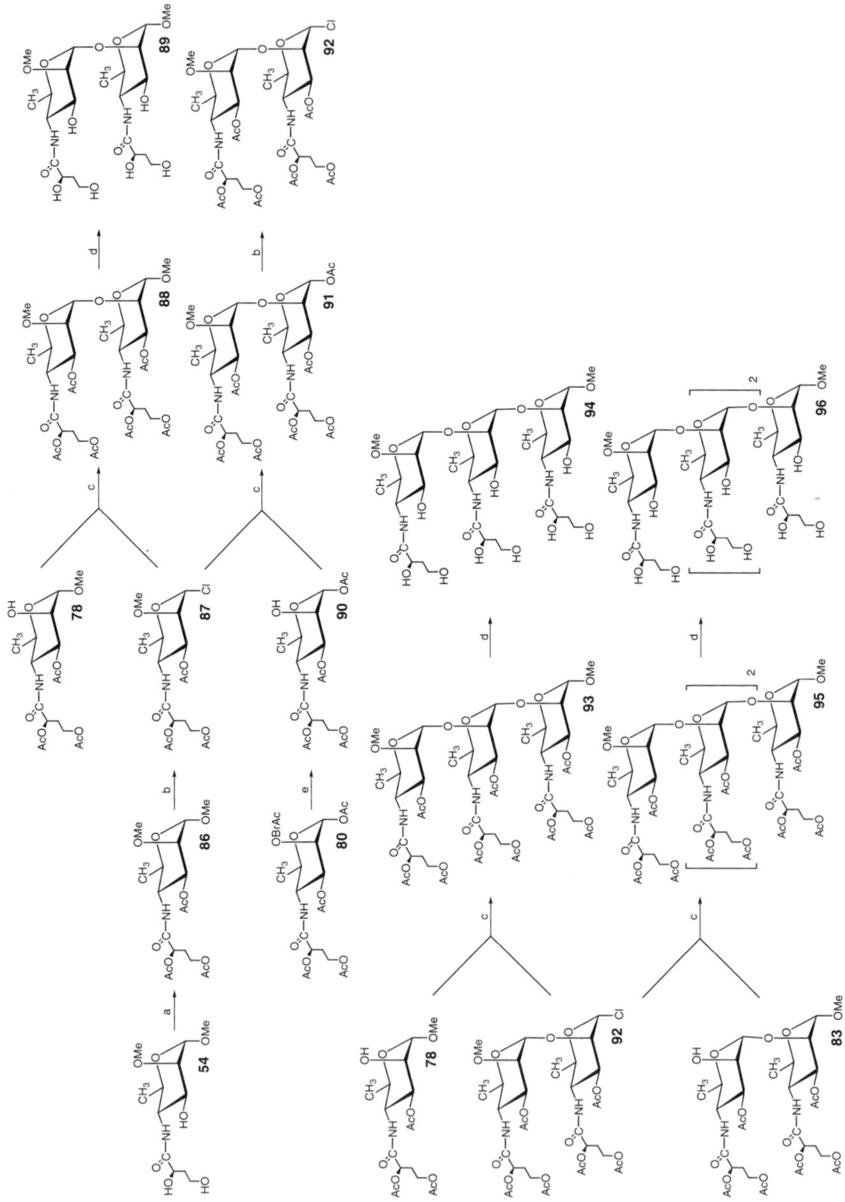

Scheme 12

a.Ac₂O; b. DCMME, ZnCl₂; c. AgOTf; d. NaOMe, MeOH; e. Thiourea.

glycosylation, the one reaction that was not scrutinized for side reactions during our preliminary work.[66] The formation of side products was not extensive, but chromatographic mobility of the by-products was very close to that of the target compounds, and isolation of pure products was difficult, mainly with the tri- and the tetrasaccharides. We concluded that practical syntheses of higher oligosaccharides would require a more stable permanent protecting group at *O*-3, instead of the acetyl group. Thus, we changed the strategy for the syntheses of methyl α-glycosides of higher oligosaccharide fragments of the O-SP of *V. cholerae* O1, serotypes Inaba and Ogawa. The new route involved the assembly of oligosaccharides from intermediates containing azido groups and having position 3 protected with the stable O-benzyl group. In addition, the thioglycoside **98**, previously developed by Bundle and co-workers,[61] was used as the key glycosyl donor, and the benzylidenated acid **40** (Scheme 6) was the *N*-acylating reagent.

The synthetic pathway from the monosaccharide azide **17**, prepared as shown in Scheme 2, to the tri- and tetrasaccharide in the Inaba series (**106** and **108**, respectively) and the di- and the trisaccharide in the Ogawa series (**110** and **112**, respectively) is shown in Scheme 13. Accordingly, diacetate **97**, product of acetolysis of glycoside **17**, was converted to two glycosyl donors, namely, glycosyl chloride **100** and thioglycoside **98**. Deacetylation of **98** afforded glycosyl acceptor **99** which, when treated with **100**, gave disaccharide thioglycoside **101**. The latter is a glycosyl donor that can be used to extend the oligosaccharide chain by two sugar moieties. In a separate sequence, reaction of **17** with thioglycoside **98** gave the fully protected disaccharide **102**, which was deacetylated to give compound **103**. This alcohol was used as the glycosyl acceptor for glycosyl donors **98** and **101**, to give the fully protected tri- and tetrasaccharides **104** and **107**, respectively. Compound **104** was then transformed to the target deprotected trisaccharide **106** by deacetylation (→**105**), reduction of the azido function to the corresponding amine, followed by *N*-acylation with 2,4-benzylidene-3-deoxy-L-*glycero*-tetronic acid (**40**, Scheme 6) and hydrogenolysis, to remove the benzyl and benzylidene protecting groups. The same sequence of reactions performed with **107** gave the Inaba tetrasaccharide **108**. The penta- and the hexasaccharide in the Inaba series were obtained in a similar way. To make the Ogawa di- and the trisaccharides (**110** and **112**, respectively), the methylation was done before the azide→amine conversion (*i.e.*, **103**→**109** and **105**→**111**). Higher Ogawa oligosaccharides were made in a similar way.[71]

3 The Mode of Binding of *Vibrio cholerae* O1 Antigens to Anti-LPS-Specific Antibodies

It has been postulated that oligosaccharides that mimic the structure of the O-SP of bacterial pathogens and bind strongly to LPS-specific immunoglobulins, when linked to a carrier protein, are likely to elicit antibodies that bind with the parent polysaccharides.[22] Therefore, we measured,[72] by fluorescence titration,[73] binding constants of synthetic mono- to hexasaccharide fragments of the Ogawa O-antigen with two murine, monoclonal, anti-*V. cholerae* O1, serotype Ogawa-specific antibodies (IgGs) (Table 1). The free energies of association for these saccharides with each of the two

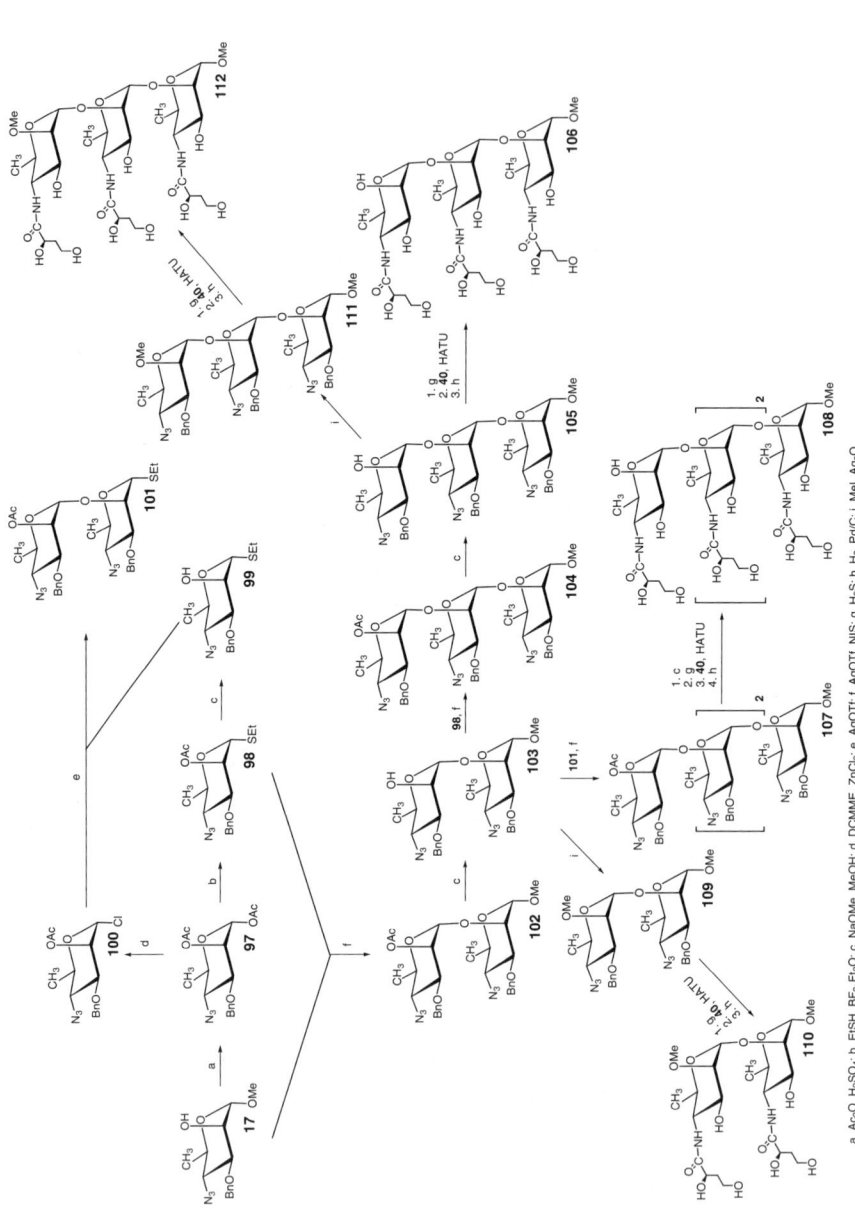

Scheme 13

a. Ac$_2$O, H$_2$SO$_4$; b. EtSH, BF$_3$·Et$_2$O; c. NaOMe, MeOH; d. DCMME, ZnCl$_2$; e. AgOTf, NIS; f. AgOTf; g. H$_2$S; h. H$_2$, Pd/C; i. MeI, Ag$_2$O.

Table 1 *The binding constants (K_a, M^{-1}) and free energy of association ($-\Delta G°$, kJ/mol) for protective monoclonal immunoglobulins G, which are specific for LPS of V. cholerae O1, serotype Ogawa, with synthetic Ogawa oligosaccharides*[*]

IgG	Synthetic oligosaccharide fragments of the Ogawa O-SP					
	Hexa-	*Penta-*	*Tetra-*	*Tri-*	*Di-*	*Mono-*
S-20-4						
$-\Delta G°$	34.9	34.5	35.4	33.8	34.7	31.9
K_a	1.3×10^6	1.1×10^6	1.6×10^6	8.5×10^5	1.2×10^6	3.9×10^5
A-20–6						
$-\Delta G°$	34.2	34.2	35.2	33.6	32.9	30.1
K_a	1.0×10^6	9.7×10^5	1.5×10^6	7.7×10^5	5.8×10^5	1.9×10^5

[*]Data taken from Ref. [72].

antibodies were found to be close, indicating that these two vibriocidal antibodies have the same fine specificity for the epitope on the Ogawa O-SP. Not surprisingly, the Ogawa hexasaccharide showed the highest association constant, but the upstream terminal Ogawa *monosaccharide* also showed strong binding, contributing ~90% of the maximal binding energy shown by the entire hexasaccharide. This finding was quite unexpected, as with carbohydrate antigens it is usually a tetra- or a higher oligosaccharide that shows a K_a of that magnitude. The two monoclonal IgGs (S-20-4 and A-20-6), specific for the LPS of *V. cholerae* O1, serotype Ogawa, exhibited similar affinities for the synthetic methyl α-glycosides of the (oligo)saccharide fragments that mimic the Ogawa O-SP but did not react with the corresponding Inaba oligosaccharides. This lack of cross-reactivity was unexpected, in view of the considerable similarity between the Inaba and Ogawa O-polysaccharides. Studies involving regular oligosaccharide fragments of the O-SP of *V. cholerae* O1 as well as their specifically deoxygenated and fluorinated analogs[72,74] provided detailed information on interactions of these O-SP mimics with their homologous antibodies. More recently, Liao *et al.*[74] documented the weak binding of the Inaba terminal monosaccharide determinant **41** (Scheme 7) with the Ogawa LPS-specific Ab S-20-4, thereby correcting our previous notion[72] regarding binding of this ligand.

Results of those solution studies were confirmed in a remarkable way when a model (Figure 4) of the complex resulting from binding of the O-SP with a homologous *V. cholerae* O1, serotype Ogawa, LPS-specific antibody was built. The model was based on the X-ray crystal structure[75] of complexes of synthetic Ogawa mono- and disaccharides with the anti Ogawa LPS-specific monoclonal antibody S-20-4. The model showed that the upstream terminal sugar moiety of the O-SP fits into a cavity in the binding site where it makes a tight contact with the immunoglobulin. Additional perosamine residues protrude outward from the Ab surface and contribute only marginally to the binding affinity and specificity.

The above results indicated that the disaccharide, or perhaps even the monosaccharide, which mimics the terminus of the O-SP could be used as the antigenic component of a potent vaccine. This would be in agreement with the notion[23] that, in theory, the antigenic oligosaccharide need not be larger than the size of the combining site.

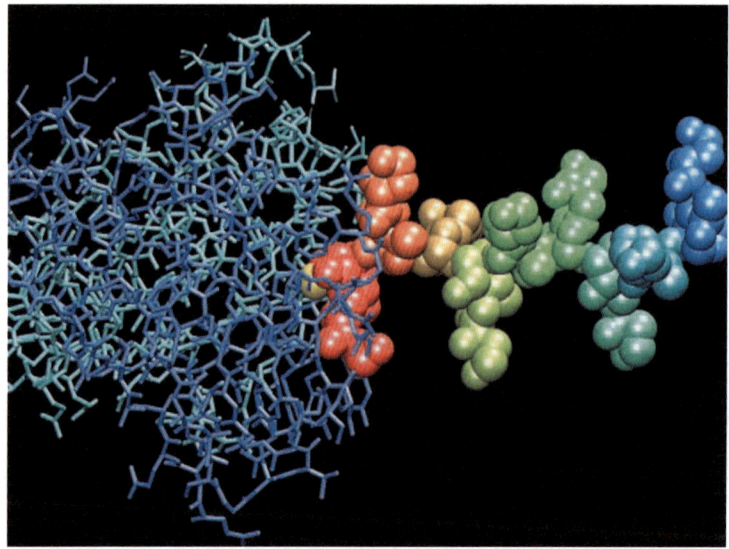

Figure 4 *Model of the Ogawa O-SP–monoclonal antibody complex. The model is based on X-ray structures of crystalline complexes of synthetic mono- and the disaccharide that mimic the upstream terminus of the O-SP of Vibrio cholerae O1, serotype Ogawa, with the Ogawa-specific monoclonal antibody. To view the model in color[75], see Ref. 75.*

Nevertheless, we decided to make our first experimental immunogens from the hexasaccharide, as some previous findings[76,77] indicate that the oligosaccharide must be sufficiently large in order to properly express the complete epitope of the native antigen.

The ability of synthetic mono- and disaccharide fragments of the O-SP of *V. cholerae* O1, serotypes Inaba and Ogawa, to inhibit interactions of anti-Ogawa antibodies with the Ogawa LPS was studied by Verez-Bencomo's laboratory.[46] Both Ogawa saccharides were potent inhibitors, but inhibition by the Inaba compounds was found negligible. This was in complete agreement with the above-discussed findings in our laboratory and, again, confirmed the dominant nature of the terminal determinant of the O-SP of *V. cholerae* O1, serotype Ogawa, which has its *O*-2 position methylated.

The Cuban group also studied[78] the conformational behavior of the trisaccharide fragments of the *V. cholerae* O1, Ogawa and Inaba serotypes, by NMR spectroscopy and molecular dynamics (MD). This was a very useful study because, in view of considerable similarity between the chemical structure of the O-SP of the Inaba and Ogawa strains, the question needed to be answered is whether the serotype specificity might lie in conformational differences between the two terminal fragments. The results[78] indicated that there are no significant conformational differences between the Inaba and Ogawa molecules. Thus, it was concluded that differences in biological activity are probably not due to conformational effects but to van der Waals and/or hydrogen bonding interactions between the antigens and the biological receptor. This finding supports the conclusions we made from the results of our solution studies.[72,74]

4 Conjugation Chemistry and Optimization of Reaction Conditions

Generally, the terminal amino groups of lysine and carboxyl groups of amino acids in carrier proteins are considered most suitable for chemical attachment of carbohydrate haptens. Therefore, synthetic oligosaccharides are normally prepared as glycosides whose aglycons (linkers, spacers) carry groups[19–21] that can be engaged in reactions with amino or carboxyl groups. Straight-chain hydroxyalkanoic acid esters are convenient linking molecules. They can be readily glycosylated, and their terminal ester groups can be further converted into, for example, carboxylic acids, acid azides, hydrazides or ω-amino amides. Lemieux and co-workers[79] first used an aglycon of this type, a linker formed from 8-(methoxycarbonyl)octanol. A similar, but more readily available alcohol is 5-(methoxycarbonyl)pentanol.[80] We have been using the latter extensively for conjugation of carbohydrates to proteins by squaric acid chemistry, and have also used a related linker molecule for conjugation of carbohydrates to proteins by reductive amination.[81–83]

During our work toward a vaccine for cholera, we not only want to make conjugate immunogens following established protocols, but also want to address some pressing problems in conjugate vaccine development. Primarily, we want to examine the effect of some fundamental variables upon conjugation and the immunogenicity of products obtained. Such parameters are, for example, the size of the oligosaccharide, the conjugation chemistry, the nature of the carrier, antigen-carrier ratio, the nature of the linker that separates the antigen from the carrier protein, and the overall geometry of the antigen. An area where synthetic vaccine development needs improvement is the methodology applied at some stages of conjugation. Two problems that we have already addressed are monitoring of the progress of the conjugation reaction and recovery of the unused antigen. Monitoring is important because we have to be able to prepare conjugates with defined and reproducible carbohydrate–protein ratio, if we want to examine the effect of antigen/carrier ratio upon immunogenicity. This requires the ability to routinely monitor the conjugation reaction, so that the reaction can be terminated when the desired hapten-carrier ratio is reached. Recovery of unchanged ligand is important because the conjugation is normally conducted with a large excess of a precious synthetic oligosaccharide, to insure a reasonable reaction rate. In view of the above, we have not pursued conjugations involving carbodiimide-type reagents because we wanted to avoid formation of cross-linked products, whose preparation in a reproducible way and characterization could be problematic. Instead, we focused on the popular, experimentally simple conjugation by reductive amination,[84–87] and conjugation by squaric acid chemistry, as the recently revived[88] interest in the latter method produced some impressive results at the time we started to work in synthetic vaccine development.

4.1 Synthesis of Mimics of the Upstream Terminus of the O-SP of *Vibrio cholerae* O1 in the Form Suitable for Conjugation

To synthesize linker-equipped hexasaccharides we had to change the synthetic strategy applied in the synthesis of methyl glycosides. Initially,[49,89] we followed the assembly protocol established by Peters and Bundle,[61] and constructed the hexasaccharide in the form of its α-2-(trimethylsilyl)ethyl (SE) glycoside from intermediates containing azido groups. The advantage of SE glycosides lies in the fact that they can be directly transformed into glycosyl chlorides.[90] Also, the aglycon can be selectively cleaved from the finished oligosaccharide leaving the interglycosidic linkages virtually intact, to give reducing sugars,[91] which are intermediates in the preparation of glycosyl donors (mainly imidates, but also glycosyl halides). The hexasaccharide was assembled from disaccharide building blocks. The two key intermediates **119** and **120** were prepared[49,92] (Scheme 14) starting from the benzylated azide **17** (Scheme 2). Good yields of 2-(trimethylsilyl)ethyl glycosides of most common sugars can be obtained conventionally, but synthesis of 2-(trimethylsilyl)ethyl α-D-mannopyranosides, such as **116**, is notoriously problematic.[91,93] Since α-D-mannopyranosylation of SEOH was observed[93] to be more efficient from a 2-*O*-benzoylated than from a fully acetylated mannopyranosyl donor, methyl glycoside **17** was first benzoylated (→**113**), and the resulting, fully protected methyl glycoside was subjected to acetolysis, to give **114**. Subsequent, sequential treatment of the latter with DCMME (→**115**) and 2-(trimethylsilyl)ethanol gave the desired compound **116**. Conventional deacylation followed by condensation of the product **117** with glycosyl chloride **100** then gave the fully protected disaccharide **118**, which was converted to the key disaccharide synthons, the disaccharide glycosyl donor and acceptor, **119** and **120**, respectively.

Having the foregoing intermediates in hand, construction of the SE α-glycoside of the hexasaccharide was accomplished (Scheme 15) by coupling of **119** with **120**, deacetylation of the formed **121** (→**122**), and repeating the elongation of the chain using the same glycosyl donor, to give the fully protected compound **123**. After deacetylation (→**124**) and methylation (→**125**), the six azido groups were reduced, and amidation of the formed **126** with tetronic acid derivative **36** (Scheme 6) gave the partially benzylated Ogawa hexasaccharide **127**. It is noteworthy that treatment of hexamine **126** with lactone **24** (Scheme 4) as for **56** (Scheme 10), did not give the desired hexasaccharide **127**.[49] Only a mixture of poorly resolved products was formed. The remaining steps from **127** to the desired hexasaccharide **133** involved (Scheme 16) debenzylation (→**128**) and exhaustive acetylation (→**129**), followed by cleavage of the SE glycoside with neat[49] TFA. The hemiacetal **130** thus obtained was then converted to the imidate glycosyl donor **131** and used for glycosylation of 5-(methoxycarbonyl)pentanol. Subsequent deacetylation of the per-*O*-acetyl derivative **132** gave finally the versatile linker-equipped hexasaccharide derivative **133**.

After evaluation of the aforementioned reaction sequence **17**→**133** (Schemes 14–16) we noted that three conversions needed to be improved, namely, the *N*-acylation of the hexasaccharide **126** (unsuccessful when effected with lactone **24**), the making of the SE glycoside **116** (giving only ~80% yield), and the glycosylation of 5-(methoxycarbonyl)pentanol with the hexasaccharide imidate **131** (~70% yield).

As already mentioned, replacement of lactone **24** with acid **36** resulted in considerable improvement of the *N*-acylation step. In anticipation of future need of larger amounts of SE glycosides of *V. cholerae* O1 oligosaccharides, we have studied in detail[94] the formation of SE α-mannopyranosides. As a result, we have been able to establish guidelines allowing to improve considerably the yield of these useful

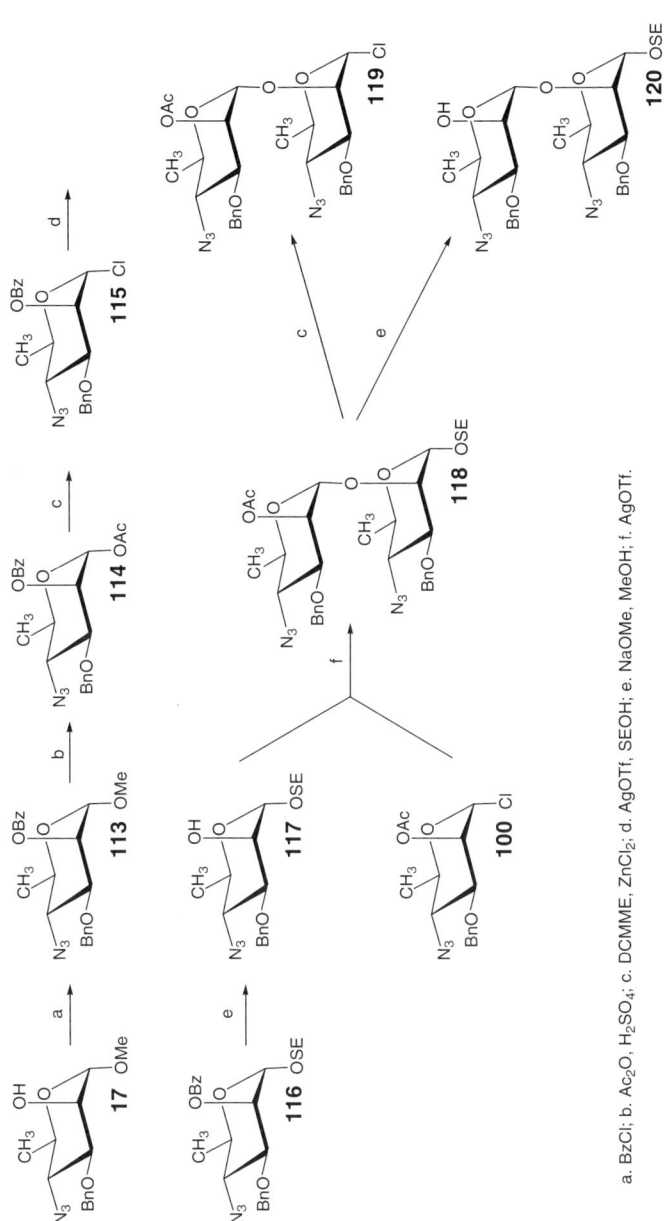

a. BzCl; b. Ac₂O, H₂SO₄; c. DCMME, ZnCl₂; d. AgOTf, SEOH; e. NaOMe, MeOH; f. AgOTf.

Scheme 14

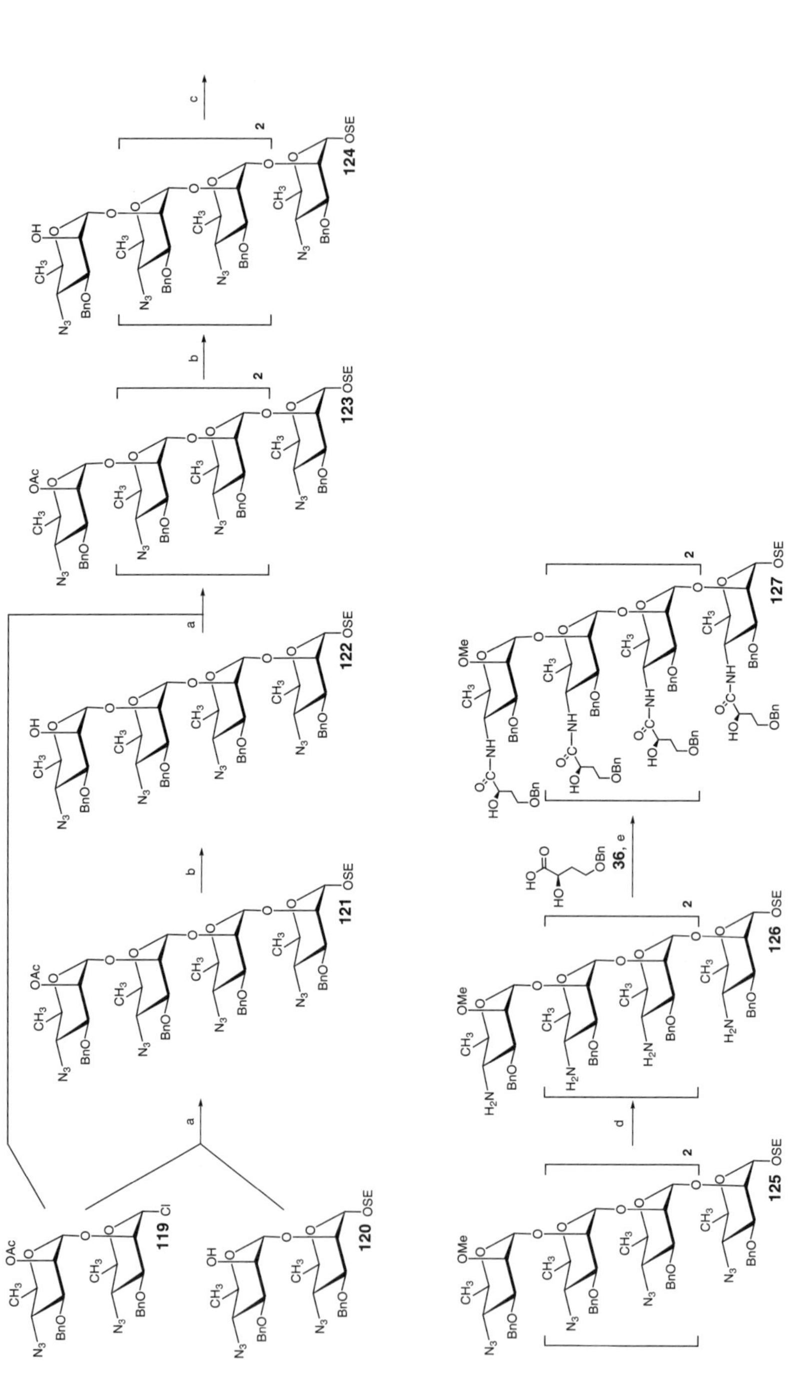

Scheme 15

a. AgOTf; b. NaOMe, MeOH; c. Ag₂O, MeI; d. H₂S; e. EDC.

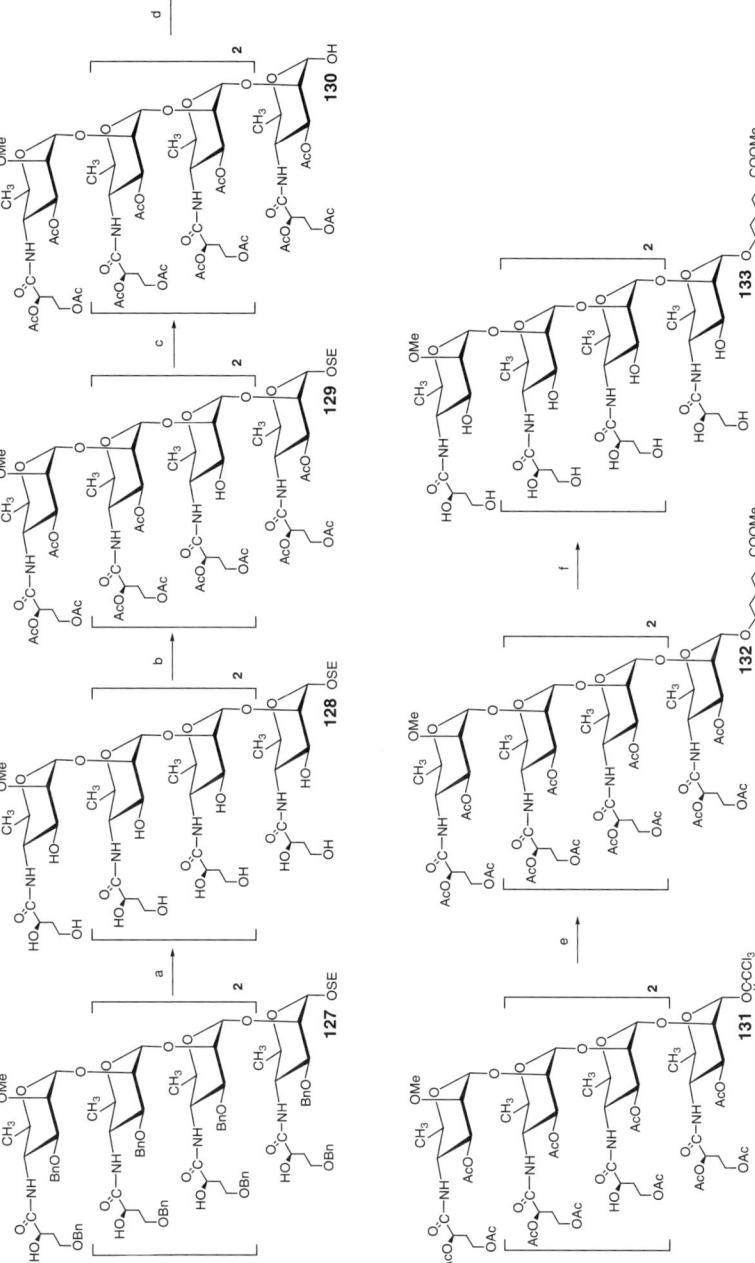

Scheme 16

a. H₂, Pd/C; b. Ac₂O, Py; c. TFA; d. TCA, DBU; e. 5-(methoxycarbonyl)pentanol, TESOTf; f. NaOMe, MeOH.

carbohydrate intermediates, applicable also to the preparation of the SE glycoside **116** (Scheme 14), which can be now obtained[94] in virtually theoretical yield from the corresponding ethyl 1-thioglycoside.

More fundamental changes in the original scheme were required to avoid the ~30% loss of product on the hexasaccharide level (**131**→**132**, Scheme 16). As shown in Scheme 17, the new synthetic strategy developed by us[95,96] involved introduction of the spacer (linker) on the monosaccharide level. Since the products of this revised synthetic scheme were intended for studies aimed at establishing the effect of the size of the antigen upon immunogenicity, the target compounds (from mono-through the pentasaccharide) were made in a stepwise manner. Also, the chain elongation of the oligosaccharides was achieved using intermediates having the amino group derivatized with 2,4-*O*-benzylidene-L-*glycero*-tetronic acid (**40**, Scheme 6). To avoid complications due to acetyl group migration, experienced during similar assembly[65,66] when the acetyl group was used as a permanent protecting group, the HO-3 in these perosamine building blocks was benzylated. Thus (Scheme 17), the key glycosyl donors **135** and **137** were made from the thioglycoside **98** (Scheme 13), whose azido group was selectively reduced, and the resulting amine **134** was *N*-acylated (**134**→**135**). To introduce the methyl group at *O*-2 and obtain the monosaccharide glycosyl donor **137**, the fully protected monosaccharide **135** was deacetylated and the alcohol **136** thus obtained was methylated. The starting glycosyl acceptor having both the *N*-tetronyl side chain and the linker already in place (**141**) was made from the linker-equipped monosaccharide **138**, accessible either from the thioglycoside **98** or directly from the anomeric acetate **97**. The pathway from **138** to **141** then followed the same reaction sequence as that from **98** to **136**. The Ogawa monosaccharide hapten **143** was obtained from **141** by methylation and deprotection by hydrogenolysis. Coupling of synthons **135** and **141** gave the fully protected disaccharide **144**. It was deacetylated, to give the glycosyl acceptor **145** suitable for extension of the oligosaccharide chain, or methylated (→**146**), to give the disaccharide hapten **147**, after final deprotection by hydrogenolysis.

Extension of the oligosaccharide chain in **145** up to the pentasaccharide **152** is shown in an abbreviated form in Scheme 18. Coupling of the former with the monosaccharide glycosyl donor **135** gave the fully protected trisaccharide **148**. Subsequent deacetylation gave alcohol **149**, which served as the glycosyl acceptor for the donor **135**. Repetition of these transformations with the product of such coupling, followed by final hydrogenolytic cleavage of the benzyl and benzylidene protecting groups, furnished the penta- and the hexasaccharide haptens in the Inaba series[95] (not shown in Scheme 18). When a methylation step preceded hydrogenolysis, the Ogawa trisaccharide **150** was obtained from the Inaba precursor **149**. The tetra- and the pentasaccharides **151** and **152**, respectively, were obtained in a similar way from the product of coupling of the glycosyl donor **135** with the glycosyl acceptor **149**.

The use of synthons **135** and **141**, having both the linker (spacer) and the tetronic side chain in place, diminished the number of chemical manipulations with the assembled oligosaccharides. Nevertheless, this was not a flawless approach. Although products of glycosylation reactions were eventually obtained in acceptable yields, these reactions were accompanied by formation of minor by-products, which were difficult to remove. Some of these side products resulted from incomplete

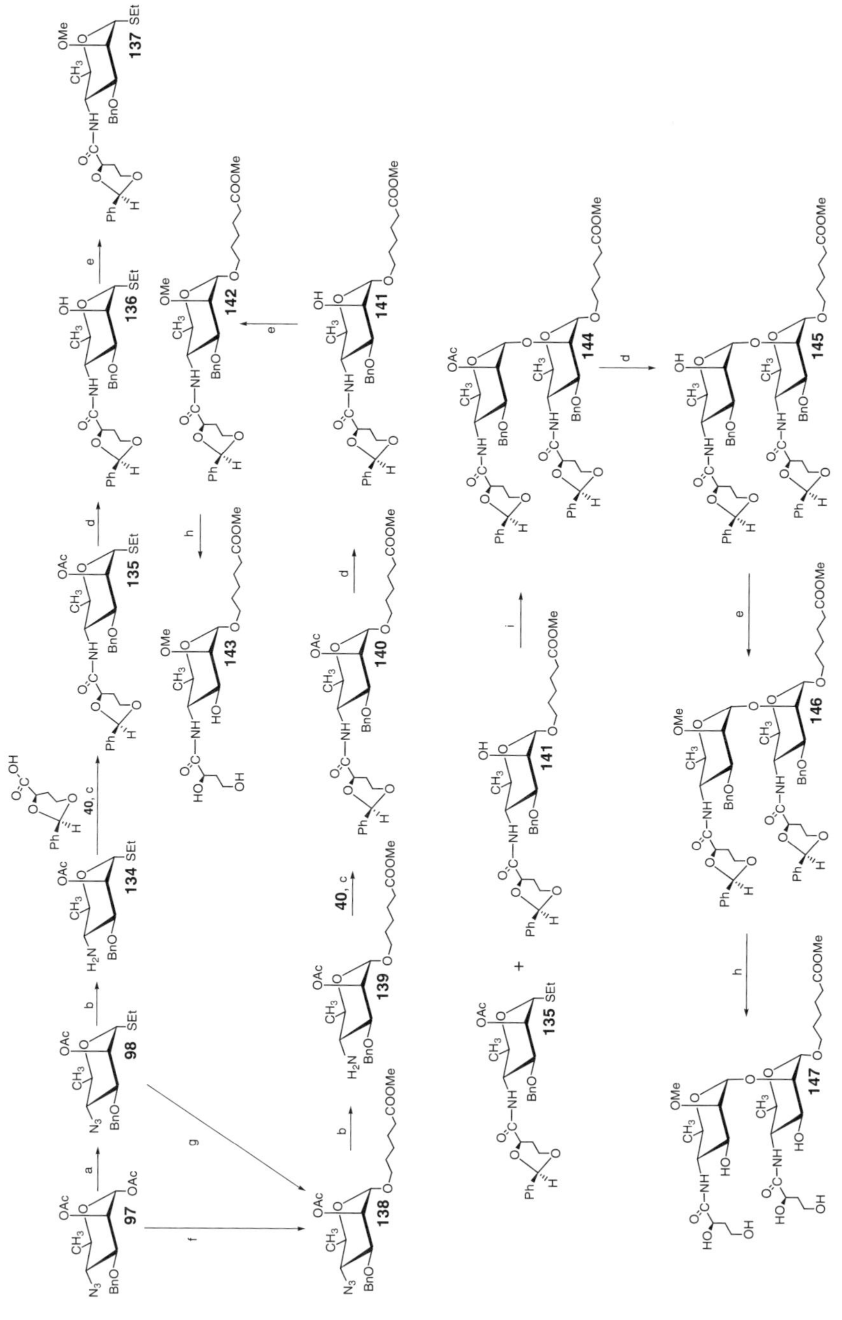

Scheme 17

a. EtSH, BF$_3$·Et$_2$O; b. H$_2$S; c. EDC; d. NaOMe, MeOH; e. MeI, Ag$_2$O; f. 5-(methoxycarbonyl)pentanol, SnCl$_4$; g. 5-(methoxycarbonyl)pentanol, AgOTf, NIS; h. H$_2$, Pd/C, i. AgOTf, NIS.

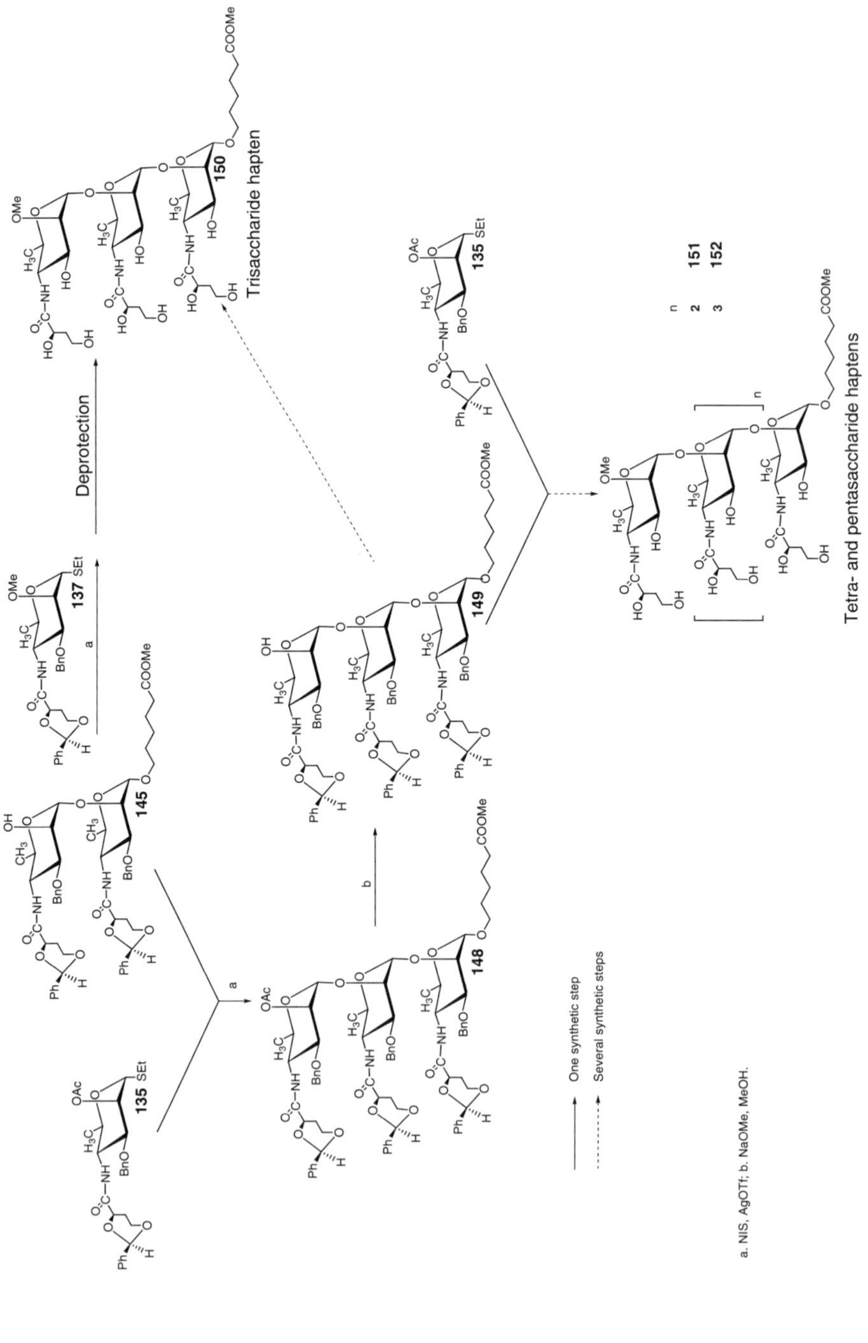
135
145
137
148
149
150
135
151
152
n
2
3
a
b
Deprotection
Trisaccharide hapten
Tetra- and pentasaccharide haptens
One synthetic step
Several synthetic steps
a. NIS, AgOTf; b. NaOMe, MeOH.

Scheme 18

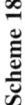

stability of the benzylidene group during glycosylation. NMR spectra of some of these materials showed a smaller ratio of benzylidene groups to the number of anomeric protons and carbons per molecule than would be expected for the desired products. Thus, we are still searching for a more efficient synthetic approach toward oligosaccharide fragments of the O-SP of *V. cholerae* O1.

4.2 Conjugation by Reductive Amination

Application of this chemistry in the carbohydrate field is based largely on the work by Gray and co-workers.[84,85,87,97] Albeit with unimpressive efficiency of conjugation (<20%), Verez-Bencomo's group successfully attached *V. cholerae* O1 mono- and disaccharides to BSA.[15] Because of the general popularity of the method in the carbohydrate field, we also have examined the utility of reductive amination as a means for conjugation of linker-equipped oligosaccharides to proteins. For this purpose, we developed the linker molecule 153 (Scheme 19). The foregoing dimethyl acetal was readily obtained[81,83] by treatment of the commercially available aminoacetaldehyde dimethyl acetal with excess (~5–10 molar equivalents) of 6-caprolactone. The latent aldehyde group in glycosyl derivatives of 153 can be unmasked by mild acid hydrolysis. We prepared glycosides of 153 from the monosaccharide and the hexasaccharide that mimic the terminus of the O-SP of *V. cholerae* O1 serotype Ogawa, and conjugated them to BSA. The monosaccharide 157 was prepared[83] (Scheme 20) from the product (154) of acetolysis of the known[66,83] Ogawa monosaccharide 54 (Scheme 9) through intermediates 155 and 156. Synthesis[98] of the corresponding hexasaccharide 167 started (Scheme 21) with the disaccharide thioglycoside 101 (Scheme 13). It was first equipped with the linker arm, and the formed 158 was deacetylated (→159). Condensation with 101 and deacetylation steps were repeated twice (→160→161→162→163), and subsequent methylation gave the hexasaccharide 164. Coupling of amine 165, obtained from 164 by reduction with hydrogen sulfide, with acid 40 (Scheme 6) gave the fully protected hexasaccharide 166. Simultaneous removal of the benzyl and benzylidene protecting groups then gave 167.

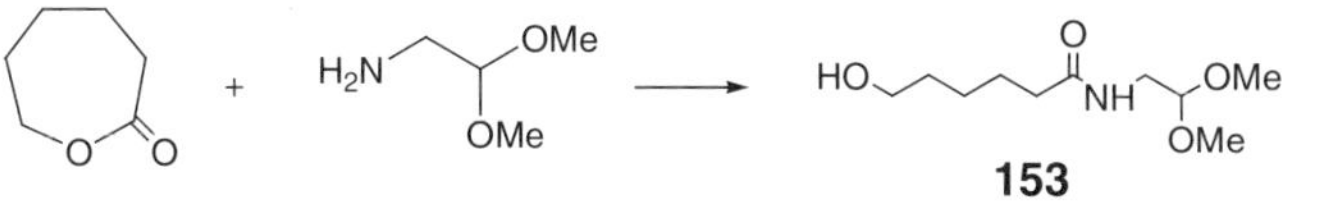

Scheme 19

Results of our studies[81,83] confirmed findings by the Cuban group[15] that the efficiency of conjugation by reductive amination is rather low when it involves synthetic oligosaccharides and high-molecular-mass proteins. The low efficiency is caused in part by the presumably low reactivity of the high-molecular-mass material and more likely, perhaps, by more important chemical factors. When carbonyl compounds, such as glycosides of carbohydrates whose aglycon contains an aldehyde group, are allowed to react with amino functions present in proteins, an unstable intermediate Schiff base (an imine) is formed through a reversible reaction. The Schiff base has to be reduced to a secondary amine, to form a stable chemical bond. This can be achieved with various reducing agents. Among those, sodium cyanoborohydride and pyridine–borane complex have been described as reagents most suitable for this purpose. However, it appears[83,99–101] that the chemospecificity[19,102] or very high chemoselectivity[20,103,104] of these reagents for reduction of Schiff base has been largely exaggerated. Clearly, any alcohol formed from the starting aldehyde, as a result of nonspecific reduction, during conjugation by reductive amination constitutes irreversible loss of the very precious synthetic material. Consequently, notwithstanding its usefulness for conjugating poly-saccharides, reductive amination is not suitable for economical conjugation of syn-thetic oligosaccharides to proteins, due to the rather poor efficiency of conjugation (expressed as a ratio of the amount of the hapten chemically attached to the carrier over the amount of the hapten added at the onset of the conjugation).

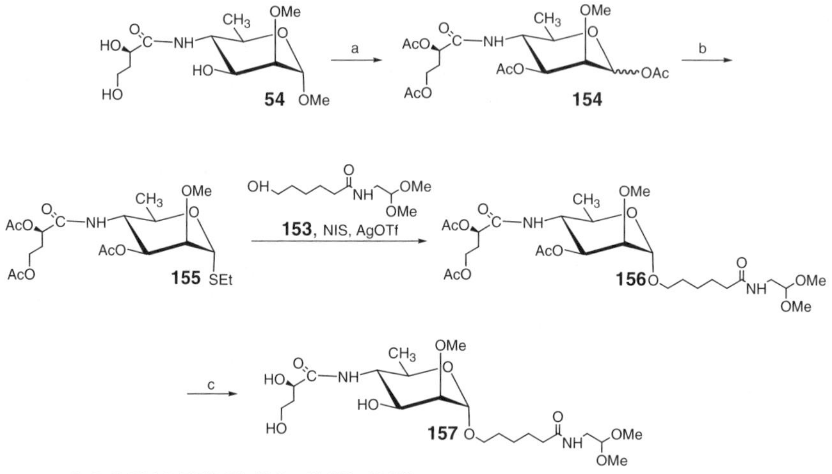

a. Ac₂O, H₂SO4; b. EtSH, BF₃.Et₂O; c. NaOMe, MeOH.

Scheme 20

4.3   Conjugation by Squaric Acid Chemistry

This method for conjugation of carbohydrates to proteins does not suffer from the above-mentioned shortcomings of reductive amination. It is based on the chemistry discovered by Tieze and co-workers.[105–107] More recently, Kamath and co-workers[88] showed that efficient conjugation of carbohydrates to proteins can be done in this way on a very small scale. The method relies on the difference in reactivity between the ester groups of a squaric acid diester and a squaric acid amide ester. At pH 7, a dialkyl squarate reacts readily with amines to give a squaric acid amide ester. Further con-version to a squaric acid diamide requires pH >9. When we compared[83,108,109] the efficiency of conjugation by this method with that of reductive amination, the conju-gation involving squaric acid chemistry was found superior. Another advantage of the conjugation with the squaric acid chemistry is that a large part of the of unchanged squaric acid derivative, which is normally added in excess at the onset of the conju-gation, can be recovered from the conjugation mixture.[110,111] Needless to say, since synthetic oligosaccharides are very expensive materials, conjugation of such sub-stances applying this methodology is particularly advantageous. Scheme 22 shows

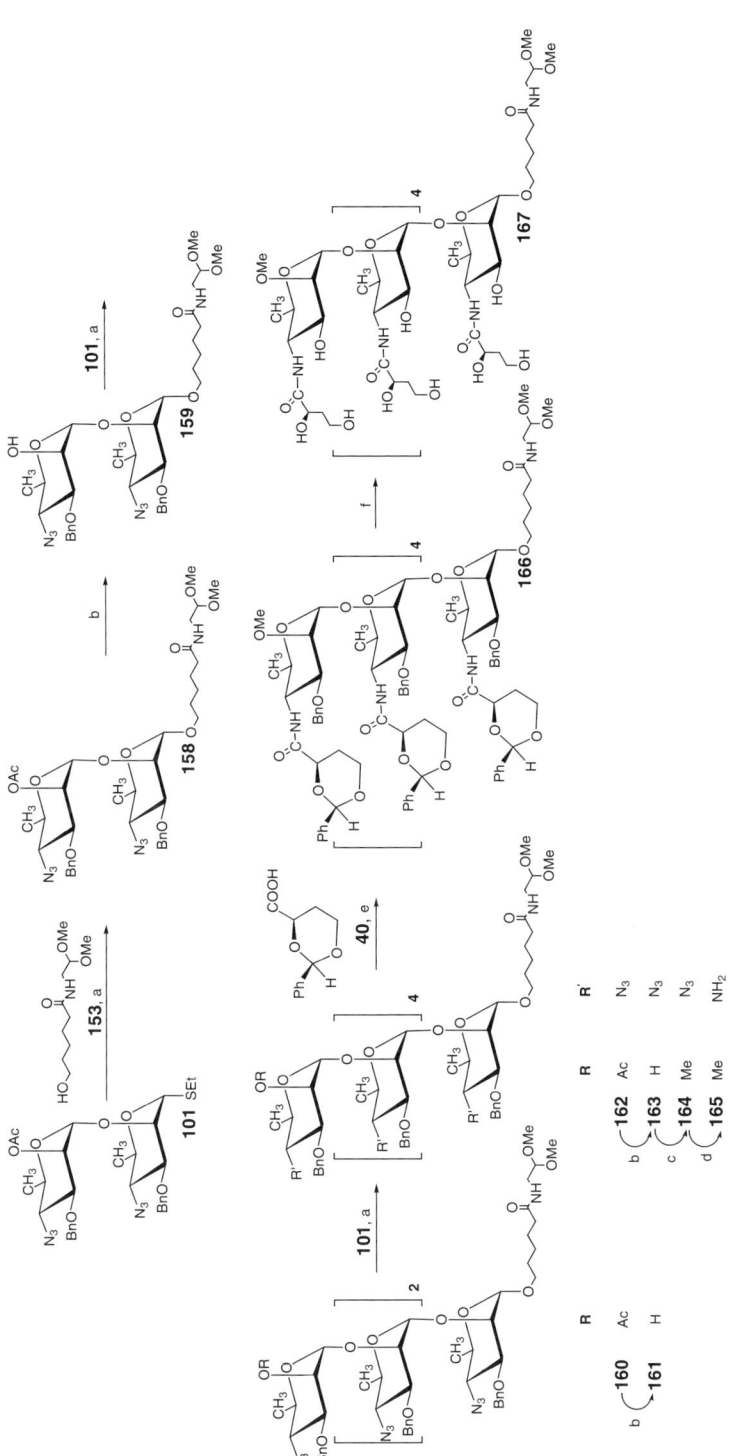

Scheme 21

conjugation by the squaric acid chemistry of a hypothetical carbohydrate hapten to an amine-containing carrier, such as a protein. The hapten is synthesized in the form of a glycoside whose aglycon contains a primary amino group (**A**). Treatment of such an amine with one of the commercially available squaric acid diesters at pH 7 gives the corresponding amide ester (**B**). When this monoester is treated at pH 9 with another amine, *e.g.*, a carrier protein, a glycoconjugate (**C**) is formed. Using this method of conjugation, we have converted mono- through the hexasaccharide fragments of the O-SP of *V. cholerae* O1 serotype Ogawa,[96,110] and also some saccharides in the Inaba series,[95,112] into a large number of neoglycoconjugates, and tested their immunogenicity.

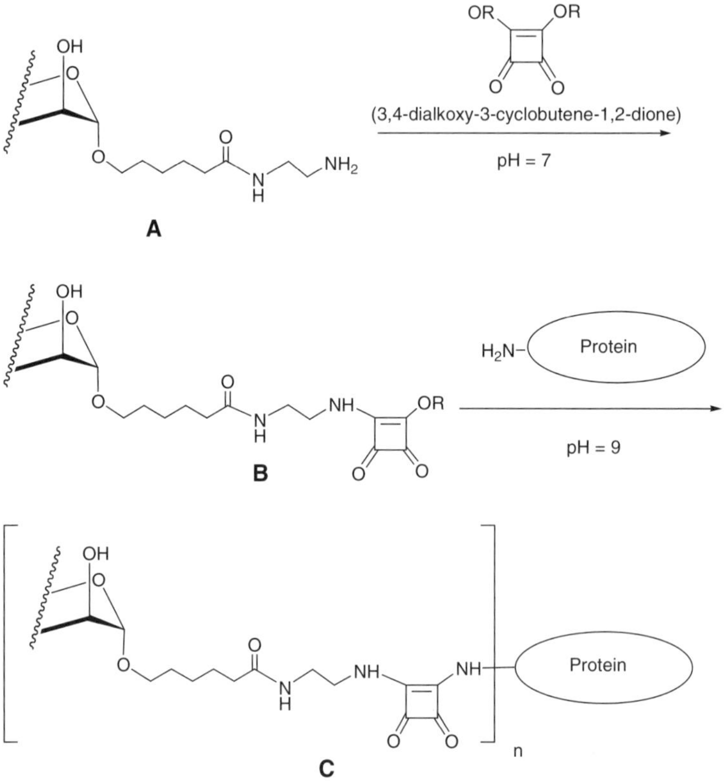

Scheme 22

5   Determination of the Carbohydrate–Protein Ratio in the Neoglycoconjugates

The number of oligosaccharide units incorporated in the carrier protein is a parameter that decisively affects the immunogenicity of the glycoconjugates. Until the last quarter of the 20th century the only means available to determine the carbohydrate–protein ratios in neoglycoconjugates were various colorimetric methods, most notably the phenol–sulfuric acid test developed by Dubois and co-workers.[113] In spite of the usefulness of the method in many areas, its accuracy is rather questionable when applied to bacterial heteropolysaccharides and neoglycoconjugates prepared from them, because a reliable calibration curve is difficult, and many times virtually impossible to construct. Also, bacterial polysaccharides often contain rare, unstable sugars that decompose nonspecifically under the conditions of analysis, which further impairs the accuracy of the determination.

A modern, objective means for determination of the carbohydrate–protein ratio in neoglycoconjugates is provided by electrospray and matrix-assisted laser desorption/ionization time of flight mass spectrometry (MALDI-TOF MS) or electron spray mass spectrometry (ES MS). Depending on the resolution power of the spectrometers, the spectra obtained can be informative not only regarding the average loading of the carrier with the carbohydrate, but also concerning the polydispersity of the conjugate.[88] Despite the high degree of sophistication the instruments have attained, the quality of mass spectra generated by these techniques depends largely on the mode of sample preparation and the nature of contaminants. Consequently, the method is far from routine, and is normally applicable only for characterization of highly purified products.

6 Development of Protocols for Efficient and Controlled Preparation of a Series of Neoglycoconjugates with Predetermined Carbohydrate–Protein Ratio

Our intended studies of the effect of the hapten density (loading) upon immunopotency required reproducible preparation of a series of neoglycoconjugates with predictable carbohydrate–protein ratio. This is not how neoglycoconjugates from synthetic oligosaccharides were made at the time we began our work on the cholera vaccine. From the analytical data reported,[88,114,115] the trial-and-error nature of those protocols is rather obvious: the reaction of the carrier with the hapten was allowed to proceed for some time, usually overnight or longer, *without any monitoring during the course of the reaction*, and the product was isolated. It had to be purified, to satisfy the requirements of mass spectroscopic methods, and the molecular mass was then determined. With a little luck, experiments conducted in this way either provided products that satisfied the needs, or provided leads for the design of subsequent experiments, targeted at glycoconjugates with different carbohydrate loading. Clearly, there was a need for a method allowing routine monitoring of the conjugation reaction. Ideally, such a method should be analogous to thin-layer chromatography, which is used as an analytical tool during syntheses of low-molecular-mass organic molecules. Thus, it should allow rapid analysis of samples withdrawn directly from the conjugation mixture and provide, within minutes, reliable information about the increasing molecular mass of the conjugate that was being formed, thereby allowing termination of the process when the desired loading is achieved. We found a tool[110] for such monitoring of the conjugation reaction in surface-enhanced laser desorption/ionization-time-of-flight mass spectrometry (SELDI-TOF MS) in combination with the ProteinChip® System, a technique which was developed for a completely different purpose. SELDI-TOF MS is closely related to MALDI-TOF MS, but the use of selectively active surfaces allows purification of the sample directly on the ProteinChip®, prior to mass spectral analysis. Due to the properties of these surfaces, the ProteinChip® system is almost ideally suited for direct monitoring of mass shift changes resulting from chemical modifications of carrier proteins. This new technology constitutes a fundamental and powerful addition to the tools available to the conjugation chemist, as it allows routine and rapid analysis of the conjugation mixture on a picomolar level, without tedious purification. Using the above technique, we can now withdraw samples from conjugation reaction mixtures, analyze them rapidly, and obtain near real-time information about the increasing molecular weight of the neoglycoconjugate. The results are obtained in the form of spectra such as shown in Figure 5. Knowing the molecular mass of the hapten and the carrier, the average number of haptens attached per one carrier molecule can be easily calculated from the difference between the molecular mass recorded for the sample of the conjugate and that of the starting carrier. The shoulder-less shape of the Gaussian curve showed in spectrum B indicates absence of underivatized BSA after 1 h of the reaction time (for comparison, see Figure 6), and an average hexasaccharide-BSA ratio of 3.8. Under the given conditions, on average, 22 hexasaccharide moieties could be attached after the reaction time of 54 h.

As already indicated by Verez-Bencomo's conjugation of a mono- and a disaccharide by reductive amination,[15] conjugation takes place more readily with small than with larger molecules. This is even more evident from our comparison of conjugation of mono- through the pentasaccharide fragments of *V. cholerae* O1, serotype Ogawa.[96] With higher oligosaccharides, the reaction rate decreases mainly at the later stages of the conjugation reaction, when these larger molecules have to penetrate deeper into the three-dimensional structure of the protein, to reach the less accessible amino groups of the protein.

When the conjugation of low-molecular-mass haptens (mass of <500 Da) is monitored by SELDI-TOF MS, it may be difficult to decide upon the right time to terminate the conjugation reaction. Because of the mechanics involved in the construction of the SELDI-TOF MS instrument, an inherent, albeit small, instrument error is involved in the reading of the molecular mass of samples, even with fine calibration. Further errors of mass reading may arise when the sample(s) and the calibration standard(s) are on a different chip, as is frequently the case. We have looked into the possibility to minimize these errors, and have found solution of the problem in using the carrier protein as an internal standard.[96] When the carrier protein is added to the sample analyzed, the laser beam reads the internal standard and the sample at the same location on the chip, and the errors resulting from the above-described phenomena are virtually eliminated. Figure 6 shows a comparison of molecular mass determination with internal and external calibration of the instrument. With internal calibration, the spectrum shows a lower mass peak, or a shoulder on the main peak, which corresponds to the internal standard, and a higher mass peak corresponding to the mass of the glycoconjugate formed. The more correct carbohydrate–protein ratio can be calculated from the difference between the two mass values.

The ability to monitor conjugation reactions and obtain near-real-time information about the size of the glycoprotein being formed has brought the process of

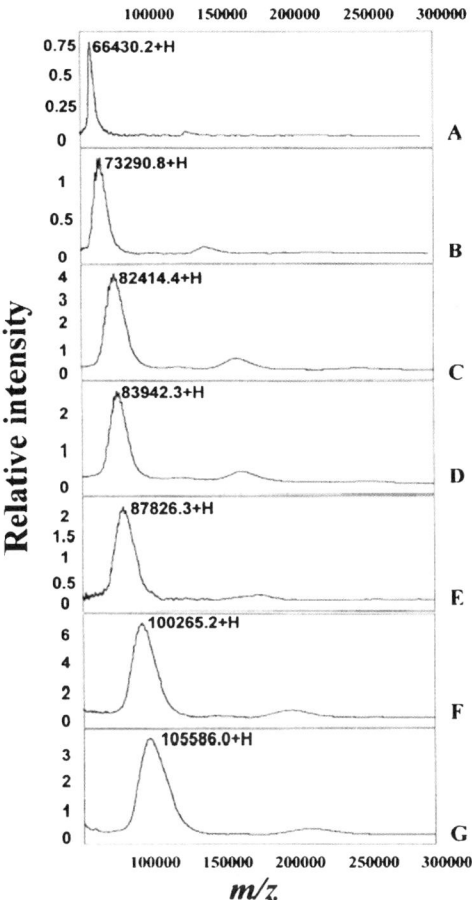

Figure 5 *The progress of conjugation of the Ogawa hexasaccharide (mol. wt. 1780.79 Da) and BSA (carrier) (mol. wt. 66,430 Da) as revealed by surface-enhanced laser desorption/ionization time of flight (SELDI-TOF) mass spectrometry. Prior to the actual measurement, the instrument was calibrated with a series of standards covering a wide range of molecular masses. These calibrants are often applied and measured on a separate ProteinChip®. For this measurement, the starting carrier BSA, serving as an additional calibrant, and individual samples were applied on different spots of the same ProteinChip®. For the way how these arrangements may affect the accuracy of molecular mass reading, see text. The conjugation was carried out at the initial hapten/BSA ratio of 80/1. Spectrum A was taken at the onset of the reaction (t = 0), and it shows the molecular mass of unchanged carrier. Spectra B through G were taken at t = 1, 3, 7, 9, 27, and 54 h, respectively, whereafter virtually no increase of molecular mass could be recorded*

making neoglycoconjugates to a new, higher level of sophistication. When the work requires a large number of conjugates with different degree of protein derivatization, our newly developed refined protocol[96] makes it possible to prepare a *series* of conjugates with a *predetermined* carbohydrate–protein ratio as a one-pot preparation.

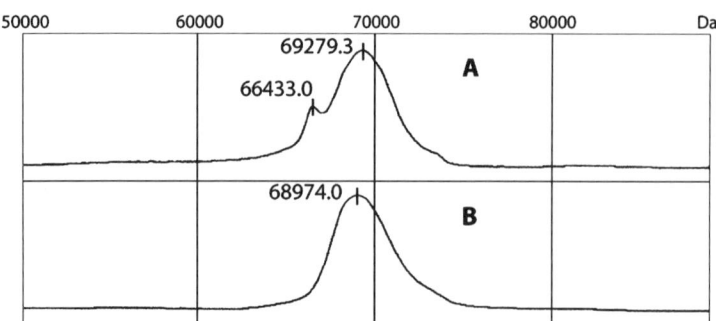

Figure 6 *A comparison of mass reading by SELDI-TOF MS of the same glycoconjugate pre-pared from the linker-equipped, terminal monosaccharide determinant of V. cholerae O1, serotype Ogawa (mol. wt. 515 Da) and BSA (mol. wt. 66,430 Da). With internal calibration (A), the determined hapten-BSA ratio is 5.5. With external calibration using a standard mixture of proteins (B) the determined hapten-BSA ratio is 4.9*

The conjugation reaction is set up with the amount of reactants sufficient to prepare several conjugates, and the reaction is monitored by SELDI-TOF MS. When the desired molecular mass is reached, a requisite amount of the mixture is withdrawn and processed, the remaining material is allowed to react until SELDI-TOF MS shows that the next desired conjugate had been formed, etc. Using the one-pot preparation protocol with the ability to monitor the conjugation reaction, the further advancement in the field of conjugate vaccines is becoming more realistic than ever before.

7   Serological Evaluation of Responses in Mice Following Immunization with Synthetic Immunogens

It has been concluded[116,117] from previous studies that infant mouse studies are predic-tive of the response of human infants to glycoconjugate immunization. The first series of neoglycoconjugates evaluated[118] for their immunogenicity and protective capacity was prepared[110] from the hexasaccharide that mimics the upstream terminus of the O-SP of *V. cholerae* O1, serotype Ogawa. Constructs with carbohydrate-carrier molar ratios of 15.5, 9.2, and 4.6 were tested for their immunogenicity in adult mice, and the protective efficacy of the resulting sera was tested in neonatal mice. The role of preim-munity to BSA and the use of adjuvant (RIBI®) in the generation of the O-SP specific antibodies, as well as protection against virulent *V. cholerae* O1 was also examined. Preimmunity to BSA did not affect the anti-Ogawa titers or the protective capacity of the antisera. Particularly encouraging was the finding that all three constructs elicited antibodies specific for LPS without addition of the adjuvant, although the latter was effective at inducing higher and earlier antibody responses. Assessing tertiary sera, we found a correlation between vibriocidal activity and protection. The protective capac-ity of antisera was evident in sera raised from all constructs, but was highest in the groups receiving the lowest substituted (hexasaccharide/BSA ~5) conjugate.

The second series of conjugates evaluated[119] was prepared from BSA and the terminal di-, tetra-, and the hexasaccharide of the O-SP of *V. cholerae* O1, serotype Inaba, having different degree of derivatization.[95,111] BALB/c mice responded serologically to the Inaba conjugates, but the serum antibodies (IgM, IgG1) were found neither vibriocidal nor protective in the infant mouse cholera model. Results of these studies raise the question as to whether immunogens prepared from mimics of the Inaba O-SP can select Inaba LPS-specific B cells that will eventually produce protective antibodies. Nevertheless, it was encouraging to find that immunization with LPS of *V. cholerae* can effectively prime the host to respond to a subsequent boost with conjugates prepared from synthetic *V. cholerae* Inaba oligosaccharides. Mice primed (ip) with two doses of Ogawa LPS and then inoculated (sq) 10 days later with oligosaccharide conjugates in QS21 adjuvant made a significant amount of anti-Ogawa LPS antibodies. A similar experiment with Inaba LPS priming at day 0, followed by a boost (day 5 or 7) with Inaba trisaccharide conjugate resulted in additive anti-Inaba titers (10 days), which were highly vibriocidal. When Inaba conjugates were used to prime on day 0 and LPS to boost on day 7, ELISA titers were found to be similar to those when LPS was delivered first, but these sera were not vibriocidal.

8 Concluding Remarks

We have been able to considerably improve the methodology of neoglycoconjugate preparation and characterization. Squaric acid chemistry is an efficient means for conjugation of labor-intensive, synthetic oligosaccharides to carrier proteins. The method also offers the possibility to recover from the conjugation mixture a portion of ligand that had been used in excess at the onset of the conjugation. Engagement of SELDI-TOF MS makes it possible to obtain real-time monitoring of the conjugation process, and allows efficient preparation of neoglycoconjugates with predetermined carbohydrate–protein ratios. We have shown that glycoconjugates prepared from fragments that mimic the upstream terminus of the O-SP of *V. cholerae* O1 can be potent in eliciting antibodies that react with the LPS, and that these antibodies have protective capacity. What remains to be accomplished is to prepare a *medically acceptable*, and *clinically useful* conjugate vaccine for cholera from synthetic carbohydrates. This is still a formidable task, as we have to find the right configuration/architecture of the conjugate. This will involve finding the right carrier protein and establish the minimum oligosaccharide size required to elicit the desired immunoresponse, when used as the antigenic component of the vaccine. We continue working toward these goals, as well as simplifying the synthetic procedures involved, to make production of conjugate vaccine for cholera also potentially amenable to industrial production. The search for a potent conjugate vaccine for cholera continues.[120]

Abbreviations

Ac	Acetyl
AcBr	Bromoacetyl
AgOTf	Silver trifluoromethanesulfonate
BMS	Borane-methyl sulfide complex
Bn	Benzyl

Boc   t-Butoxycarbonyl
BTMA   Borane-trimethylamine complex
Bz   Benzoyl
CAN   Ceric ammonium nitrate
DBU   1,8-Diazabicyclo[5,4,0]-undec-7-ene
DCMME   α,α-Dichloromethyl methyl ether
DDQ   2,3-Dichloro-5,6-dicyano-1,4-benzoquinone
DMP   2,2-Dimethoxypropane
DMSO   Dimethyl sulfoxide
EDC   1-(3-Dimethylaminopropyl)-3-ethylcarbodiimide hydrochloride
EEDQ   2-ethoxy-1-ethoxycarbonyl-1,2-dihydroquinoline
HATU   O-(7-azobenzotriazol-1-yl)-1,1,3,3-tetramethyluronium
             hexafluorophosphate
MBn   p-methoxybenzyl
Me   Methyl
Ms   Methanesulfonyl (Mesyl)
Py   Pyridine
SE   2-(Trimethylsilyl)ethyl
SEOH   2-(Trimethylsilyl)ethanol
TBAB   Tetrabutylammonium bromide
TBI   2,4,5-Tribromo-1H-imidazole
TCA   Trichloroacetonitrile
TEA   Triethylamine
TESOTf   Triethylsilyl trifluoromethanesulfonate
Tf   Trifluoromethanesulfonyl (Triflyl)
TFA   Trifluoroacetic acid
TII   2,4,5-Triiodo-1H-imidazole
TPP   Triphenylphosphine

References

1. S. Chatterjee and K. Chaudhuri, *Biochem. Biophys. Acta*, 2003, **1639**, 65.
2. P.A. Manning, U.H. Stroeher and R. Morona, in *Vibrio cholerae and Cholera: Molecular to Global Perspectives*, I.K. Wachsmuth, P.A. Blake and O. Olsvik (eds), American Society for Microbiology, Washington, DC, 1994, p. 77.
3. E.B. Steinberg, K.D. Greene, C.A. Bopp, D.N. Cameron, J.G. Wells and E.D. Mintz, *J. Infect. Dis.*, 2001, **184**, 799.
4. M.L. Bennish, in *Vibrio cholerae and Cholera: Molecular to Global Perspectives*, I.K. Wachsmuth, P.A. Blake and O. Olsvik (eds), American Society for Microbiology, Washington, DC, 1994, pp. 229–255.
5. E.T. Rietschel, L. Brade, B. Lindner and U. Zähringer, in *Bacterial Endotoxic Lipopolysaccharides*, Vol. 1, D.C. Morrison and J.L. Ryan (eds), CRC Press, Boca Raton, FL, 1992, p. 3.
6. U. Zähringer, B. Lindner and E.T. Rietschel, *Adv. Carbohydr. Chem. Biochem.*, 1994, **50**, 211.
7. J.B. Robbins, R. Schneerson, S.C. Szu and V. Pozsgay, *Pure Appl. Chem.*, 1999, **71**, 745.

8. P. Anderson, M.E. Pichichero and R.A. Insel, *J. Clin. Invest.*, 1985, **76**, 52.
9. O.T. Avery and W.F. Goebel, *J. Exp. Med.*, 1929, **50**, 533.
10. W.F. Goebel and O.T. Avery, *J. Exp. Med.*, 1929, **50**, 521.
11. S. Kabir, *J. Med. Microbiol.*, 1987, **23**, 9.
12. R.K. Gupta, S.C. Szu, R.A. Finkelstein and J.B. Robbins, *Infect. Immun.*, 1992, **60**, 3201.
13. J.A. Benitez, A.J. Silva, B.L. Rodriguez, R. Fando, J. Campos, A. Robert, H. Garcia, L. Garcia and J.L. Perez, *Arch. Med. Res.*, 1996, **27**, 275.
14. N.I. Smirnova, L.F. Livanova, G.V. Chekhovskaya, G.A. Eroshenko, Y.V. Lazovsky and T.L. Zakharova, *Zh. Mikrobiol. Epidemiol. Immunobiol.*, 2000, 47.
15. A. Ariosa-Alvarez, A. Arencibia-Mohar, O. Madrazo-Alonso, L. Garcia-Imia, G. Siera-Gonzalez and V. Verez-Bencomo, *J. Carbohydr. Chem.*, 1998, **17**, 1307.
16. V. Verez-Bencomo, V. Fernández-Santana, E. Hardy, M.E. Toledo, M.C. Rodríguez, L. Heynngnezz, A. Rodriguez, A. Baly, L. Herrera, M. Izquierdo, A. Villar, Y. Valdés, K. Cosme, M.L. Deler, M. Montane, E. Garcia, A. Ramos, A. Aguilar, E. Medina, G. Torano, I. Sosa, I. Hernandez, R. Martínez, A. Muzachio, A. Carmenates, L. Costa, F. Cardoso, C. Campa, M. Diaz and R. Roy, *Science*, 2004, **305**, 522.
17. J.M. Cruse and R.E. Lewis (eds), *Conjugate Vaccines*, Karger, New York, 1989.
18. I.K. Wachsmuth, P.A. Blake and O. Olsvik (eds), *Vibrio cholerae and Cholera*, ASM Press, Washington, DC, 1994.
19. G.T. Hermanson, *Bioconjugate Techniques*, Academic Press, New York, 1996.
20. Y.C. Lee and R.T. Lee, *Neoglycoconjugates: Preparation and Application*, Academic Press, New York, 1994.
21. S.S. Wong, *Chemistry of Protein Conjugation and Cross-linking*, CRC Press, Boca Raton, FL, 1993.
22. W.E. Dick Jr. and M. Beurret, in *Conjugate Vaccines*, Vol. 10, J.M. Cruse and R.E. Lewis Jr. (eds), Krager, Basel, 1989, p. 48.
23. K.E. Stein, *Int. J. Technol. Assess. Health Care*, 1994, **10**, 167.
24. C.P. Stowell and Y.C. Lee, *Adv. Carbohydr. Chem. Biochem.*, 1980, **37**, 225.
25. S.M. Dimick, Ph.D. Thesis, Duke Univ., Durham, N.C., p. 185, 1999.
26. R.T. Lee and Y.C. Lee, *Glycoconjugate J.*, 2001, **17**, 543.
27. R. Roy, *Curr. Opin. Struct. Biol.*, 1996, **4**, 692.
28. R.Z. Dintzis, *Pediatr. Res.*, 1992, **32**, 370.
29. L. Kenne, B. Lindberg, P. Unger, B. Gustafsson and T. Holme, *Carbohydr. Res.*, 1982, **100**, 341.
30. McNaught, *Carbohydr. Res.*, 1997, **297**, 1.
31. K. Hisatsune, S. Kondo, Y. Isshiki, T. Iguchi and Y. Haishima, *Biochem. Biophys. Res. Commun.*, 1993, **190**, 302.
32. Y. Isshiki, S. Kondo, Y. Haishima, T. Iguchi and K. Hisatsune, *J. Endotoxin Res.*, 1996, **3**, 143.
33. J.W. Redmond, *Biochim. Biophys. Acta*, 1978, **542**, 378.
34. C.L. Stevens, R.P. Glinski, K.G. Taylor, P. Blumberg and S. K. Gupta, *J. Am. Chem. Soc.*, 1970, **92**, 3160.

35. C.-H. Lee and C.P. Schaffner, *Tetrahedron Lett.*, 1966, **7**, 5837.
36. L. Kenne, P. Unger and T. Wehler, *J. Chem. Soc. Perkin Trans. 1*, 1988, 1183.
37. C.L. Stevens, R.P. Glinski, K.G. Taylor, P. Blumberg and F. Sirokman, *J. Am. Chem. Soc.*, 1966, **88**, 2073.
38. S. Hanessian, *Chem. Commun.*, 1966, 796.
39. J.S. Brimacombe, O.A. Ching and M. Stacey, *J. Chem. Soc. (C)*, 1969, 1270.
40. M.J. Eis and B. Ganem, *Carbohydr. Res.*, 1988, **176**, 316.
41. D.R. Bundle, M. Gerken and T. Peters, *Carbohydr. Res.*, 1988, **174**, 239.
42. M. Gotoh, C.L. Barnes and P. Kováč, *Carbohydr. Res.*, 1994, **260**, 203.
43. M. Gotoh and P. Kováč, *J. Carbohydr. Chem.*, 1994, **13**, 1193.
44. J.W.E. Glattfeld and F.V. Sander, *J. Am. Chem. Soc.*, 1921, **43**, 2675.
45. P. Brewster, F. Hiron, E.D. Hughes, C.K. Ingold and P.A.D.S. Rao, *Nature*, 1950, **166**, 178.
46. A. Arencibia-Mohar, A. Ariosa-Alvarez, O. Madrazo-Alonso, E.G. Abreu, L. Garcia-Imia, G. Sierra-Gonzalez and V. Verez-Bencomo, *Carbohydr. Res.*, 1998, **306**, 163.
47. P.-s. Lei, Y. Ogawa, J.L. Flippen-Anderson and P. Kováč, *Carbohydr. Res.*, 1995, **275**, 117.
48. P.-s. Lei, Y. Ogawa and P. Kováč, *J. Carbohydr. Chem.*, 1996, **15**, 485.
49. Y. Ogawa, P.-s. Lei and P. Kováč, *Carbohydr. Res.*, 1996, **293**, 173.
50. M. Fouquey, J. Polonsky and E. Lederer, *Bull. Soc. Chim. Biol.*, 1958, **40**, 315.
51. A.H.C. Chang, D. Horton and P. Kováč, *Tetrahedron: Asymmetry*, 2000, **11**, 595.
52. E.J. Corey and B. Samuelsson, *J. Org. Chem.*, 1984, **49**, 4735.
53. P.J. Garegg and B. Samuelsson, *Carbohydr. Res.*, 1978, **67**, 267.
54. M. Gotoh and P. Kováč, *J. Carbohydr. Chem.*, 1993, **12**, 981.
55. T. Ogawa and M. Matsui, *Carbohydr. Res.*, 1978, **62**, C1.
56. M.A. Nashed, *Carbohydr. Res.*, 1978, **60**, 200.
57. P. Fugedi and P. Nanasi, *J. Carbohydr. Nucleosides, Nucleotides*, 1981, **8**, 547.
58. P. Kováč, in *Handbook of Derivatives for Chromatography*, K. Blau and J.M. Halket (eds), Wiley, Chichester, 1993, p. 109.
59. P.-M. Aberg, L. Blomberg, H. Lonn and T. Norberg, *J. Carbohydr. Chem.*, 1994, **13**, 141.
60. E. Eichler, J. Kihlberg and D.R. Bundle, *Glycoconjugate J.*, 1991, **8**, 69.
61. T. Peters and D.R. Bundle, *Can. J. Chem.*, 1989, **67**, 491.
62. T. Peters and D.R. Bundle, *Can. J. Chem.*, 1989, **67**, 497.
63. J. Kihlberg, E. Eichler and D.R. Bundle, *Carbohydr. Res.*, 1991, **211**, 59.
64. P. Kováč, in *Modern Methods in Carbohydrate Synthesis*, S.H. Khan and R.A. O'Neill (eds), Harwood Academic, Amsterdam, 1996, p. 55.
65. P.-s. Lei, Y. Ogawa and P. Kováč, *Carbohydr. Res.*, 1996, **281**, 47.
66. P.-s. Lei, Y. Ogawa and P. Kováč, *Carbohydr. Res.*, 1995, **279**, 117.
67. P. Kováč and C.P.J. Glaudemans, *Carbohydr. Res.*, 1985, **140**, 313.
68. P. Kováč, H.C.J. Yeh and C.P.J. Glaudemans, *Carbohydr. Res.*, 1985, **140**, 277.
69. T. Ziegler, *Liebigs Ann. Chem.*, 1990, 1125.

70. When referring to individual moieties in oligosaccharides, sugar residues are serially numbered beginning with the one bearing the aglycon, and are identified by a Roman numeral superscript.

71. J. Zhang and P. Kováč, *Carbohydr. Res.*, 1997, **300**, 329.

72. J. Wang, J. Zhang, C.E. Miller, S. Villeneuve, Y. Ogawa, P.-s. Lei, P. Lafaye, F. Nato, A. Karpas, S. Bystrický, S.C. Szu, J.B. Robbins, P. Kováč, J.-M. Fournier and C.P.J. Glaudemans, *J. Biol. Chem.*, 1998, **273**, 2777.

73. C.P.J. Glaudemans and M.E. Jolley, in *Methods In Carbohydrate Chemistry*, Vol. 8, R.L. Whistler and J.N. BeMiller (eds), Academic Press, New York, 1980, p. 145.

74. X. Liao, E. Poirot, A.H.C. Chang, X. Zhang, J. Zhang, F. Nato, J.-M. Fournier, P. Kováč and C.P.J. Glaudemans, *Carbohydr. Res.*, 2002, **337**, 2437.

75. S. Villeneuve, H. Souchon, M.M. Riottot, J.C. Mazie, P.-s. Lei, C.P.J. Glaudemans, P. Kováč, J.M. Fournier and P.M. Alzari, *Proc. Natl. Acad. Sci. USA*, 2000, **97**, 8433.

76. L.C. Paoletti, D.L. Kasper, F. Michon, J. DiFabio, H.J. Jennings, T.D. Tosteson and M.R. Wessels, *J. Clin. Invest.*, 1992, **89**, 203.

77. C.A. Laferriere, R.K. Sood, J.-M. De Muys, F. Michon and H. J. Jennings, *Infect. Immun.*, 1998, **66**, 2441.

78. L. Gonzalez, J.L. Asensio, A. Ariosa-Alvarez, V. Verez-Bencomo and J. Jimenez-Barbero, *Carbohydr. Res.*, 1999, **321**, 88.

79. R.U. Lemieux, D.R. Bundle and D.A. Baker, *Glycoside-Ether-Ester Compounds*, 1979, US Patent No. 4,137,4011.

80. S. Sabesan and J.C. Paulson, *J. Am. Chem. Soc.*, 1986, **108**, 2068.

81. P. Kováč and J. Zhang, *Linking Compounds Useful for Coupling Carbohydrates to Amine Containing Carriers*, US Patent No. 5,952,454.

82. J. Zhang and P. Kováč, *Tetrahedron Lett.*, 1998, **39**, 1091.

83. J. Zhang, A. Yergey, J. Kowalak and P. Kováč, *Tetrahedron*, 1998, **54**, 11783.

84. G.R. Gray, in *Methods in Enzymology*, Vol. 50, V. Ginsburg (ed), Academic Press, New York, 1978, p. 155.

85. G.R. Gray, B.A. Schwartz and B.J. Kamicker, *Prog. Clin. Biol. Res.*, 1978, **23**, 583.

86. B.J. Kamicker, B.A. Schwartz, R.M. Olson, D.C. Drinkwitz and G.R. Gray, *Arch. Biochem. Biophys.*, 1977, **183**, 393.

87. B.A. Schwartz and G.R. Gray, *Arch. Biochem. Biophys.*, 1977, **181**, 542.

88. V.P. Kamath, P. Diedrich and O. Hindsgaul, *Glycoconjugate J.*, 1996, **13**, 315.

89. Y. Ogawa, P.-s. Lei and P. Kováč, *Bioorg. Med. Chem. Lett.*, 1995, **5**, 2283.

90. K. Jansson, G. Noori and G. Magnusson, *J. Org. Chem.*, 1990, **55**, 3181.

91. K. Jansson, S. Ahlfors, T. Frejd, J. Kihlberg, G. Magnusson, J. Dahmen, G. Noori and K. Stenwall, *J. Org. Chem.*, 1988, **53**, 5629.

92. Y. Ogawa, P.-s. Lei and P. Kováč, *Carbohydr. Res.*, 1996, **288**, 85.

93. V. Pozsgay, *Tetrahedron Lett.*, 1993, **34**, 7175.

94. R. Saksena, J. Zhang and P. Kováč, *J. Carbohydr. Chem.*, 2002, **21**, 453.

95. X. Ma, R. Saksena, A. Chernyak and P. Kováč, *Org. Biomol. Chem.*, 2003, **1**, 775.

96. R. Saksena, X. Ma and P. Kováč, *Carbohydr. Res.*, 2003, **338**, 2591.

97. G.R. Gray, *Arch. Biochem. Biophys.*, 1974, **163**, 426.

98. J. Zhang and P. Kováč, *Carbohydr. Res.*, 1999, **321**, 157.

99. M. Dubber and T.K. Lindhorst, *Synthesis*, 2001, 327.

100. R. Roy, E. Katzenellenbogen and H.J. Jennings, *Can. J. Biochem. Cell Biol.*, 1984, **62**, 270.

101. Chernyak, A. and Kováč, P., unpublished results.

102. N. Jentoft and D.G. Dearborn, *J. Biol. Chem.*, 1979, **254**, 4359.

103. R.F. Borch, M.D. Bernstein and H.D. Durst, *J. Am. Chem. Soc.*, 1971, **93**, 2897.

104. C.F. Lane, in *Selections from the Aldrichimica Acta, 1968–1982*, Aldrich Chemical Company, Inc., Aldrich, 1984, p. 67.

105. K.-H. Glüsenkamp, W. Drosdziok, G. Eberle, E. Jähde, and M.F. Rajewsky, *Z. Naturforsch., C, Bioscience*, 1991, **46**, 498.

106. L.F. Tietze, M. Arlt, M. Beller, K.-H. Glüsenkamp, E. Jähde and M.F. Rajewsky, *Chem. Ber.*, 1991, **124**, 1215.

107. L.F. Tietze, C. Schröter, S. Gabius, U. Brinck, A. Goerlach-Graw and H.-J. Gabius, *Bioconjugate Chem.*, 1991, **2**, 148.

108. J. Zhang and P. Kovac, *Bioorg. Med. Chem. Lett.*, 1999, **9**, 487.

109. J. Zhang, A. Yergey, J. Kowalak and P. Kováč, *Carbohydr. Res.*, 1998, **313**, 15.

110. A. Chernyak, A. Karavanov, Y. Ogawa and P. Kováč, *Carbohydr. Res.*, 2001, **330**, 479.

111. R. Saksena, A. Chernyak, E. Poirot and P. Kováč, in *Methods in Enzymology*, Vol. 362, Y.C. Lee and R. Lee (eds), Academic Press, New York, 2003, p. 140.

112. R. Saksena, A. Chernyak, A. Karavanov and P. Kováč, in *Methods in Enzymology*, Vol. 362, Y.C. Lee and R. Lee (eds), Academic Press, New York, 2003, p. 125.

113. M. Dubois, K.A. Gilles, J.K. Hamilton, P.A. Rebers and F. Smith, *Anal. Chem.*, 1956, **28**, 350.

114. V. Pozsgay, E. Dubois and L. Pannell, *J. Org. Chem.*, 1997, **62**, 2832.

115. V. Pozsgay, *J. Org. Chem.*, 1998, **63**, 5983.

116. K.E. Stein, D.A. Zopf, C.B. Johnson, C.B. Miller and W.E. Paul, *J. Immunol.*, 1982, **128**, 1350.

117. K.E. Stein, *J. Infect. Dis.*, 1992, **165**(suppl. 1), S49.

118. A. Chernyak, S. Kondo, T.K. Wade, M.D. Meeks, P.M. Alzari, J.-M. Fournier, R.K. Taylor, P. Kováč and W.F. Wade, *J. Infect. Dis.*, 2002, **185**, 950.

119. M.D. Meeks, R. Saksena, X. Ma, T.K. Wade, R.K. Taylor, P. Kováč and W.F. Wade, *Infect. Immun.*, 2004, **72**, 4090.

120. R.K. Taylor, T.J. Kirn, N. Bose, E. Stonehouse, S.A. Tripathi, P. Kováč and W.F. Wade, *Chem. Biodiv.*, 2004, **1**, 1036.

CHAPTER 11

Carbohydrate Microarrays for High-Throughput Analysis of Carbohydrate–Protein Interactions

INJAE SHIN

Department of Chemistry, Yonsei University, Seoul 120-749, Korea

1 Introduction

Carbohydrate–protein interactions are involved in a wide variety of physiological and pathogenic processes in living organisms.[1–6] Biophysical approaches have been mainly employed to elucidate these interactions for biological research and biomedical applications. For instance, isothermal titration calorimetry (ITC) and surface plasmon resonance (SPR) spectroscopy have been utilized to determine the binding affinities between glycans and proteins.[7–10] NMR spectroscopy and X-ray crystallography have been also used for investigating binding modes between glycans and proteins.[11–13] As an alternative, specifically modified synthetic sugars provide useful tools for understanding the molecular basis of carbohydrate–protein interactions.[14] However, these well-established biophysical and biochemical approaches are not suitable for high-throughput analysis of these biomolecular interactions.

During the last decade, microarray-based technologies have been widely exploited for fast, quantitative, and simultaneous analyses of a large number of biomolecular interactions. DNA microarray technology, which was first developed, is a good example for high-throughput studies of DNA–RNA (or DNA–DNA) interactions. This technology has been applied to understanding gene mutations, tracking the activities of many genes at the same time, and studying changes in the pattern of gene expression in disease.[15–20] Protein microarrays, which were developed after DNA microarrays, have been used for the profiling of protein expression in normal and diseased states and high-throughput analysis of protein–protein interactions.[21–26] As these microarray technologies are used for genomic, transcriptomic, and proteomic studies in a high-throughput manner, carbohydrate microarrays have the potential to serve as valuable tools for high-throughput analysis of carbohydrate–protein interactions in glycomic research.

In general, carbohydrate-binding proteins recognize monovalent carbohydrates weakly but exhibit a strong binding affinity to multivalent carbohydrates because of the cluster effects.[27–29] The carbohydrates immobilized on the solid surface with proper spacing and orientation may display multivalency and act as cell surface carbohydrates. Therefore, carbohydrate microarrays are ideal for elucidating recognition events between glycans and proteins in a high-throughput fashion.[30–36] In this chapter, the immobilization technique of maleimide-linked carbohydrates on the thiol-coated glass slides and applications of the fabricated carbohydrate microarrays are described. Furthermore, this chapter also includes a brief summary of other immobilization methods for the preparation of carbohydrate microarrays and their potential for biological research and biomedical applications.

2   Strategy for Immobilization of Carbohydrate Probes on the Solid Surface

Efficient immobilization techniques of carbohydrates on the solid surface are essential for the successful construction of carbohydrate microarrays. Functional groups connected to carbohydrate ligands should be specifically reacted with functional groups derivatized on the solid surface. Chemoselective reaction between maleimide and thiol groups has been widely used for the preparation of versatile bioconjugates.[37] This ligation reaction is highly selective even in the presence of other potent nucleophiles, such as amines, alcohols, and carboxylates, at pH 6.5–7.5. For example, neoglycopeptides and neoglycoproteins were produced by coupling of maleimide-linked sugars to cysteine-containing peptides and proteins.[38,39] This methodology was applied to the attachment of maleimide-linked carbohydrates to thiol-coated glass slides through stable thioether linkages (Figure 1).[40,41]

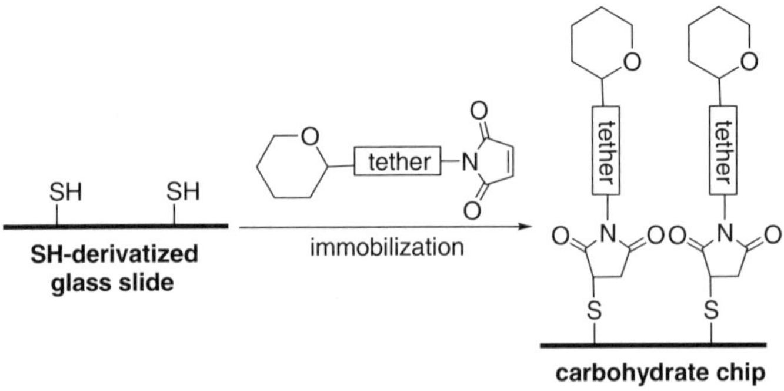

Figure 1 *Immobilization of carbohydrates by chemoselective ligation between maleimide-conjugated carbohydrates and thiols derivatized on the glass slides*

3   Synthesis of Maleimide-Linked Carbohydrates

The lengths of tethers inserted between maleimide and carbohydrate moieties are important for the strong binding of proteins to the glycan ligands immobilized on the surface. If the length of a tether is too short, a solid surface may prevent protein binding to the immobilized carbohydrate ligands. To examine effect of tethers on protein binding, carbohydrates conjugated by linkers of various lengths (**S**, **L1**, **L2**, and **L3**) were prepared. The required linkers were obtained from glycine or 6-aminohexanoic acid (**1**) according to the procedure shown in Scheme 1.[40] The amine group was converted to maleimide by reacting with maleic anhydride and a subsequent cyclization with hexamethyldisilazide (HMDS) and zinc chloride (ZnCl$_2$).[38,39,42] The maleimide-containing acid was then reacted with pentafluorophenol (Pfp-OH), diphenyl chlorophosphate, and *N*-ethylmorpholine (NEM) to provide a pentfluorophenyl ester **S** or **L1**.[43] Linker **L2** was synthesized by coupling **L1** to **1**, followed by esterification with Pfp-OH and *N*-ethyl-*N*-(3-dimethylaminopropyl)carbodiimide·HCl (EDC). Repetition of this procedure with **L2** afforded **L3**.

The N-linked maleimide-conjugated sugars were prepared by a one-pot amination and a subsequent coupling with the linkers (Scheme 2). Reactions of mono- and disaccharides with saturated ammonium bicarbonate for 24 h at 45 °C gave β-glycosyl amines.[44,45] The resulting β-anomers were then coupled to the linkers (**S** or **L1–L3**) to produce the N-linked maleimide-conjugated sugars shown in Figure 2. This synthetic method successfully produced various carbohydrate probes but is limited to providing only β-anomeric carbohydrate probes.

Alternatively, the allylation of carbohydrates was performed to synthesize O-linked α- and β-anomeric maleimide-conjugated carbohydrates as well as to investigate the

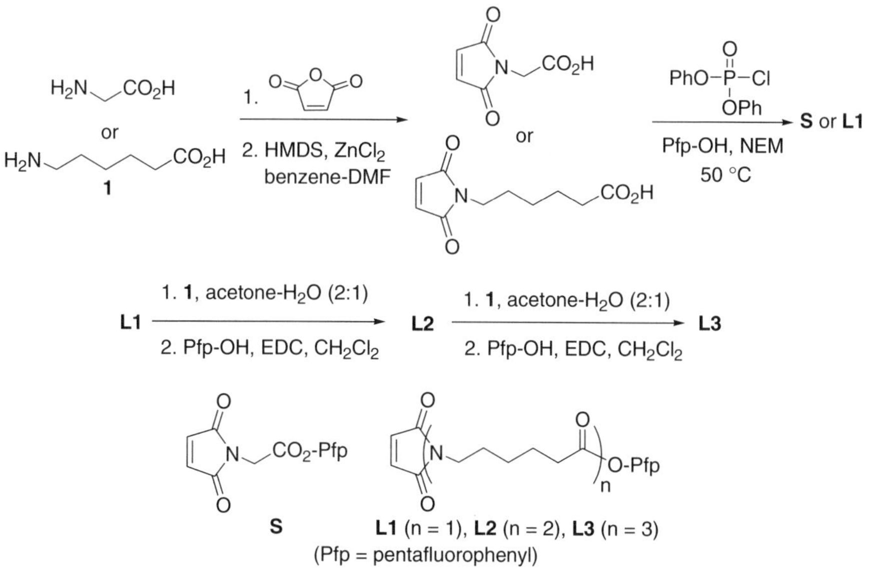

Scheme 1 *Synthesis of linkers S, L1, L2, and L3*

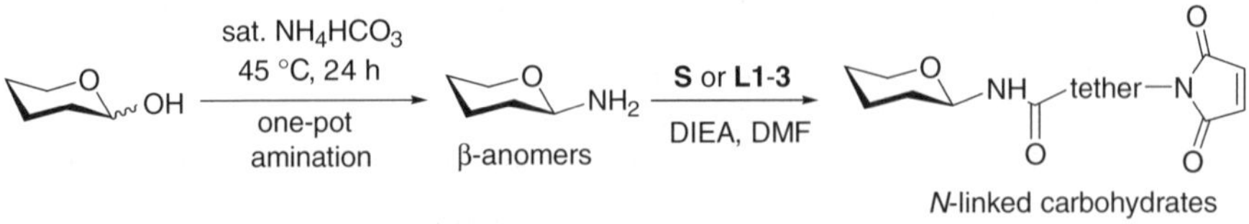

Scheme 2 *Synthesis of N-linked maleimide-conjugated carbohydrates*

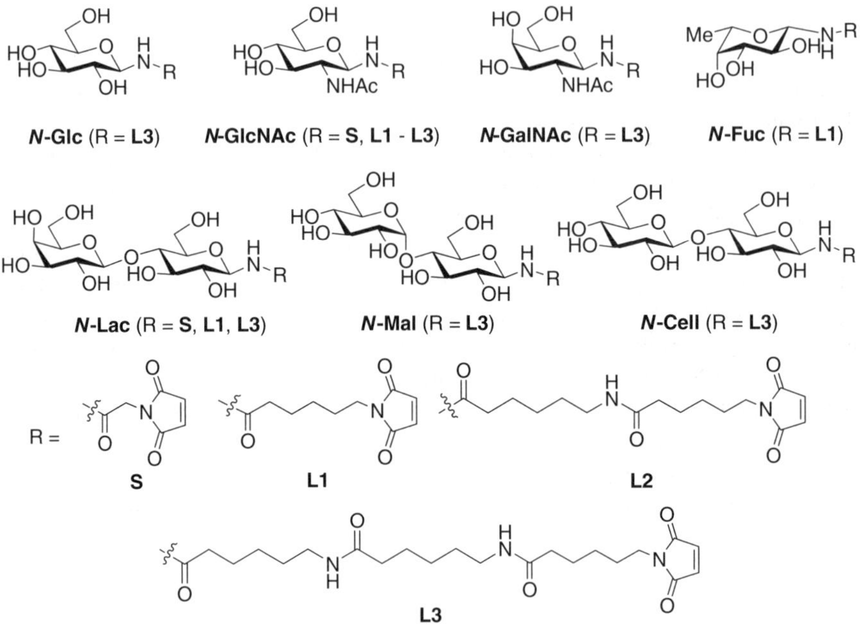

Figure 2 *Structures of synthesized N-linked maleimide-conjugated carbohydrates*

difference of binding affinities between N- and O-linked carbohydrates to lectins (Scheme 3). The α- and β-allyl glycosides prepared from known procedures were treated with cysteamine,[46–50] and then the resulting amines were reacted with **L2** to produce O-linked maleimide-conjugated sugars (Figure 3). Moreover, the sialic acid probe (**NeuAc**) was also prepared from *N*-acetylneuraminic acid (Scheme 4). Esterification of *N*-acetylneuraminic acid followed by peracetylation and chlorination provided **2**.[51] The chloro group in **2** was replaced by thioacetyl group with potassium thioacetate to give **3**. Selective deacetylation of the thioacetate in **3** with 1 equiv. NaOMe and a subsequent coupling of the resulting thiol to bromoacetylated compound **4** produced **5**. Removal of all the protecting groups in **5** followed by coupling to **L2** gave the sialic acid probe (**NeuAc**).

Carbohydrate microarrays containing a variety of carbohydrate probes are more useful for rapidly and systematically studying carbohydrate–protein interactions. Several researchers have developed combinatorial synthesis of carbohydrate libraries to prepare diverse carbohydrates.[52–57] As an effort to synthesize more

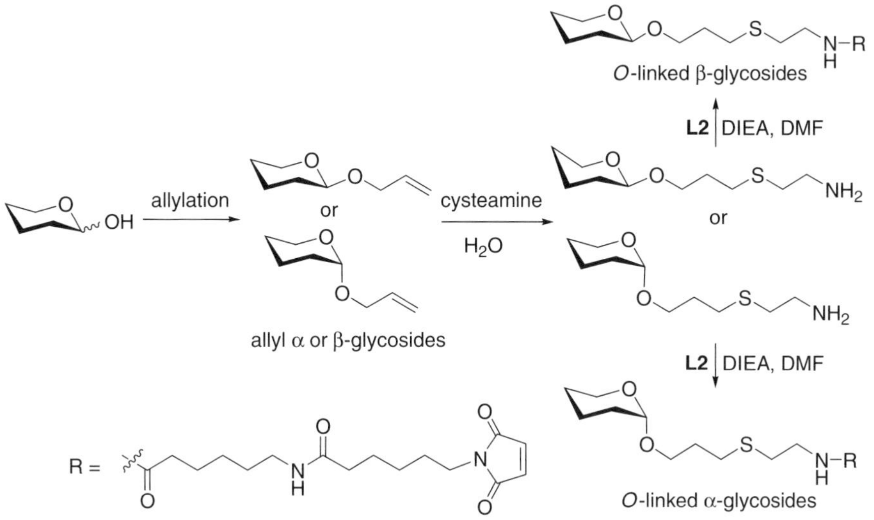

Scheme 3 *Synthesis of O-linked maleimide-conjugated carbohydrates*

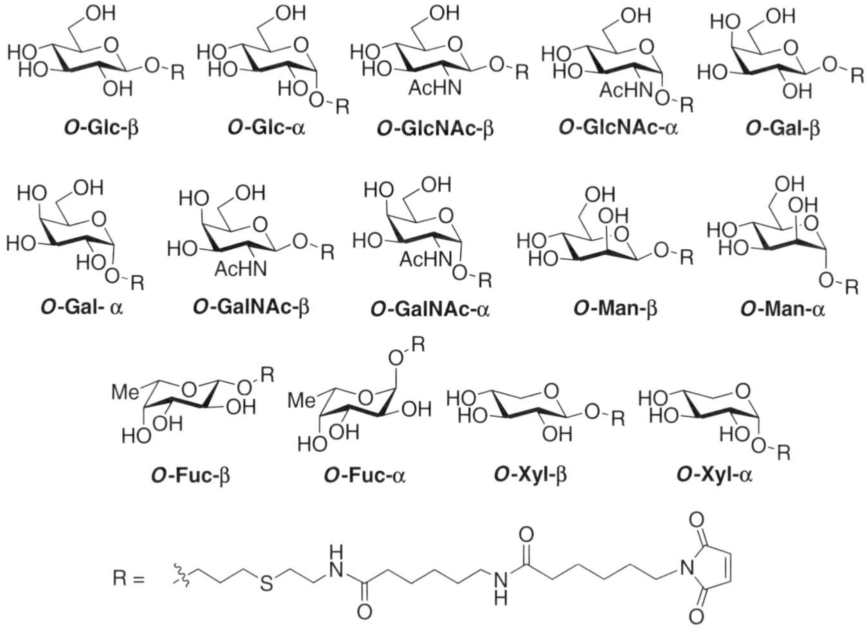

Figure 3 *Structures of synthesized O-linked maleimide-conjugated carbohydrates*

versatile maleimide-linked carbohydrate probes, three disaccharides (**Glcβ1,6Man**, **Galβ1,6Man**, and **Manα1,6Man**) were prepared by solution-phase parallel glycosylation (Scheme 5). The 2-bromoethyl α-mannoside (**6**) was in parallel glycosylated with three glycosyl bromides (**7–9**) in the presence of silver triflate (AgOTf) as

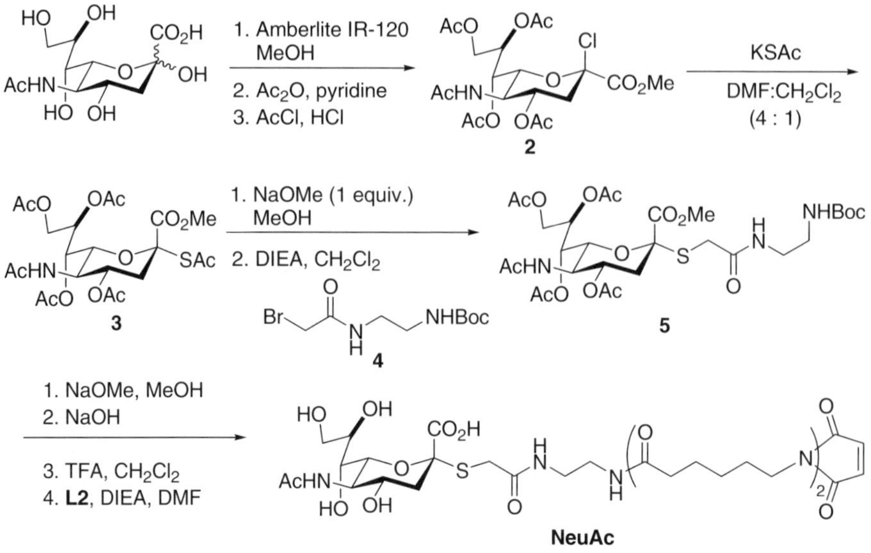

Scheme 4 *Synthesis of a maleimide-conjugated sialic acid probe*

an activator to provide the 1,6-linked disaccharides (**10–12**).[54] The bromoethyl group was used as an anomeric substituent because it could be readily converted to other functional groups.[54,58] The three disaccharides (**10–12**) were transformed into maleimide-conjugated disaccharides (**Glcβ1,6Man**, **Galβ1,6Man**, and **Manα1,6Man**) by treatment with Boc-protected cysteamine, followed by removal of all the protecting groups and coupling of the resulting amines to **L2**.

4   Fabrication of Carbohydrate Microarrays

For the construction of carbohydrate microarrays, solutions of maleimide-linked carbohydrates were prepared by dissolving in phosphate-buffered saline (PBS) containing 40–50% glycerol (Figure 4). The glycerol suppresses undesired evaporation of the nanodroplets during spotting and ligation reaction. The solutions of carbohydrate probes in a 384-microtiter plate were printed in the predetermined places on a thiol-coated glass slide by using a high-precision pin-type microarrayer (spotting volume, 1 nL; spot size, ~100 μm in diameter). After 5-h immobilization in a humid chamber, the glass slide was washed with water and then treated with 1% N-ethyl-maleimide (NEM) in water to remove unreacted thiol groups. This capping process prevents oxidative disulfide bond formation between surface thiols and cysteine residues of proteins used in the next incubation step. A blocked plastic film (0.1–0.2 mm thickness), which is coated by adhesive at one side, was then carefully attached to the microspotted slide. Use of this plastic film facilitates compartmentalization that is required for simultaneous treatment with several proteins and determination of IC_{50} values. The prepared carbohydrate microarrays can be stored in a desiccator for several months without any problem.

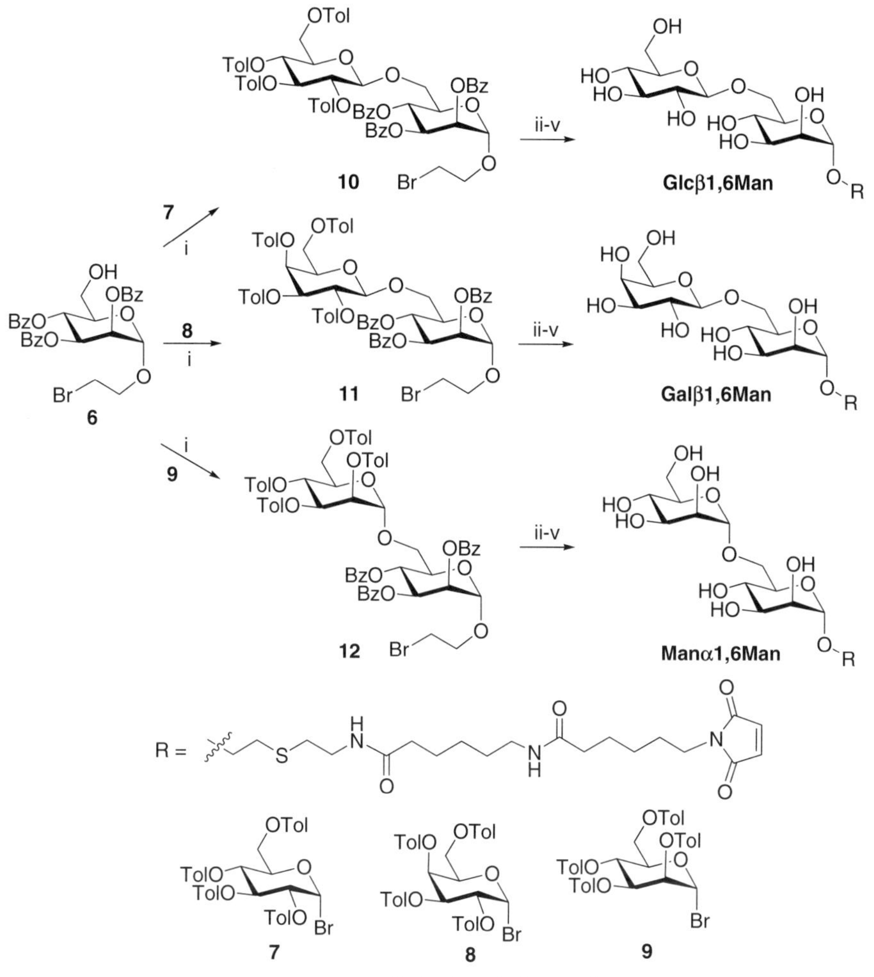

Scheme 5 *Solution-phase parallel synthesis of disaccharides. Conditions: (i) AgOTf, CH₂Cl₂; (ii) HSCH₂CH₂NHBoc, TEA, DMF, 50 °C; (iii) NaOMe, MeOH; (iv) TFA; (v) **L2**, DIEA, DMF*

Before incubation of the slides with fluorophore-labeled proteins, the carbohydrate microarrays were preincubated with a solution of 3% bovine serum albumin (BSA) in PBS buffer containing 0.2% Tween 20. The treatment of carbohydrate microarrays with BSA reduces background fluorescence resulting from nonspecific binding of fluorophore-labeled proteins. The slides were then probed with fluorophore-labeled proteins in PBS buffer containing 0.1% Tween 20 for 1 h. After extensive washing of the protein-probed slides with the same buffer and removal of a plastic film, protein binding was visualized and/or quantitated by using a fluorescence scanner.

It was worth noting that immobilization reaction for more than 5 h did not improve protein binding to carbohydrates immobilized on the surface. This result indicated that

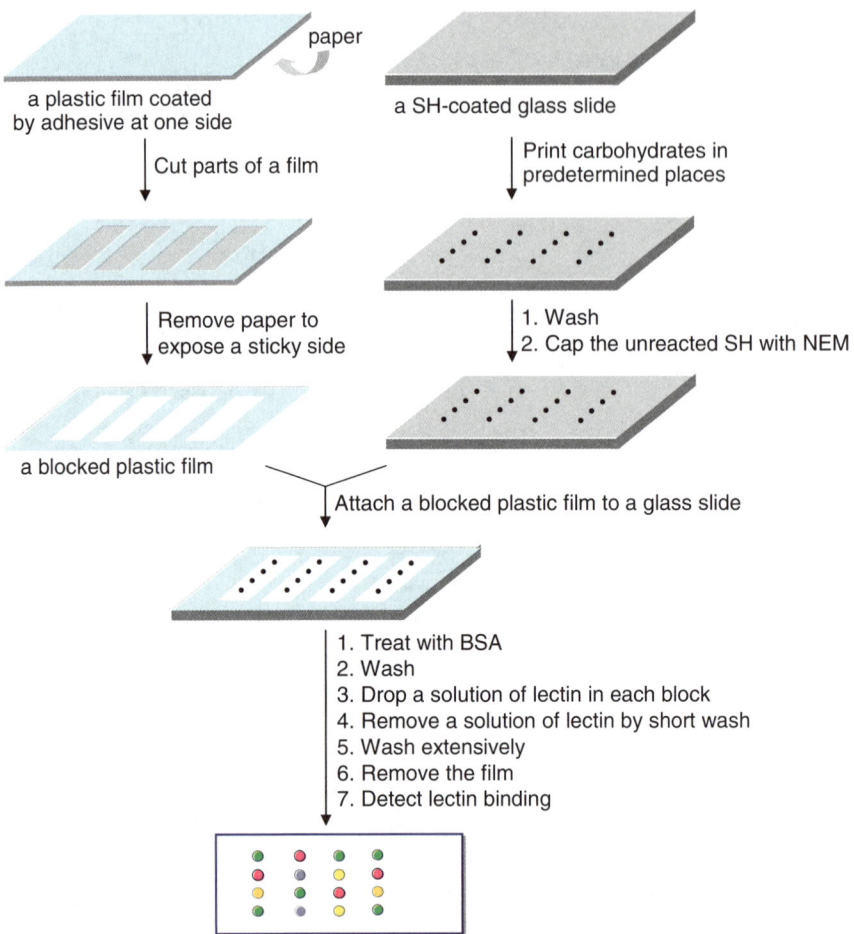

Figure 4 *Procedure for the fabrication of carbohydrate microarrays*

5-h immobilization was suitable for efficient attachment of carbohydrates to the thiol-coated glass slides. However, thorough washing of the slides after incubation with fluorophore-labeled proteins was critical in reducing background fluorescent signal. For example, short washes (three for 1 min each) of the slide exhibited high background fluorescence, whereas three washes for 5–10 min consistently showed a low background.

5 Applications

5.1 High-Throughput Analysis of Carbohydrate–Protein Interactions

Before embarking on applications of carbohydrate microarrays, the effects of immobilization concentration of carbohydrates and length of linkers on protein binding were examined to establish optimal conditions. For these studies, various concentrations of **N-GlcNAc** and **N-Lac** (5.0, 1.0, 0.5, and 0.1 mM) connected by **S** or **L1–L3** were printed on the thiol-derivatized glass slide. The microspotted slide was then probed with fluorescein-labeled *Triticum vulgaris* (FITC-TV, a GlcNAc-binding lectin, also known as wheat germ agglutinin) and *Erythrina cristagalli* (FITC-EC, a GalNAc/Gal binding lectin).[59,60] The binding affinity of TV and EC to the corresponding carbohydrates on the slide depended on both the immobilization concentration of carbohydrate probes and the length of linkers (Figure 5). Carbohydrates connected by the shortest linker (**S**) were weakly recognized by the lectins at less than 5 mM concentration. However, the lectins strongly interacted with carbohydrates tethered to longer linkers (**L1–L3**) at > 0.5 mM concentration. It was noteworthy that fluorescent signals of immobilized carbohydrates were retained even after extensive washing. This suggested that carbohydrates immobilized on the glass slide bound to lectins strongly and may display multivalency.

On the basis of these observations, 22 mono- and disaccharides conjugated by the proper length of linkers (1 mM) shown in Figure 6 were microspotted on the thiol-coated slides. The fabricated carbohydrate microarrays were then incubated with five FITC-labeled lectins (1–10 µg mL^{-1}), including FITC-TV, FITC-*A. aurantia* (FITC-AA), FITC-EC, FITC-*Concanavalin A* (FITC-ConA), and FITC-*N. pseudonarcissus* (FITC-NPA).[59,60] The carbohydrate microarray probed with FITC-TV (a GlcNAc-binding lectin) showed strong binding of **O-GlcNAc-α** and lower binding of **O-GlcNAc-β** (Figure 7a). This lectin interacted with **O-GlcNAc-β** and **O-GalNAc-α** with the similar binding affinity but recognized **O-GalNAc-β** and **N-GalNAc** with the reduced binding affinity.[61–63] Interestingly, microspots of N-linked β-GlcNAc (**N-GlcNAc**) showed weaker fluorescence intensity than those of O-linked β-GlcNAc (**O-GlcNAc-β**). This indicated that the nature of the anomeric linkage in monosaccharides had influence on their binding affinities to the lectin.

When the carbohydrate microarray was treated with FITC-AA (a Fuc binding lectin), the lectin bound strongly to O-linked Fuc (**O-Fuc-α** and -**β**) but weakly to N-linked Fuc (**N-Fuc**) (Figure 7b). This lectin recognized N-linked β-Fuc (**N-Fuc**) less tightly than O-linked β-Fuc (**O-Fuc-β**). Unexpectedly, AA also bound to **O-Man-β**

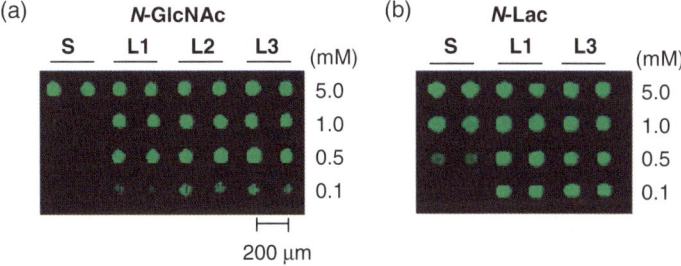

Figure 5 *Effects of immobilization concentration of carbohydrate probes and length of linkers connected to the carbohydrates on the lectin binding. Fluorescence images of carbohydrate microarrays containing (a) **N-GlcNAc** and (b) **N-Lac** probed with FITC-TV and FITC-EC, respectively*

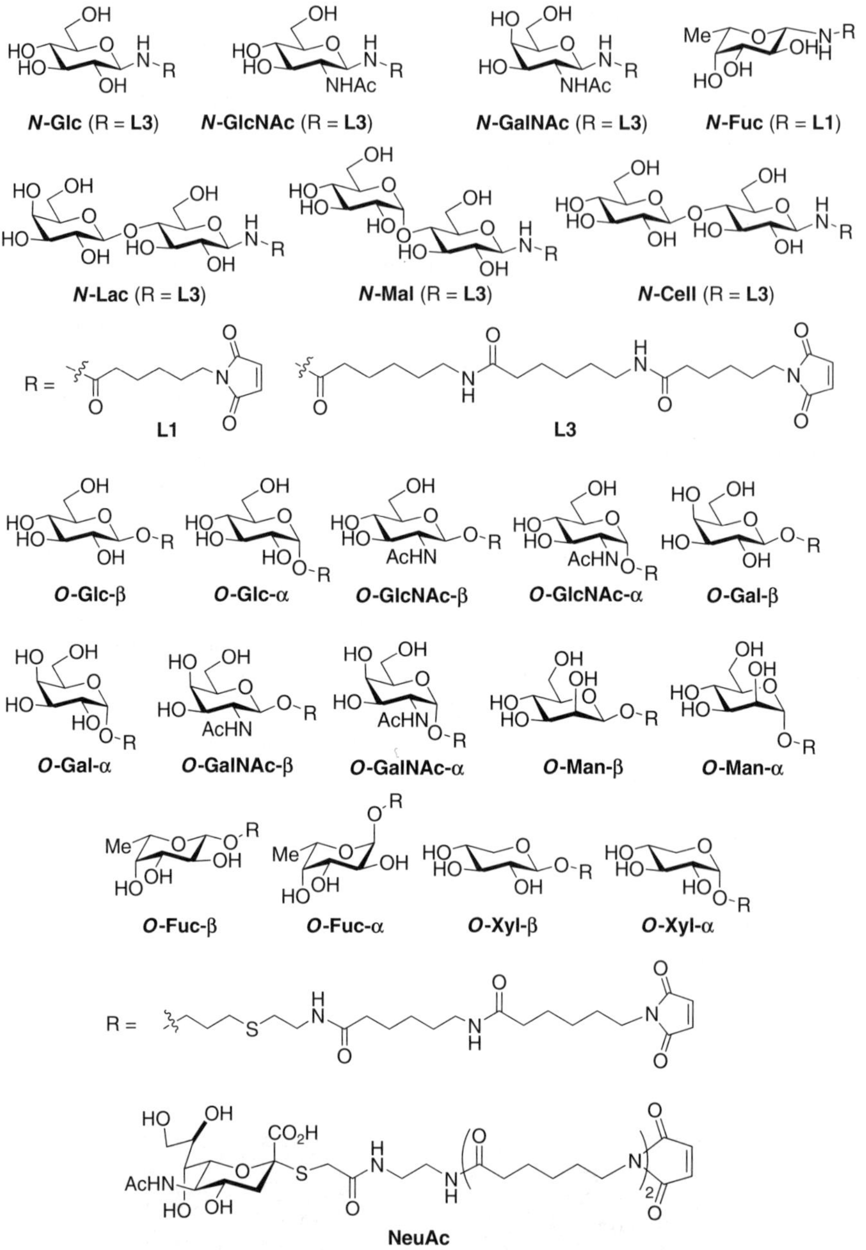

Figure 6 *Structure of 22 carbohydrate probes used for the fabrication of carbohydrate microarrays*

with a similar affinity as it did to **N-Fuc**. On the other hand, the carbohydrate microarray probed with FITC-EC (a GalNAc/Gal binding lectin) exhibited fluorescent signals in the regions of Gal, GalNAc, and lactose (Figure 7c). The lectin bound to **O-GalNAc-α**, **N-GalNAc**, and **N-Lac** with similar binding affinities. However,

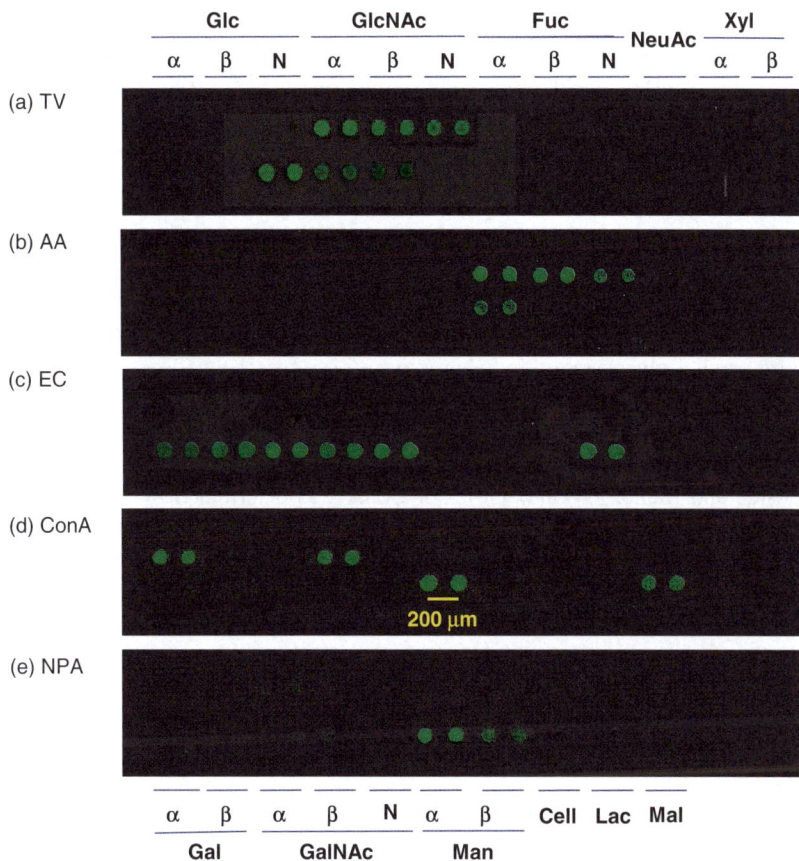

Figure 7 *Fluorescence images of carbohydrate microarrays containing 22 carbohydrates probed with (a) FITC-TV, (b) FITC-AA, (c) FITC-EC, (d) FITC-ConA, (e) FITC-NPL. The α and β symbols denote O-linked α- and β-anomeric maleimide-conjugated sugars and N denotes N-linked maleimide-conjugated sugars. (Reprinted from Ref. 40 with permission from American Chemical Society)*

O-linked Gal (**O-Gal-α** and -**β**) and **O-GalNAc-β** were recognized by the lectin with the reduced binding affinities. Microspots of N-linked β-GalNAc (**N-GalNAc**) showed stronger fluorescence intensity than those of O-linked β-GalNAc (**O-GalNAc-β**). The relative binding affinities of EC with the carbohydrate microarray were similar to those observed for solution-based assays.[64]

When the carbohydrate microarray was incubated with FITC-ConA (an α-Man/α-Glc binding lectin), the lectin bound to **O-Man-α** strongly, **O-Glc-α** with reduced affinity, and **N-Mal** and **O-GlcNAc-α** weakly (Figure 7d). These results were consistent with binding tendencies assessed by ITC.[65] Dissociation constants (K_d) of methyl α-mannoside, methyl α-glucoside, maltose, and methyl α-GlcNAc for ConA were determined by ITC to be 0.12, 0.51, 0.76, and 0.93 mM, respectively. However, β-anomeric carbohydrates (**O-Man-β**, **O-Glc-β**, **O-GlcNAc-β**, and **N-Cell**) were not recognized by this lectin. This showed that the anomeric configuration of glucose, mannose, and GlcNAc on the solid surface governed lectin binding. The carbohydrate microarray treated with FITC-NPA (a Man binding lectin which does not bind to Glc) exhibited fluorescent signals only in the region of ***O*-Man-α** and -**β** (Figure 7e). Fluorescence intensity of O-linked α-Man (***O*-Man-α**) was greater than that of O-linked β-Man (***O*-Man-β**).[66] According to the relative fluorescence intensity of ***O*-Man-α** for ConA and NPA, ConA interacted with this carbohydrate more tightly than NPA. It was reported that K_d's of Manα 1,3Man for ConA and NPA determined by ITC were 0.07 and 2.00 mM, respectively.[65,67] Therefore, both the carbohydrate microarray and the ITC method showed that ConA recognized Man more strongly than NPA.

Three disaccharide probes (**Glcβ1,6Man**, **Galβ1,6Man**, and **Manα1,6Man**) synthesized by parallel glycosylation in solution were also microspotted on the thiol-coated slide along with four monosaccharides (***O*-Fuc-β**, ***O*-GlcNAc-α**, ***O*-GalNAc-β**, and ***O*-Man-α**). The ConA-treated slide exhibited that this lectin recognized **Manα1,6Man** and ***O*-Man-α** with similar binding affinities (Figure 8). K_d of both methyl α-mannoside and Manα1,6Man-α-OMe for ConA was determined by ITC to be 0.12 mM.[65] Thus, the relative binding affinities assessed by carbohydrate microarrays were consistent with those determined by ITC. However, another mannose-binding lectin, NPA, interacted with **Manα1,6Man** more tightly than ***O*-Man-α**. The EC-probed slide showed stronger fluorescence intensity in the region of ***O*-GalNAc-β** than **Galβ1,6Man**. As anticipated, TV and AA recognized only the corresponding monosaccharides.

High-density carbohydrate microarrays are more useful for their applications to biological and biomedical research. To demonstrate the feasibility of this immobilization method for the fabrication of high-density carbohydrate microarrays, three of ***O*-Fuc-α**, ***O*-Man-α**, and ***O*-GlcNAc-β** were alternately printed on the thiol-coated slide to generate 12,000 microspots (60 × 200). The fabricated carbohydrate

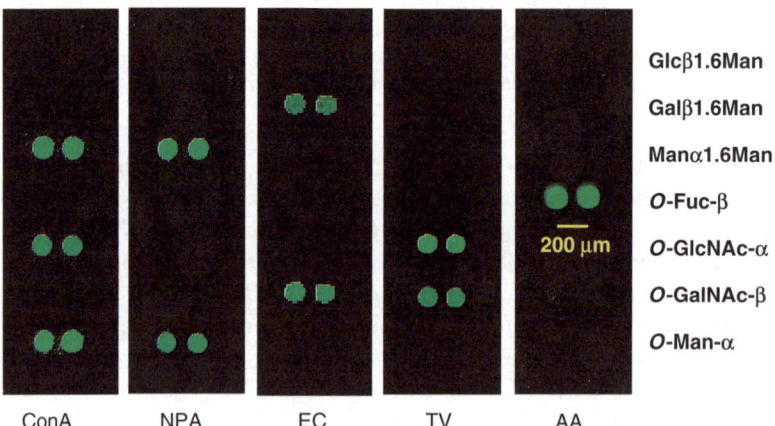

Figure 8 *Fluorescence images of carbohydrate microarrays containing mono- and disaccharides probed with FITC-labeled lectins. (Reprinted from Ref. 40 with permission from American Chemical Society)*

microarray was then incubated with a mixture of FITC-ConA, Cy3-TV, and Cy5-AA. The monosaccharides on the surface were selectively recognized by the corresponding lectins (Figure 9). The lectin binding studies described above clearly demonstrated that carbohydrate microarray technology was ideal for investigating recognition events between carbohydrates and proteins in a high-throughput manner.

5.2   Quantitative Analysis of Protein-Binding Affinities

Carbohydrate microarrays were also applied to quantitatively analyzing protein-binding affinities by determining IC_{50} values against soluble inhibitors. For example, the difference of binding affinity between ***O*-GlcNAc-α** and ***O*-GalNAc-α** for TV was examined by determining concentrations (IC_{50}) of methyl *N*-acetyl-α-glucosaminide (α-GlcNAc-OMe) to inhibit 50% of TV binding to ***O*-GlcNAc-α** and ***O*-GalNAc-α** on the solid surface. The carbohydrate microarray containing ***O*-GlcNAc-α** and ***O*-GalNAc-α** was treated with a series of mixtures of Cy3-TV (1 μg mL^{-1}) and α-GlcNAc-OMe (~10 μM to 0.4 M) as an inhibitor. After thorough washing of the slide, the amount of bound lectin was quantitated by determining fluorescence intensity. The IC_{50} values of soluble α-GlcNAc-OMe for ***O*-GalNAc-α** and ***O*-GlcNAc-α** immobilized on the surface were determined to be 3.5 and 8.9 mM, respectively (Figure 10a and 10b). This showed that ***O*-GlcNAc-α** competed more efficiently with α-GlcNAc-OMe for TV binding than ***O*-GalNAc-α**.[68]

Furthermore, binding affinity of ***O*-Man-β** immobilized on the surface to AA was also assessed by using carbohydrate microarrays. In this case, methyl α-fucoside (α-Fuc-OMe) and allyl α-fucoside (α-Fuc-OAllyl) were used as soluble inhibitors. Microspots of ***O*-Man-β** were incubated with a series of mixtures of Cy5-AA (1 μg mL^{-1}) and α-Fuc-OMe or α-Fuc-OAllyl (1 nM–0.01 M). Then, the amount of bound lectin was analyzed by measuring fluorescence intensity. The IC_{50} values of α-Fuc-OAllyl and α-Fuc-OMe that inhibited 50% of AA binding to the ***O*-Man-β** were

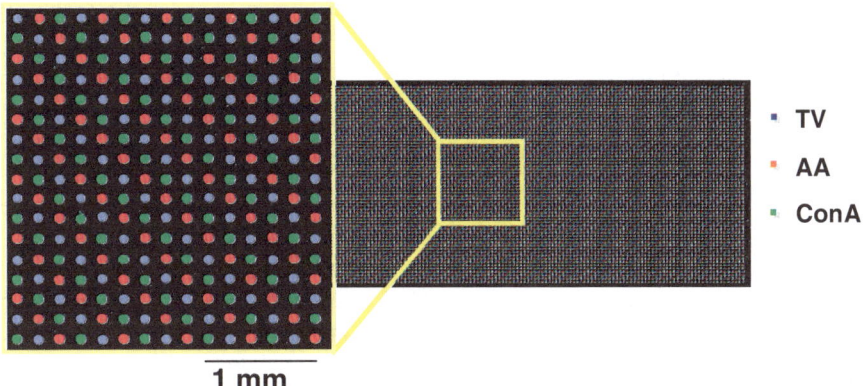

1 mm

Figure 9 *Fluorescence image of 12,000 microspots (60 × 200) consisting of **O-GlcNAc-β**, **O-Man-α**, and **O-Fuc-α** probed with a mixture of Cy3-TV, Cy5-AA, and FITC-ConA. (Reprinted from Ref. 40 with permission from American Chemical Society)*

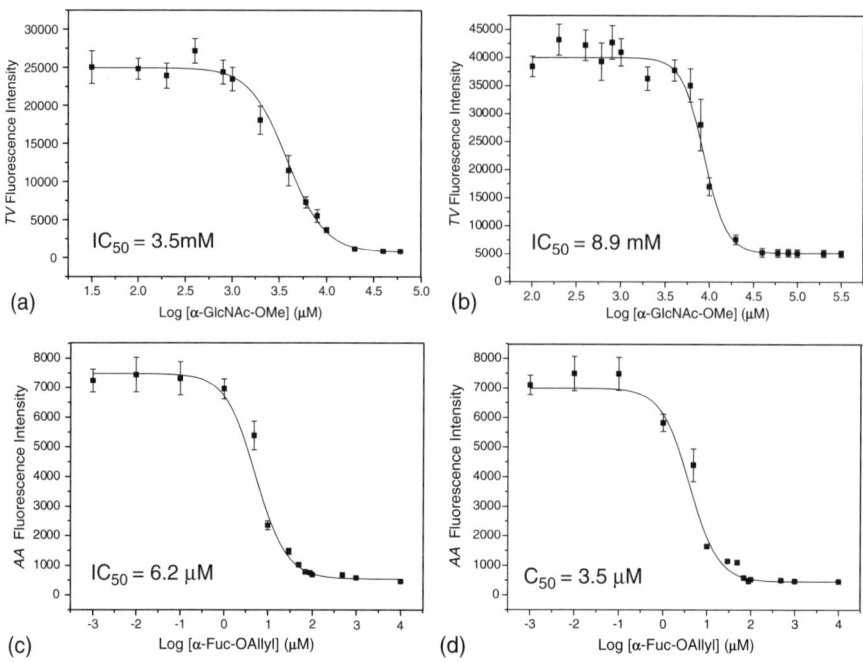

Figure 10 *Determination of IC$_{50}$ values using carbohydrate microarrays. Concentrations of soluble α-GlcNAc-OMe to inhibit 50% of TV binding to (a)* **O-GalNAc-α** *and (b)* **O-GlcNAc-α** *on the surface, and concentrations of (c) soluble α-Fuc-OAllyl and (d) α-Fuc-OMe to inhibit 50% of AA binding to* **O-Man-β** *on the surface. (Reprinted from Ref. 40 with permission from American Chemical Society)*

determined to be 6.2 and 3.5 μM, respectively (Figure 10c and 10d). Previous studies showed that AA did not bind to Man up to 50 mM concentration.[69,70] However, it seems that AA weakly binds to Man immobilized on the surface perhaps because of the cluster effects. The quantitative studies of AA binding showed that **O-Man-β** was a poor ligand for AA and α-Fuc-OMe was a slightly stronger ligand for the lectin than α-Fuc-OAllyl.

5.3 Characterization of Carbohydrate-Processing Enzymes

Another application of the carbohydrate microarrays is to characterize the substrate specificity or enzymatic activity of carbohydrate-processing enzymes. To demonstrate this, two monosaccharides (**N-GlcNAc** and **O-Fuc-α**) were alternately printed on the thiol-coated glass slide to produce 100 microspots (5 × 20). The slide was then incubated with β-1,4-galactosyltransferase (GalT) and UDP-Gal for 15 h at 37 °C in a humid chamber. The fluorescence image of the carbohydrate microarray probed with a mixture of Cy5-AA and FITC-EC showed that only GlcNAc was selectively converted to LacNAc by GalT as in solution (Figure 11). This study suggested that the carbohydrate microarray technology could be useful for characterizing novel carbohydrate-processing enzymes in the postgenomic era.

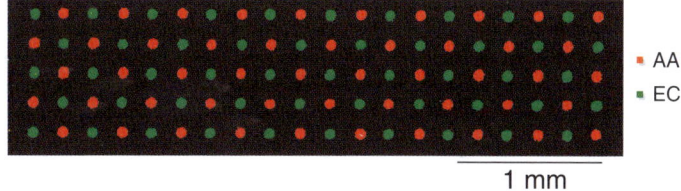

1 mm

Figure 11 *Fluorescence image of the carbohydrate microarray containing **N-GlcNAc** and **O-Fuc-α** after incubation with β-1,4-galactosyltransferase and UDP-Gal followed by probing with Cy5-AA and FITC-EC. (Reprinted from Ref. 40 with permission from American Chemical Society)*

On the other hand, this enzymatic glycosylation can be applied to preparing complex oligosaccharides on carbohydrate microarrays. For example, sialyl Le[x] was synthesized from GlcNAc immobilized on the solid surface by using three glycosyltransferases consecutively (Figure 12). The **N-GlcNAc** was microspotted on the thiol-coated glass slide and then treated with GalT in the presence of UDP-Gal under the glycosylation conditions given above. Successful enzymatic galactosylation was proved by probing with Cy5-EC. The GalT-incubated slide was further treated with α-2,3-sialyltransferase (SialT) and CMP-NeuAc for 15 h at 37 °C in a humid chamber.[71] The resulting microarray was then incubated with Cy5-EC. No fluorescence was observed, which indicated that LacNAc was converted to NeuNAcα2,3LacNAc. Finally, the carbohydrate microarray containing NeuNAcα2, 3LacNAc was incubated with α-1,3-fucosyltransferase (FucT) and GDP-Fuc for 15 h at 37 °C in a humid chamber. After rinsing with PBS three times, the enzymatic fucosylation was repeated under the same conditions to bring about more fucosylation of NeuNAcα2,3LacNAc.[72] The slide was then incubated sequentially with mouse antisialyl Le[x] antibody and goat Cy5-anti-antibody. Microspots of the enzymatically synthesized sialyl Le[x] exhibited visible fluorescence. It was noteworthy that a single fucosylation resulted in poor glycosylation based on low fluorescence intensity. These enzymatic glycosylations could be used for the preparation of carbohydrate microarrays containing more complex and perhaps more diverse carbohydrate probes.

6   Other Examples of Carbohydrate Chips and Their Applications

6.1   Other Immobilization Methods

Several immobilization methods other than the technique described above have been also developed for the construction of carbohydrate microarrays. Wang and co-workers[73] prepared microbial polysaccharide microarrays by attaching a variety of unmodified microbial polysaccharides to nitrocellulose-coated glass slides (Figure 13). The polysaccharide probes were noncovalently and site-nonspecifically immobilized on the surface mainly by hydrophobic interactions. The immobilization efficiency of this method was considerably affected by the size of carbohydrate probes. Whereas

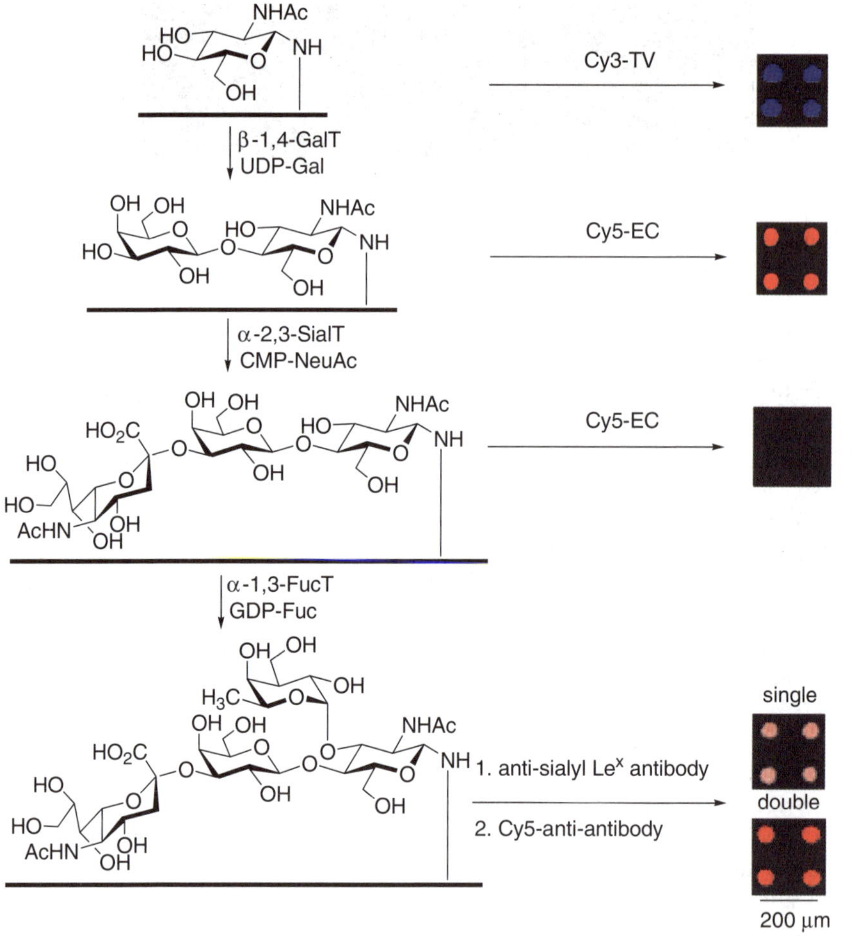

Figure 12 *Enzymatic synthesis of sialyl Le^x from **N-GlcNAc** immobilized on the surface by using three glycosyltransferases consecutively. (Reprinted from Ref. 40 with permission from American Chemical Society)*

polysaccharides of 3.3–2000 kDa were efficiently adsorbed on the solid surface, smaller carbohydrates were less retained on the surface after extensive washing. However, the binding experiments with monoclonal antibodies indicated that the recognition properties of noncovalently and site-nonspecifically immobilized polysaccharides were preserved. Willats and co-workers[74] fabricated microarrays containing diverse carbohydrate probes using black polystyrene slides, prepared from injection moulding of black polystyrene and an oxidative surface modification (Figure 13). The carbohydrate samples, including polysaccharides, proteoglycans, neoglycoproteins, and plant cell extracts, were noncovalently and site-nonspecifically immobilized on the slides through hydrogen bonding, ionic bonding, and hydrophobic interactions. Since the carbohydrate microarrays were prepared with the hydrophobic polystyrene slide, a relatively high signal-to-noise ratio was observed after probing with fluorophore-labeled

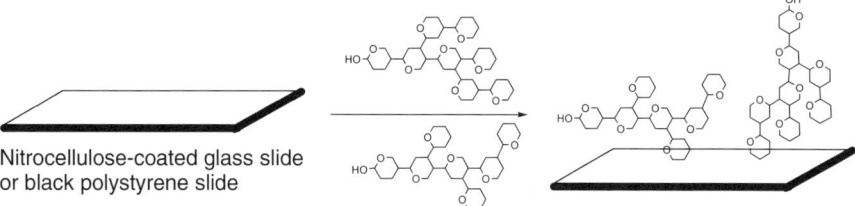

Figure 13 *Noncovalent and site-nonspecific attachment of chemically unmodified polysaccharides on the underivatized solid surface*

proteins. The above-mentioned two strategies require neither a chemically modified surface nor chemical linking techniques, and thus researchers can easily fabricate glycan microarrays, especially polysaccharide microarrays, by using these methods. However, these strategies are not suitable for fabricating mono- and oligosaccharide microarrays because of easy detachment of the probes from the surface during washing.

Another interesting method to noncovalently immobilize carbohydrate probes was developed by Fukui and co-workers.[75] The modified oligosaccharide probes, neoglycolipids (NGLs), were noncovalently and site-specifically attached to nitrocellulose or polyvinylidene fluoride (PVDF) membrane to prepare oligosaccharide microarrays (Figure 14). The required NGLs were produced by reductive amination of oligosaccharides with 1,2-dihexadecyl-*sn*-glycero-3-phosphoethanolamine. The oligosaccharides were obtained by chemical or enzymatic methods by using glycoproteins, glycolipids, proteoglycans, polysaccharides, and whole organs, or from chemical synthesis. The immobilization efficiency of the NGLs on nitrocellulose membrane was better than that observed on PVDF. Interestingly, NGLs were efficiently immobilized on nitrocellulose membrane irrespective of the size of carbohydrates.

More general methods to construct microarrays containing simple carbohydrates and oligosaccharides are to site-specifically and covalently immobilize modified carbohydrates on the properly derivatized surface. These strategies need both a modified surface and chemical-linking techniques, described in Section 2. However, the advantage of these methods is that both simple carbohydrates and oligosaccharides are efficiently immobilized on the surface. Houseman and Mrksich[76] fabricated carbohydrate chips by immobilizing cyclopentadiene-linked carbohydrates on the gold surface coated by benzoquinone group via a Diels–Alder reaction (Figure 15a). The gold surface was initially modified by immersing the gold plate into a mixture of alkanethiols with (1%) and without (99%) appended hydroquinone groups. This procedure generated self-assembled monolayers of hydroquinone and penta(ethylene glycol) groups. The hydroquinone groups were then converted to benzoquinone groups by chemical or electrochemical oxidation. The monosaccharides conjugated by cyclopentadiene groups were covalently attached to the gold surface through the Diels–Alder reaction.

Wong and co-workers[77] fabricated microtiter plate-type carbohydrate arrays by attaching azide-linked sugars to alkynylated lipids noncovalently immobilized on the microtiter plates (Figure 15b). Preliminary studies showed that lipid components of 13–15 carbons in length were tightly attached to the surface of a microtiter plate and

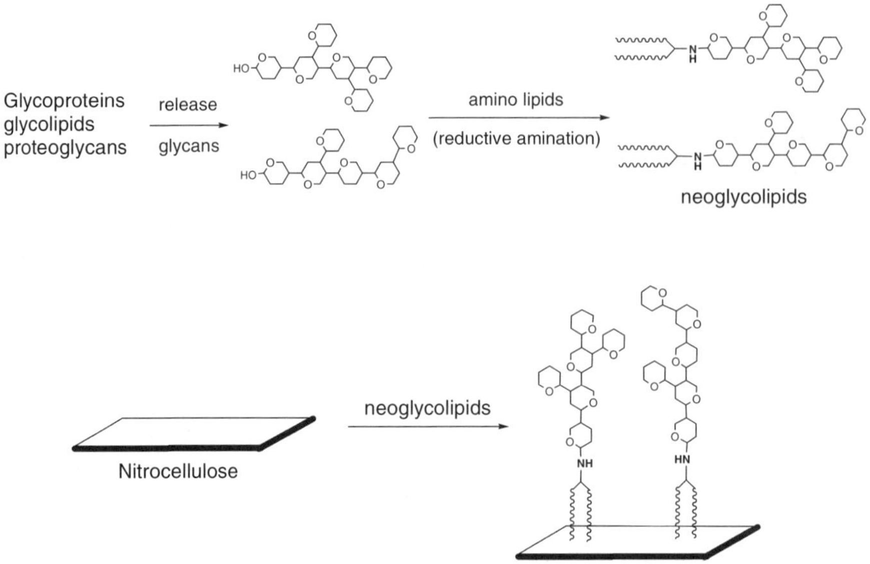

Glycoproteins
glycolipids   →   release
proteoglycans   glycans

amino lipids
(reductive amination)

neoglycolipids

Nitrocellulose

neoglycolipids →

Figure 14 *Noncovalent and site-specific attachment of neoglycolipids (NGLs) on the underivatized solid surface*

were not removed by repeated washing with citrate or bicarbonate buffers.[78] On the basis of these observations, the alkyne containing a 14-carbon lipid component was first noncovalently adsorbed on the plate by hydrophobic interactions. Then, a series of azide-linked sugars were covalently immobilized by employing copper(I)-accelerated regiospecific 1,3-dipolar cycloaddition reactions between the alkyne and azide groups. Waldmann and co-workers[79] prepared small molecule microarrays including carbohydrates by using Staudinger reactions between azide-linked substances and phosphane-derivatized glass slides (Figure 15c). The glass surface was derivatized with fourth-generation polyamidoamine (PAMAM) dendrimers to increase reactive sites on the surface. After reacting PAMAM-derivatized glass slides with glutaric anhydride to introduce terminal carboxylic acid groups, azide-reactive phosphane moieties were introduced by treating with 2-(diphenylphosphinyl)phenol and diisopropylcarbodiimide (DIC). The azide-linked substances were then covalently immobilized onto the phosphane-coated slides through the Staudinger ligation. The required azide-linked substances were easily prepared on the solid support by using the safety-catch linker strategy.

Schwarz *et al.*[80] developed glycan arrays by covalently immobilizing various *p*-aminophenyl glycosides on cyanuric chloride-modified glass slides patterned with a hydrophobic Teflon mask (Figure 15d). Seeberger and co-workers fabricated carbohydrate microarrays by reacting thiol-linked carbohydrates on maleimide-functionalized BSA-coated glass slides.[81,82] Aldehyde-derivatized glass slides were immersed into PBS buffer containing 1% BSA and then treated with a bifunctional crosslinker, succinimidyl 4-(*N*-maleimidomethyl)cyclohexane-1-carboxylate, to introduce maleimide groups on the protein surface. The thiol-containing carbohydrates were attached to the

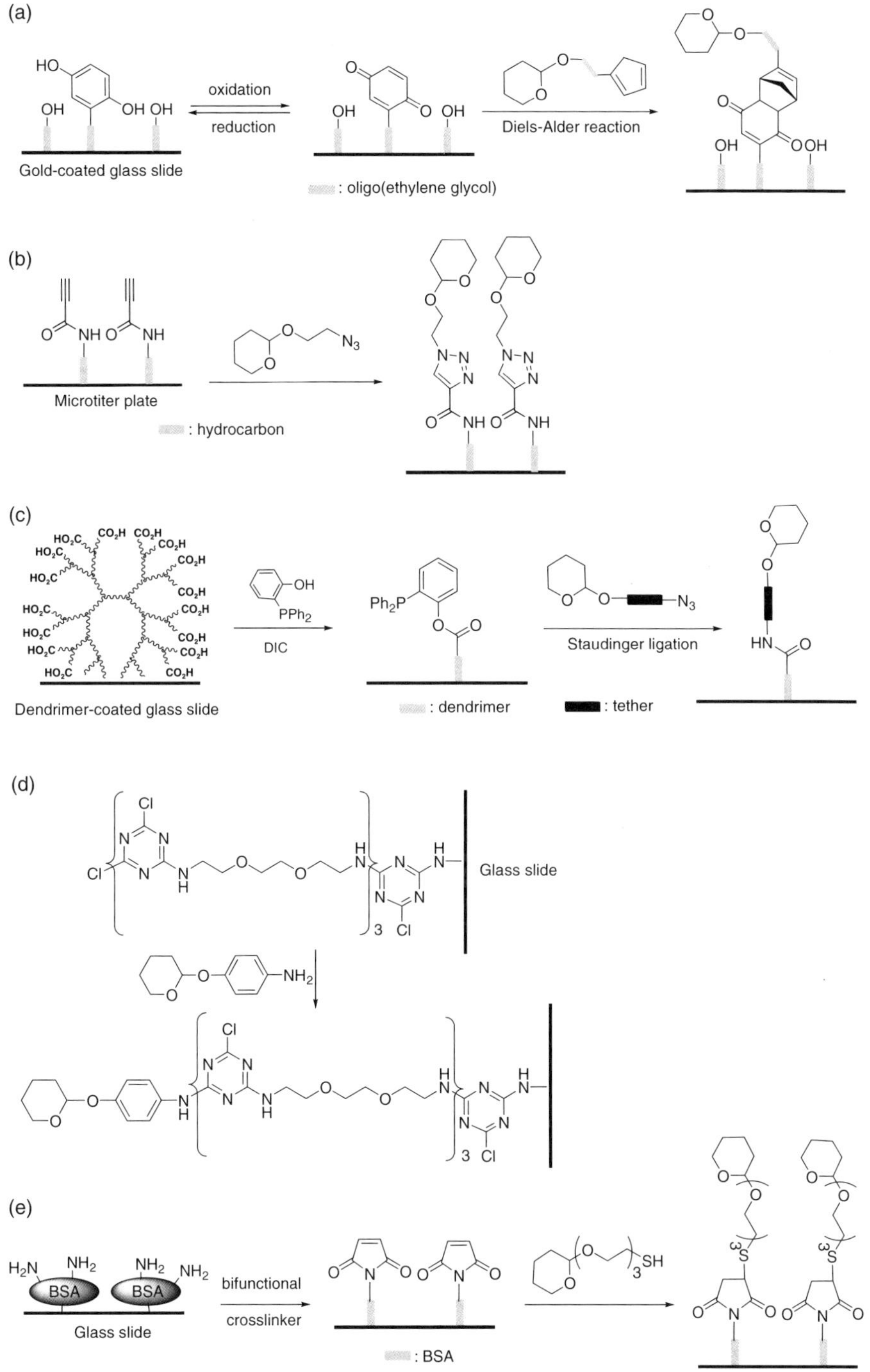

Figure 15 *Covalent and site-specific immobilization of chemically modified carbohydrates on the properly derivatized solid surface*

maleimide-derivatized BSA-coated glass slides by chemoselective reaction between thiol and maleimide groups (Figure 15e).

6.2   Other Applications

As described in Section 5, carbohydrate microarrays can be used for high-throughput analysis of carbohydrate–protein interaction, quantitative analysis of protein-binding affinities, and characterization of carbohydrate-processing enzymes. Furthermore, this technology can be applied to assessing the binding specificities of antibodies, studies of carbohydrate-mediated cell recognition events and the detection of pathogens for diagnosis, deciphering the carbohydrate code (structures of glycans), and the discovery of novel inhibitors of carbohydrate-binding proteins and carbohydrate-processing enzymes (Figure 16).

Carbohydrate microarrays can be utilized for analyzing carbohydrate–antibody interactions and detecting specific carbohydrate-binding antibodies for the diagnosis of diseases (Figure 16a). For example, binding specificities of human antibodies were assessed with microarrays containing 48 microbial polysaccharides by using only a limited amount of human serum (1 µL).[73] The antibody-binding experiments demonstrated that antibodies specifically recognized the corresponding polysaccharides on the surface. Moreover, unexpected antibody specificities were found and previously unknown cellular markers (Dex-Ids) were discovered by using this microarray. Most pathogens contain specific polysaccharides on their surface. Once pathogens infect humans, antibodies that bind to the pathogenic polysaccharides are produced. Therefore, the microbial polysaccharide microarray can be used for the diagnosis of pathogen infection by using human serum samples.

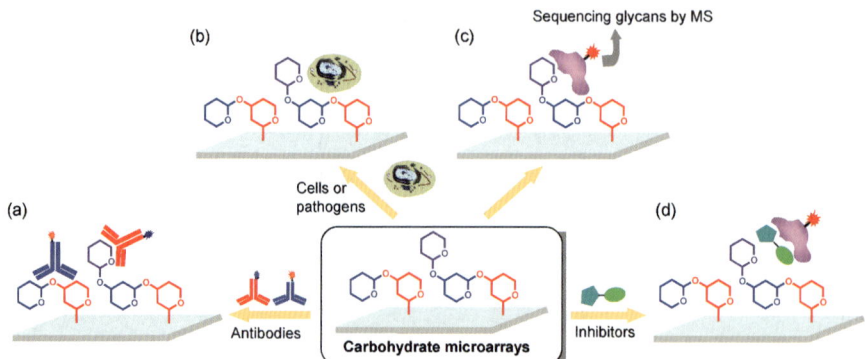

Figure 16 *Applications of carbohydrate microarrays. (a) Profiling of carbohydrate–antibody interactions and detection of specific carbohydrate-binding antibodies for diagnosis of diseases, (b) characterization of carbohydrate-mediated cell recognition events and detection of pathogens for diagnosis, (c) deciphering the oligosaccharide code in a glycome, and (d) high-throughput screening of inhibitors of carbohydrate-binding proteins and carbohydrate-processing enzymes*

Another potential application of carbohydrate microarrays for diagnosis is the detection of tumor cells in humans, because they frequently express tumor-associated carbohydrates.

This microarray technology can also be employed for characterizing carbohydrate-mediated cell recognition events and the detection of pathogens (Figure 16b). Primary chicken hepatocytes expressing GlcNAc-specific lectin on their surface were monitored by using carbohydrate microarrays.[83] For these studies, the glycan arrays were incubated with dye-labeled primary chicken hepatocytes. The results showed that intact cells adhered to GlcNAc but not to Gal or GalNAc immobilized on the solid surface. This technology was also applied to examining adhesion of human CD4+ T-cell to carbohydrates on the solid surface. CD4+ cells adhered to sialyl Le[x], maybe through cell surface L-selectin, but rarely recognized the nonfucosylated form. Furthermore, the carbohydrate microarrays were applied to the direct detection of pathogens that expressed specific carbohydrate-binding proteins on their surfaces. The carbohydrate microarray composed of Man, Glc, GlcNAc, Gal, and Fuc was incubated with *Escherichia coli* ORN178 that had been stained with cell-permeable fluorescence dye.[84] This *E. coli* strain expresses mannose-binding proteins on the cell surfaces and thus bound only to microspots of mannose. However, a mutant of this strain, ORN209, which exhibits a reduced binding affinity to mannose, showed much lower fluorescence intensity than ORN178. This microarray was also used to screen inhibitors of carbohydrate–cell interactions. Three inhibitors (mannose, *p*-nitrophenyl-α-D-mannopyranoside, and a mannose-functionalized polymer) were preincubated with *E. coli* cells, and then the mixture was poured onto the slide. It was found that a mannose-functionalized polymer was a much more efficient inhibitor than the other two compounds. These experiments demonstrated that carbohydrate microarrays were suitable for the diagnosis of pathogen infection and screening of antiadhesion reagents.

Carbohydrate structures in glycoproteins or glycolipids can be mapped by using a combination of neoglycolipid microarray technology and mass spectrometry (MS; Figure 16c).[32,75] To this end, oligosaccharides are initially released from glycoproteins, glycolipids, or proteoglycans by chemical or enzymatic methods. The isolated oligosaccharides are converted to NGLs by coupling to an aminolipid by reductive amination. After immobilizing NGLs on nitrocellulose and probing with proteins, sequences of oligosaccharides that are recognized by specific proteins are determined by MS. Ultimately, this application can be extended for deciphering the oligosaccharide code in a glycome.

A potential application of this technology is high-throughput screening of inhibitors for carbohydrate-processing enzymes that correspond to the biosynthesis of the disease-related carbohydrates and modulators that disrupt carbohydrate–protein interactions (Figure 16d). Earlier, there was no example for this application using carbohydrate microarrays. However, Wong and co-workers[85] reported the screening of fucosyltransferase (FucT) inhibitors by using microtiter-type carbohydrate arrays. LacNAc (a substrate for FucT) immobilized on the microtiter was incubated with 85 synthetic compounds in the presence of FucT and GDP-Fuc. fter probing with peroxidase-coupled *T. purpureas* (a fucose specific lectin), four inhibitors with nanomolar K_i's were discovered. Although microtiter type carbohydrate arrays were successfully used for screening inhibitors of carbohydrate-processing enzymes, more work should be needed to apply carbohydrate microarrays for drug discovery. However, the microarray technology could be eventually used for developing new inhibitors of carbohydrate-processing enzymes and carbohydrate-binding proteins.

7 Detection Methods

Sensitive detection is important for the wide applications of carbohydrate microarrays. In most cases, fluorescence detection using fluorophore-labeled proteins has been extensively employed because of its high sensitivity. For this method, fluorophore-labeled proteins are required. However, protein labeling with fluorophores sometimes causes protein denaturation and/or interference with carbohydrate ligand binding. In order to prevent protein denaturation during labeling with fluorophores, C-terminal dye-labeled proteins were prepared by *in vitro* expression in the presence of fluorophore-containing puromycin derivatives.[86] Since the required conjugated proteins are generated during protein expression, an extra protein labeling procedure can be omitted. Moreover, C-terminal modification of proteins may not interfere with their binding to carbohydrate ligands. However, detection methods that do not use labeled proteins are still more useful. One of detection methods that are suitable for this purpose is SPR spectroscopy. This method does not require labeled proteins, and its high sensitivity allows the detection of low-affinity binding. The limitation of SPR is that this technique is inappropriate for high-throughput analysis of carbohydrate–protein interactions. To overcome this limitation, SPR imaging technology has been exploited. This technology has been applied to the detection of RNA–DNA, protein–DNA, and carbohydrate–protein interactions.[87–89] Alternatively, MS was also employed for characterizing the modification of carbohydrates on glycan microarrays.[90,91] For instance, enzymatic reactions by carbohydrate-processing enzymes on carbohydrate microarrays were assessed by using MALDI-TOF MS.[91] Furthermore, this method was also used to characterize the time-dependent enzymatic glycosylation.

8 Conclusion

Conventional biochemical and biophysical approaches have mainly been used to elucidate recognition events between carbohydrates and proteins. However, in the postgenomic era, a pressing need exists to develop high-throughput analytic tools for use in functional glycomic research. Carbohydrate microarray technology may be the best choice to achieve this goal. Although this technology is at an early stage of development, it is clear that it has a high potential in the context of biological research as well as biomedical applications, as described above. In the near future, carbohydrate microarrays will become more practicable for the functional studies of carbohydrates, the diagnosis of pathogen-infected diseases or cancers, and the discovery of novel drugs.

References

1. J. Roth, *Chem. Rev.*, 2002, **102**, 285.
2. G.E. Ritchie, B.E. Moffatt, R.B. Sim, B.P. Morgan, R.A. Dwek and P.M. Rudd, *Chem. Rev.*, 2002, **102**, 305.
3. N.E. Zachara and G.W. Hart, *Chem. Rev.*, 2002, **102**, 431.
4. C.R. Bertozzi and L.L. Kiessling, *Science*, 2001, **291**, 2357.
5. R.A. Dwek, *Chem. Rev.*, 1996, **96**, 683.
6. A. Varki, *Glycobiology*, 1993, **3**, 97.
7. W.B. Turnbull, B.L. Precious and S.W. Homans, *J. Am. Chem. Soc.*, 2004, **126**, 1047.
8. T.K. Dam and C.F. Brewer, *Chem. Rev.*, 2002, **102**, 387.
9. R.J. Green, R.A. Frazier, K.M. Shakesheff, M.C. Davies, C.J. Roberts and S.J.B. Tendler, *Biomaterials*, 2000, **21**, 1823.
10. D.A. Mann, M. Kanai, D.J. Maly and L.L. Kiessling, *J. Am. Chem. Soc.*, 1998, **120**, 10575.
11. M.P. Wormald, A.J. Petrescu, Y.-L. Pao, A. Glithero, T. Elliott and R.A. Dwek, *Chem. Rev.*, 2002, **102**, 371.
12. D.F. Wyss, J.S. Choi, J. Li, M.H. Knoppers, K.J. Willis, A.R.N. Arulanandam, A. Smolyar, E.L. Reinherz and G. Wagner, *Science*, 1995, **269**, 1273.
13. N.K. Vyas, *Curr. Opin. Struct. Biol.*, 1991, **1**, 732.
14. Y.C. Lee, R.T. Lee, K. Rice, Y. Ichikawa and T.-C. Wong, *Pure Appl. Chem.*, 1991, **63**, 499.
15. C.M. Perou, T. Sørlie, M.B. Eisen, M. van de Rijn, S.S. Jeffrey, C.A. Rees, J.R. Pollack, D.T. Ross, H. Johnsen, L.A. Akslen, Ø. Fluge, A. Pergamenschikov, C. Williams, S.X. Zhu, P.E. Lønning, A.L. Børresen-Dale, P.O. Brown and D. Botstein, *Nature*, 2000, **406**, 747.
16. M. Bittner, P. Meltzer, Y. Chen, Y. Jiang, E. Seftor, M. Hendrix, M. Radmacher, R. Simon, Z. Yakhini, A. Ben-Dor, N. Sampas, E. Dougherty, E. Wang, F. Marincola, C. Gooden, J. Lueders, A. Glatfelter, P. Pollock, J. Carpten, E. Gillanders, D. Leja, K. Dietrich, C. Beaudry, M. Berens, D. Alberts, V. Sondak, N. Hayward and J. Trent, *Nature*, 2000, **406**, 536.
17. A.A. Alizadeh, M.B. Eisen, R.E. Davis, C. Ma, I.S. Lossos, A. Rosenwald, J.C. Boldrick, H. Sabet, T. Tran, X. Yu, J.I. Powell, L. Yang, G.E. Marti, T. Moore, J. Hudson, Jr., L. Lu, D.B. Lewis, R. Tibshirani, G. Sherlock, W.C. Chan, T.C. Greiner, D.D. Weisenburger, J.O. Armitage, R. Warnke, R. Levy, W. Wilson, M.R. Grever, J.C. Byrd, D. Botstein, P.O. Brown and L.M. Staudt, *Nature*, 2000, **403**, 503.
18. L. Wodicka, H. Dong, M. Mittmann, M.-H. Ho and D.J. Lockhart, *Nat. Biotechnol.*, 1997, **15**, 1359.
19. J.L. DeRisi, V.R. Lyer and P.O. Brown, *Science*, 1997, **278**, 680.
20. M. Chee, R. Yang, E. Hubbell, A. Berno, X.C. Huang, D. Stern, J. Winkler, D.J. Lockhart, M.S. Morris and S.P.A. Fodor, *Science*, 1996, **274**, 610.
21. D.A. Hall, H. Zhu, X. Zhu, T. Royce, M. Gerstein and M. Snider, *Science*, 2004, **306**, 482.

22. H. Zhu, M. Bilgin, R. Bangham, D. Hall, A. Casamayor, P. Bertone, N. Lan, R. Jansen, S. Bidlingmaier, T. Houfek, T. Mitchell, P. Miller, R.A. Dean, M. Gerstein and M. Snyder, *Science*, 2001, **293**, 2101.

23. G. MacBeath and S.L. Schreiber, *Science*, 2000, **289**, 1760.

24. H. Zhu, J.F. Klemic, S. Chang, P. Bertone, A. Casamayor, K.G. Klemic, D. Smith, M. Gerstein, M.A. Reed and M. Snyder, *Nat. Genet.*, 2000, **26**, 283.

25. M.F. Templin, D. Stoll, M. Schrenk, P.C. Traub, C.F. Vöhringer and T.O. Joos, *Trends Biotechnol.*, 2002, **20**, 160.

26. S.R. Weinberger, E.A. Dalmasso and E.T. Fung, *Curr. Opin. Chem. Biol.*, 2001, **6**, 86.

27. J.J. Lundquist and E.J. Toone, *Chem. Rev.*, 2002, **102**, 555.

28. M. Mannen, S.-K. Choi and G.M. Whitesides, *Angew. Chem. Int. Ed.*, 1998, **37**, 2754.

29. Y.C. Lee and R.T. Lee, *Acc. Chem. Res.*, 1995, **28**, 321.

30. I. Shin, S. Park and M.-r. Lee, *Chem. Eur. J.*, 2005, **11**, 2894.

31. I. Shin, J.W. Cho and D.W. Boo, *Combin Chem High Throughput Screen*, 2004, **7**, 565.

32. T. Feizi and W. Chai, *Nat. Rev. Mol. Cell Biol.*, 2004, **5**, 582.

33. D. Wang, *Proteomics*, 2003, **3**, 2167.

34. T. Feizi, F. Fazio, W. Chai and C.-H. Wong, *Curr. Opin. Struct. Biol.*, 2003, **13**, 637.

35. C. Ortiz Mellet and J.M. Garcia Fernández, *ChemBioChemistry*, 2002, **3**, 819.

36. K.R. Love and P.H. Seeberger, *Angew. Chem. Int. Ed.*, 2002, **41**, 3583.

37. G.T. Hermanson, *Bioconjugate Techniques*, Academic Press, San Diego, 1996, 148.

38. I. Shin, H.-j. Jung and M.-r. Lee, *Tetrahedron Lett.*, 2001, **42**, 1325.

39. I. Shin, H.-j. Jung and J. Cho, *Bull. Korean Chem. Soc.*, 2000, **21**, 845.

40. S. Park, M.-r. Lee, S.-J. Pyo and I. Shin, *J. Am. Chem. Soc.*, 2004, **126**, 4812.

41. S. Park and I. Shin, *Angew. Chem. Int. Ed.*, 2002, **41**, 3180.

42. P.Y. Reddy, S. Kondon, S. Fujita and T. Toru, *Synthesis*, 1998, **1998**, 999.

43. P. Pöchlauer and W. Hendel, *Tetrahedron*, 1998, **54**, 3489.

44. D. Vetter and M.A. Gallop, *Bioconjugate Chem.*, 1995, **6**, 316; D. Vetter, E.M. Tate and M.A. Gallop, *Bioconjugate Chem.*, 1995, **6**, 319.

45. E. Meinjohanns, M. Meldal, H. Paulsen, R.A. Dwek and K. Bock, *J. Chem. Soc., Perkin Trans. 1*, 1998, 549.

46. R.T. Lee and Y.C. Lee, *Carbohydr. Res.*, 1974, **37**, 193.

47. T.C. Wong, T.R. Townsend and Y.C. Lee, *Carbohydr. Res.*, 1987, **170**, 27.

48. J.-P. Utille and B. Priem, *Carbohydr. Res.*, 2000, **329**, 431.

49. P.J. Garegg and T. Norberg, *Carbohydr. Res.*, 1976, **52**, 235.

50. H.J. Rosenberg, A.M. Riley, R.D. Marwood, V. Correa, C.W. Taylor and B.V.L. Potter, *Carbohydr. Res.*, 2001, **332**, 53.

51. W. Fitz, P.B. Rosenthal and C.-H. Wong, *Bioorg. Med. Chem.*, 1996, **4**, 1349.

52. L.A. Marcaurelle and P.H. Seeberger, *Curr. Opin. Chem. Biol.*, 2002, **6**, 289.

53. P.H. Seeberger and W.-C. Haase, *Chem. Rev.*, 2000, **100**, 4349.

54. T. Takahashi, M. Adachi, A. Matsuda and T. Doi, *Tetrahedron Lett.*, 2000, **41**, 2599.

55. M.J. Sofia, *Mol. Divers.*, 1998, **3**, 75.
56. R. Liang, L. Yan, J. Loebach, M. Ge, Y. Uozumi, K. Sekanina, N. Horan, J. Gildersleeve, C. Thompson, A. Smith, K. Biswas, W.C. Still and D. Kahne, *Science*, 1996, **274**, 1520.
57. K.M. Koeller and C.-H. Wong, *Chem. Rev.*, 2000, **100**, 4465.
58. C. Kieburg, K. Sadalapure and T.K. Lindhorst, *Eur. J. Org. Chem.*, 2000, **2000**, 2035.
59. I.E. Liener, N. Sharon and I.J. Goldstein, *The Lectins: Properties, Functions, and Applications in Biology and Medicine*, Academic Press, Orlando, 1986.
60. E.J.M. Van Damme, W.J. Psumans, A. Pusztai and S. Bardocz, *Handbook of Plant Lectins: Properties and Biomedical Applications*, Wiley, Chichester, UK, 1998.
61. M. Monsigny, A.-C. Roche, C. Sene, R. Maget-Dana and F. Delmotte, *Eur. J. Biochem.*, 1980, **104**, 147.
62. C.S. Wright, *J. Mol. Biol.*, 1980, **141**, 267.
63. V.P. Bhavanandan and A.W. Kaltlic, *J. Biol. Chem.*, 1979, **254**, 4000.
64. P.M. Kaladas, E.A. Kabat, J.L. Iglesias, H. Lis and N. Sharon, *Arch. Biochem. Biophys.*, 1982, **217**, 624.
65. D.K. Mandal, N. Kishore and C.F. Brewer, *Biochemistry*, 1994, **33**, 1149.
66. H. Kaku, E.J.M. Van Damme, W.J. Peumans and I.J. Goldstein, *Arch. Biochem. Biophys.*, 1990, **279**, 298.
67. M.K. Sauerborn, L.M. Wright, C.D. Reynolds, J.G. Grossmann and P.J. Rizkallah, *J. Mol. Biol.*, 1999, **290**, 185.
68. I.J. Goldstein, S. Hammarström and G. Sundblad, *Biochim. Biophys. Acta*, 1975, **405**, 63.
69. N. Kochibe and K. Furukawa, *Biochemistry*, 1980, **19**, 2841.
70. D. Sudakevitz, A. Imberty and N. Gilboa-Garber, *J. Biochem.*, 2002, **132**, 353.
71. H.J.M. Gijsen, L. Qiao, W. Fitz and C.-H. Wong, *Chem. Rev.*, 1996, **96**, 443.
72. O. Seitz and C.-H. Wong, *J. Am. Chem. Soc.*, 1997, **119**, 8766.
73. D. Wang, S. Liu, B.J. Trummer, C. Deng and A. Wang, *Nat. Biotechnol.*, 2002, **20**, 275.
74. W.G.T. Willats, S.E. Rasmussen, T. Kristensen, J.D. Mikkelsen and J.P. Knox, *Proteomics*, 2002, **2**, 1666.
75. S. Fukui, T. Feizi, C. Galustian, A.M. Lawson and W. Chai, *Nat. Biotechnol.*, 2002, **20**, 1011.
76. B.T. Houseman and M. Mrksich, *Chem. Biol.*, 2002, **9**, 443.
77. F. Fazio, M.C. Bryan, O. Blixt, J.C. Paulson and C.-H. Wong, *J. Am. Chem. Soc.*, 2002, **124**, 14397.
78. M.C. Bryan, O. Plettenburg, P. Sears, D. Rabuka, S. Wacowich-Sgarbi and C.-H. Wong, *Chem. Biol.*, 2002, **9**, 713.
79. M. Köhn, R. Wacker, C. Peters, H. Schröder, L. Soulère, R. Breinbauer, C.M. Niemeyer and H. Waldmann, *Angew. Chem. Int. Ed.*, 2003, **42**, 5830.
80. M. Schwarz, L. Spector, A. Gargir, A. Shtevi, M. Gortler, R.T. Altstock, A.A. Dukler and N. Dotan, *Glycobiology*, 2003, **13**, 749.
81. D.M. Ratner, E.W. Adams, J. Su, B.R. O'Keefe, M. Mrksich and P.H. Seeberger, *ChemBioChemistry*, **2004**, **5**, 379.

82. B.T. Houseman, E.S. Gawalt and M. Mrksich, *Langmuir*, **2003**, **19**, 1522.

83. L. Nimrichter, A. Gargir, M. Gortler, R.T. Altstock, A. Shtevi, O. Weisshaus, E. Fire, N. Dotan and R.L. Schnaar, *Glycobiology*, 2004, **14**, 197.

84. M.D. Disney and P.H. Seeberger, *Chem. Biol.*, 2004, **11**, 1701.

85. M.C. Bryon, L.V. Lee and C.-H. Wong, *Bioorg. Med. Chem. Lett.*, 2004, **14**, 3185.

86. Y. Kawahashi, N. Doi, H. Takashima, C. Tsuda, Y. Oishi, R. Oyama, M. Yonezawa, E. Miyamoto-Sato and H. Yanagawa, *Proteomics*, 2003, **3**, 1236.

87. E.A. Smith, W.D. Thomas, L.L. Kiessling and R.M. Corn, *J. Am. Chem. Soc.*, 2003, **125**, 6140.

88. E.A. Smith, M.G. Erickson, A.T. Ulijasz, B. Weisblum and R.M. Corn, *Langmuir*, 2003, **19**, 1486.

89. B.P. Nelson, T.E. Grimsrud, M.R. Liles, R.M. Goodman and R.M. Corn, *Anal. Chem.*, 2001, **73**, 1.

90. M.C. Bryan, F. Fazio, H.-K. Lee, C.-Y. Huang, A. Chang, M.D. Best, D.A. Calarese, O. Blixt, J.C. Paulson, D. Burton, I.A. Wilson and C.-H. Wong, *J. Am. Chem. Soc.*, 2004, **126**, 8640.

91. J. Su and M. Mrksich, *Angew. Chem. Int. Ed.*, 2002, **41**, 4715.

Subject Index